COPING WITH BIOLOGICAL GROWTH ON STONE HERITAGE OBJECTS

Methods, Products, Applications, and Perspectives

COPING WITH BIOLOGICAL GROWTH ON STONE HERITAGE OBJECTS

Methods, Products, Applications, and Perspectives

Daniela Pinna

CRC Press
Taylor & Francis Group
Boca Raton London New York

CRC Press is an imprint of the
Taylor & Francis Group, an **informa** business

First published 2017 by Apple Academic Press Inc.

Published 2019 by CRC Press
Taylor & Francis Group
6000 Broken Sound Parkway NW, Suite 300
Boca Raton, FL 33487-2742

ISBN-13: 978-1-77188-532-4 (hbk)
ISBN-13: 978-1-315-36551-0 (ebk)

Library and Archives Canada Cataloguing in Publication

Pinna, Daniela, author
Coping with biological growth on stone heritage objects: methods, products, applications, and perspectives / Daniela Pinna.

Includes bibliographical references and index.
Issued in print and electronic formats.

ISBN 978-1-77188-532-4 (hardcover).--ISBN 978-1-315-36551-0 (PDF)
1. Antiquities--Collection and preservation--Case studies.
2. Biodegradation--Case studies. 3. Biodegradation--Prevention--Case studies. I. Title.

CC135.P55 2017 930.1028 C2017-900917-6 C2017-900918-4

Library of Congress Cataloging-in-Publication Data

Names: Pinna, Daniela, author.
Title: Coping with biological growth on stone heritage objects : methods, products, applications, and perspectives / Daniela Pinna.
Description: Toronto ; [Waretown, New Jersey] : Apple Academic Press, 2017. |
Includes bibliographical references and index.
Identifiers: LCCN 2017003495 (print) | LCCN 2017010685 (ebook) | ISBN 9781771885324 (hardcover : acid-free paper) | ISBN 9781315365510 (ebook) |
Subjects: LCSH: Antiquities--Collection and preservation--Handbooks, manuals, etc. | Cultural property--Protection--Handbooks, manuals, etc. | Stone--Biodegradation--Handbooks, manuals, etc. | Monuments--Conservation and restoration--Handbooks, manuals, etc. | Stone carving--Conservation and restoration--Handbooks, manuals, etc. | Excavations (Archaeology)--Handbooks, manuals, etc. | Caves--Handbooks, manuals, etc.
Classification: LCC CC135 .P57 2017 (print) | LCC CC135 (ebook) | DDC 363.6/9--dc23
LC record available at https://lccn.loc.gov/2017003495

Visit the Taylor & Francis Web site at
http://www.taylorandfrancis.com

and the CRC Press Web site at
http://www.crcpress.com

ABOUT THE AUTHOR

Daniela Pinna

Daniela Pinna has worked as a biologist at the Italian Cultural Heritage Ministry since 1987. Since 2011, she has been lecturing on "Biodeterioration and degradation of bioarcheological materials" at Bologna University (International Degree Course Science for the Conservation-Restoration of Cultural Heritage). She was involved in the European Projects EU-ARTECH (Access Research and Technology for the Conservation of the European Cultural Heritage, 2004–2009) and CHARISMA (Cultural Heritage Advanced Research Infrastructures: Synergy for a Multidisciplinary Approach to Conservation/Restoration, 2009–2014).

She has published two books and several articles on those subjects. She is editor of the book *Scientific Examination for the Investigation of Paintings: A Handbook for Conservators-Restorers* (Centro Di, Firenze, 2009). Moreover, she is involved in the activity of CEN/TC 346. CEN is the European Committee for Standardization and TC346 is in charge for standards related to conservation of cultural heritage. Her areas of specialization include biodeterioration of heritage objects, prevention and treatment of biodeteriogens, testing and evaluation of conservation products, and the control of treatment efficacy and durability in the field of stone, mortar, mural paintings, and polychrome sculpture conservation. She graduated in Biology from Padua University, Italy.

CONTENTS

LIST OF ABBREVIATIONS

a.i.	active ingredient
a.s.	active substance
AAC	autoclaved aerated concrete
AFNOR	Association Française de Normalization
AHAS	acetohydroxyacid synthase
Al_2O_3	aluminum oxide
ALS	acetolactate synthase
At	astatine
ATP	adenosine triphosphate
BCF	bacterial cell wall fraction
BPR	biocidal product regulation
Br	bromine
BSE	backscattered electrons
BSI	British Standards Institution
$CaC_2O_4 \cdot 2H_2O$	weddellite
$CaC_2O_4 \cdot H_2O$	whewellite
CaO	calcium oxide
CFU	colony-forming unit
CICPs	complex inorganic colored pigments
Cl	chlorine
CLSM	confocal laser scanning microscopy
CNT	carbon nanotubes
D-ALA	D-aminolevulinic acid
DDT	dichloro diphenyl trichloroethane
DGGE	denaturing gradient gels
DI	danger index
DIN	Deutsches Institut fur Normung
DNA	deoxyribonucleic acid
EC	electrical conductivity
EC_{50}	effective concentration at 50%
EDDS	ethylenediaminedisuccinate
EDDS	ethylenediaminedisuccinate
EDTA	ethylenediaminetetraacetic acid

EDX	energy dispersive X-ray analysis
ENs	European Standards
EPA	Environmental Protection Agency
EPS	extracellular polymeric substances
EPSP	5-enolpyruvylshikimate 3-phosphate
ERY	erythrosine
ESEM	environmental SEM
EU	European Union
F	fluorine
FISH	fluorescence in situ hybridization
GOx	glucose oxidase
H_2PO_4	phosphoric acid
HEPA	high efficiency particulate arrestants
I	iodine
IAA	indole-3-acetic acid
IARC	International Agency for Research on Cancer
LC_{50}	lethal concentration
LD_{50}	lethal dose
LECA	light expanded clay aggregate
LEDs	light-emitting diodes
LPBA	lichen potential biodeteriogenic activity
MgO	magnesium oxide
MIC	minimum inhibitory concentration
MMOs	mixed metal oxides
mRNAs	messenger RNAs
Na_2CO_3	sodium carbonate
NPs	nanoparticles
PAM	pulse amplitude modulation
PAS	periodic acid-Schiff
PCP	pentachlorophenol
PCR	polymerase chain reaction
PEI	polyethyleneimine
PNA	peptide nucleic acid
POEA	polyethoxylated tallow amine
QACs	quaternary ammonium compounds
QS	quorum sensing
RNA	ribonucleic acid
ROS	reactive oxygen species

rRNAs	ribosomal RNAs
SEM	scanning electron microscope
SiF_4	silicon tetrafluoride
SiF_4	silicon tetrafluoride
SiO_2	silicon dioxide
SMA	shape memory alloys
TEOS/TMOS	*n*-octyltriethoxysilane/tetramethylorthosilicate
tRNAs	transfer RNAs
UNI	Ente di Unificazione Italiano
UV	ultraviolet
ZnO–NPs	zinc oxide nanoparticles

PREFACE

From bacteria to plants, biological agents pose serious risks to the preservation of cultural heritage. In an effort to save heritage objects, buildings, and sites, conservators' activity aims to arrest, mitigate, and prevent the damages caused by bacteria, algae, fungi, lichens, plants, and birds. Although a lot has been learned about these problems, information is scattered across meeting proceedings and journals of all sorts of fields that often are not available to restorers and conservators. Therefore, there is a gap in the existing literature that needs to be filled. This book provides a comprehensive selection and examination of international papers published in the last 15 years, focusing on the methods and on the more suitable techniques and products that are useful for the prevention and removal of micro- and macroorganisms that grow on artificial and natural stone works of art, including wall paintings. Results on new substances with antimicrobic properties and alternative methods for the control of biological growth are presented as well.

This book offers hands-on guidance for facing the specific challenges involved in conserving monuments, sculptures, archaeological sites, and caves colonized by micro and macroorganisms. It provides many case studies of removal of biological growth with practical advices for making adequate choices and presents detailed updated information related to biocides and to alternative substances, feature that will be valuable to the audience. However, it is not the purpose of this text to tell the reader what to do. Rather, the envisioned goal is to provide access to information and offer the conceptual framework needed to understand complex issues, so that the reader can comprehend the nature of conservation problems and formulate her/his own views.

Another goal of the book is to emphasize issues on bioreceptivity of stones and the factors influencing biological growth. It also includes an outline of the various organisms able to develop on stones, a discussion on the bioprotection of stones by biofilms and lichens, a review of the main analytical techniques and a section on bioremediation.

Cultural heritage conservators and restores, scientists, heritage-site staff involved in conservation and maintenance of buildings, archaeological sites, parks, caves will find this book of great value to enable them to make informed decisions for the prevention and eradication of biological growth.

This text fits well within the programs of university courses on conservation and restoration of cultural heritage. It can be used either a primary or as a supplemental text, depending on the programs of the courses. In most countries, the first step of training into the professions connected to conservation is a full-time academic course. Although training in conservation of cultural heritage has traditionally taken the form of an apprenticeship, in recent years recognized conservation courses at university have become the norm in many countries.

It is a great hope of the author to contribute a useful tool for anybody with an interest in conservation of cultural heritage.

ACKNOWLEDGMENTS

I owe a special debt of gratitude to the Getty Conservation Institute of Los Angeles (USA), especially to the Director Tim Whalen and to Giacomo Chiari, former head of GCI Science. Without my scholarship period at the Getty some years ago, this book would not have been possible. While at the GCI, I reviewed many international papers on the biodeterioration of artificial and natural stone objects, and on methods, techniques and products for the removal of biodeteriogens. The invaluable benefit was the access to the huge amount of conservation literature and related research resources provided by the GCI. Furthermore, the skill and kindness of the GCI staff was of relevant importance. Without their precious help and advices, my stay would have not been as fruitful as it was.

I owe huge thanks to José Delgado Rodrigues for providing excellent support and advice. He read, wrote, and offered precious comments.

I am deeply indebted to Giulia Caneva and Ornella Salvadori who read some chapters and gave me many valuable suggestions for improving the manuscript. Giulia, Ornella, José, Antonella Pomicetti, Barbara Salvadori, Isetta Tosini, Mauro Tretiach, and Clara Urzì provided some of the figures used in this book.

I am grateful to my daughter Livia for help with graphic design of some figures.

I would like to express my gratitude to my friend Maria Ruggieri who offered perceptive comments and helpful advices.

I want to thank Editor Sandra Jones Sickels for her professional guidance and assistance.

I am grateful to my husband Antonio who supported and encouraged me.

Finally, I would like to mention my mentor Raffaella Rossi Manaresi who was a teacher to me through research, practice, and training.

FOREWORD

Daniela Pinna is a reputable biologist with an extensive practice in the field of conservation of Cultural Heritage. To tackle the subject "Coping with biological growth on stone objects" in a very extensive and comprehensive way is at reach to a very limited number of people, and among those, Daniela Pinna is certainly the best placed to make a useful "State of the art and perspectives of methods and products applied for the prevention and control of biological growth on artificial and natural stone heritage objects."

Essentially, the book is a very detailed state of the art, but it contains also a personal perspective about the addressed topics, helping the reader better move among these often complex subjects. Circa 500 references were collected, and many of them were summarized to support the author's leading ideas or to illustrate the complexity and sometimes the contradictions that can be found within such a multifaceted theme.

The book analyses this topic under the scientific point of view quoting and commenting research papers, but it also tackles the practical issues connected with biodeterioration, its occurrences, consequences, and possible remediation.

The agents of biodeterioration and their ecology are dealt with in Chapters 2 and 3, while Chapters 4 and 5 deal with control, eradication, and prevention of biological growth. Chapter 6 addresses biological growth from its positive side, as a potential help for conservation, and Chapter 7 illustrates some specific aspects of scientific examinations applied to the study of these subjects.

For the extensive list of references and the detailed analyses made out of them, the book is of interest to biologists, especially for those newcomers in the field, but I found it particularly interesting for non-biologists, as is my case. I read with high personal interest Chapter 3 "Outline of biodeterioration of stone objects", and found Section 4.3 "Biocides" extremely relevant and useful for any person dealing with biodeterioration in Cultural Heritage. The known categories of biocides are mentioned and their acting mechanisms explained, which turns this section an updated and very relevant working tool for professionals working in the field. The extensive list of products, and the respective properties and suppliers is a very handy contribution to conservator-restorers and authorities.

The book is easy to read, although it might be used essentially as a consulting manual, for conservation scientists searching for identifying the problems, their causes, and solutions, as well as for practitioners searching for a feasible solution for the problem at hand.

The book has large potential to be a reference in the shelf of any conservation scientist and conservator-restorer working in the field of Cultural Heritage.

—**José Delgado Rodrigues**
Geologist, former Principal Research Officer LNEC,
Consultant in Stone Conservation

INTRODUCTION

The cultural attitude toward flora and microflora that grow on stone monuments and artifacts went through substantial transformations in the past. In the 19th century, the wild flora on archaeological ruins, abbeys, and castles was appreciated for its esthetical value adding a picturesque quality to them.[1] The cultural sensibility of that period much valued the ruined nature of the sites and antiquities considering biological growth as an essential complement of them. For example, the covering of walls in ivy was regarded as a genuine enhancement.[2] Many artists took inspiration from the "melancholy contemplation" of archaeological ruins over-run by greenery like those of the Forum in Rome.[2] A conference on *American Ruins and Antiquities in the Long 19th Century* (The Huntington, San Marino, California, US, 2010) explored the "necessity for ruins" in the new American nation. Native American ruins and antiquities, and urban ruins of their own making both reassured Americans of their antiquity and helped them cope with the dislocations of modernity.

This view did not permit any analytical approach to conservation aspects connected to the interaction between vegetation and stone substrate. Even the scientific community was not interested in the preservation of monuments from harmful vegetation so that some scientists even denied the damage caused by spontaneous vegetation.

In a short time, this attitude turned into a distinction between the vegetation to be protected for its naturalistic value and the one harmful for the monuments. The esthetic aspects assumed an important role as well because vegetation is an impediment to the site's presentation to the audience when it compromises the legibility of a monument.[2] Conversely, the microflora (algae, fungi, lichens) that grow on stones, though appreciated in the 19th century, did not evoke the same esthetical meaning as the vegetation did, and usually, it was removed as some documents confirm. In the case of the Cathedral and Monumental Cemetery of Pisa, the archive papers mention the use of very aggressive treatments to "eradicate the lichen," for example, "potassium hydroxide and a solution of water and hydrochloric acid,"[3] not acceptable anymore today.

At present, the importance of biodeterioration process on historical objects of art has reached growing attention of people in charge with the conservation of cultural heritage. Many studies increasingly have documented and discussed the interaction between biological growth and cultural heritage.

This book highlights issues on bioreceptivity of stones and the factors influencing biological growth. It outlines the various organisms able to develop on stones and their effect in terms of degradation, including a discussion on an important topic, the bioprotection of stones by biofilms and lichens.

This text focuses on the methods and on the more suitable techniques and products that are useful for the prevention and removal of micro and macroorganisms that grow on artificial and natural stone works of art, including wall paintings. It details the different strategies performed for the prevention of biological growth on indoor and outdoor stone heritage objects. It includes also the revision of recent studies on bioremediation applied to conservation of cultural heritage involving the use of microorganisms to remove various kinds of materials present on the surfaces.

The analytical techniques—both noninvasive and micro-invasive—suitable to assess the weathering effects of biological growth on materials and the efficacy of control methods are described and discussed.

Furthermore, the appendix reports information about some biocides providing data on chemical and physical properties, toxicity, potential health effects, along with the manufacturers and/or suppliers of trade products containing the active ingredients.

This book covers a comprehensive selection of international articles on these issues published in the last 15 years and has been inspired by the book *Plant Biology for Cultural Heritage* (The Getty Conservation Institute, Los Angeles, 2008) edited by my colleagues and friends Giulia Caneva, Maria Pia Nugari, and Ornella Salvadori. The outstanding and complete publication is a sort of starting point for this book. Although focused on the literature published after 2005, year of the first Italian publication of the book *Plant Biology for Cultural Heritage*, the present publication, includes some of the more relevant information contained in that book. Comparing the recent results with those published more than 15 years ago, we can easily appreciate the progress in this field. For example, there has been tremendous progress on analytical methods applied to the examination of the interaction between microorganisms and substrates.

Regarding the removal of biological growth, the book provides updated results on mechanical, physical, and chemical methods. The reviewed literature included not only papers and books connected to cultural heritage conservation but also publications dealing with the control of biodeterioration on other kinds of objects. Results deriving from studies in other fields—civil buildings and their structures, water systems, plastics, etc.—can bring to cultural heritage conservation field new and interesting applications. On the other hand, the biocides used in conservation field were originally developed for medical and agricultural applications.

The successful treatment techniques adopted, as well as failures, have been taken into consideration. Moreover, the book reports nonconventional methods such as use of laser, microwaves, and nanoparticles. Laser technology has recently offered new solutions to the problems of lichens' removal from stone artifacts, despite the complexity of the physicochemical risks associated with the ablation of photosensitive substrates and the limited opportunities for testing on real cases. The use of microwaves and nanoparticles appears particularly notable for further developments.

Large space is devoted to the chemical methods using biocide formulations. They are worth studying more in depth than other methods as they undoubtedly are the most used in conservation practice. Several aspects are relevant when using biocides. The evaluation of their effectiveness, toxicity, environmental impact, chemical stability, long-term effect, best application methods, and harmlessness toward substrates have been the topics of many studies. Toxicity and environmental impact are nowadays important causes for concern in the selection of biocides.

A description of the results as well as a view on possible future developments are included in this book. Topics associated with chemical biocides include methods of biocides' application on stones, types of microbicides and herbicides, mechanisms of antimicrobial action, resistance of organisms to biocides, toxicity and legislative regulations, recolonization after treatment, novel biocides and alternative methods for the control of biological growth. Information about major active agents of biocide formulations applied on stone objects are provided. This book contains also up-to-date information on legislation and regulations governing the use of biocides in the European Union.

In situ monitoring of the performance of methods and products for the control of biodeteriogens is discussed. The concept of planned maintenance has reached recently growing consideration, and several techniques and diagnostic tools have been developed for the detection of possible risks, thus facilitating maintenance and management decisions.

CHAPTER 1

BASIC PRINCIPLES OF BIOLOGY*

CONTENTS

*The data and information sources have been the books:
J. B. Reece, L. A. Urry, M. L. Cain, S. A. Wasserman, V. P. Minorsky, R. B. Jackson. *Campbell Biology*. Benjamin Cummings. San Francisco, 2014.
D. L. Nelson, M. M. Cox. *Lehninger Principles of Biochemistry*. W.H. Freeman and Company, New York, 2012.
R. F. Evert, S. E. Eichhorn. *Raven Biology of Plants*. W.H. Freeman/Palgrave Macmillan, New York, 2013.

ABSTRACT

The smallest organisms consist of single cells and are microscopic. Multicellular organisms contain many different types of cells, which vary in size, shape, and specialized function. Cells are the structural and functional units of all living organisms. The structural features shared by cells of all kinds are described. The organisms are divided according to how they obtain the energy and carbon they need for synthesizing cellular material.

The chapter provides some fundamentals of biochemistry. Biochemistry describes in molecular terms the structures, mechanisms, and chemical processes shared by all living organisms. Thousands of different molecules constitute the cell's intricate internal structures. Diverse living organisms share common chemical features. Besides water and electrically charged ions, the cell is composed of organic molecules, that is, of molecules that contain carbon—carbohydrates, lipids, proteins, and nucleid acids. The chapter includes an outline of the cell structure, and of the various micro-and macroorganisms able to develop on stones.

As the reader goes through the book, she/he may find it helpful to refer to this chapter at intervals to refresh the memory of this background material.

1.1 THE CELL

Some characteristics distinguish any forms of life:

The cell, the basic structural and functional unit of all living organisms;
A high degree of chemical complexity and microscopic organization;
The presence of systems for extracting, transforming, and using energy from the environment;
Defined functions for each component and regulated interactions among them;
The presence of mechanisms for sensing and responding to alterations in their surroundings;
The ability to grow and to reproduce;
A capacity to change over time by gradual evolution.

Cells are the smallest unit of life that can replicate independently. The cell membrane, or plasma membrane, surrounds the cytoplasm of a cell. It serves to separate and protect it from the free passage of inorganic ions and other charged or polar compounds. Plasma membranes are permeable to water. This permeability is due largely to protein channels in the membrane that selectively permit the passage of water. Plasma membrane mostly consists in a bilayer of phospholipids that are amphipathic: One end of the molecule is hydrophobic (insoluble in water), the other is hydrophilic (soluble in water) (Fig. 1.1). A phospholipid is made of a molecule of glycerol with a phosphate added to one end, and two side chains of fatty acids attached at the other end. In the cell membrane, the glycerol and phosphate part of the molecule hangs out at the surface with the long side chains sandwiched in the middle (Fig. 1.2).

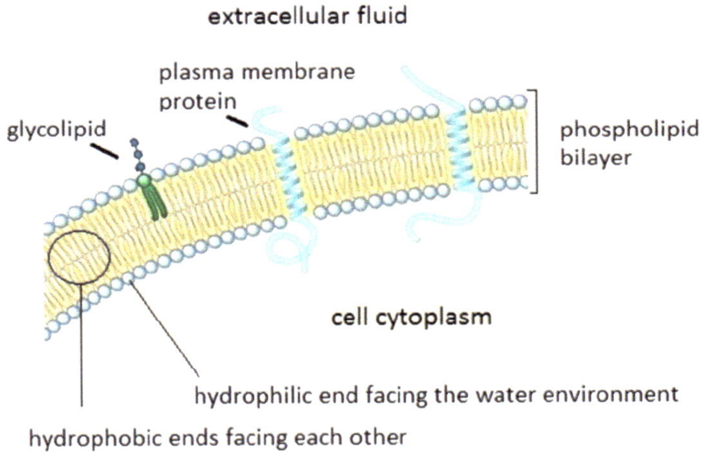

FIGURE 1.1 Structure of the cell membrane, composed of the phospholipid bilayer with embedded proteins. It is involved in a variety of cellular processes such as cell adhesion, ion conductivity and cell signaling. It is a barrier for most molecules and serves as the location for the transport of molecules into the cell.

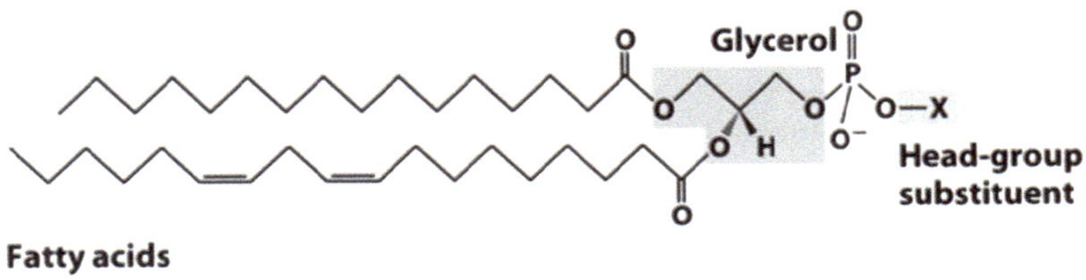

FIGURE 1.2 A phospholipid molecule is made of two fatty acids linked to a glycerol molecule. The third carbon of glycerol is linked to a phosphate group. The letter "X" indicates the atom or group of atoms that makes up the "rest of the molecule." The phospholipid tail is nonpolar and uncharged and is thus hydrophobic, while the polar head, which contains the phosphate and X groups, is hydrophilic.

The lipid bilayer creates an effective chemical barrier around the cell. The cell membrane includes transport proteins that allow the passage of certain ions and molecules, receptor proteins that transmit signals into the cell, and membrane enzymes that participate in some reaction pathways. It contains also glycolipids and sterols. Sterols account for approximately 25% of the weight of the cell membrane. The type and content of sterols differ among plant, animal, and fungal cells. Whereas plant cell membranes contain primarily stigmasterol and animal cell membranes cholesterol, ergosterol is the predominant sterol in many fungi. Cells of archaea, bacteria, algae, fungi, and plants have also a cell wall around the membrane, a semi rigid layer that provides cells with structural support and protection. It plays key roles in cell differentiation, intercellular communication, water movement, and defense environment. It acts as a filtering mechanism as well. The plant cell wall is the primary source of cellulose, the most abundant and useful biopolymer on Earth.

Within the cell membrane, there is the cytoplasm, composed of an aqueous solution, the cytosol, and a variety of suspended particles with specific functions. It contains ribosomes, the complex processing molecules that assemble proteins for the cell. It also contains the genome—the complete set of genes composed of DNA (deoxyribonucleic acid). DNA and proteins are the major components of chromosomes. In the cytoplasm, there are also structures called organelles that carry out specific functions, from providing energy to producing hormones and enzymes.

Cells are too small to be seen without magnification. They range in size from 1 to 100 μm (Fig. 1.3).

There are two primary types of cells: prokaryotic cells and eukaryotic cells.

In prokaryotic cells (Greek *pro*, "before," and *karyon*, "nucleus"), the nucleoid contains a single circular molecule of DNA that is not separated by a membrane from the rest of the cell. The cytoplasm contains also small, circular segments of DNA called plasmids. In bacteria, the plasmids can be transmitted from one cell to another. For this ability, genes, like those for antibiotic resistance, may be spread very rapidly through bacterial populations. All prokaryotic organisms are unicellular, meaning that the entire organism is only one cell. Most of them reproduce through a process called binary fission where the cell just splits in half after copying its DNA. Prokaryotic cells contain a few organelles that are not membrane-bound. Archaea and bacteria have prokaryotic cells (Fig. 1.4).

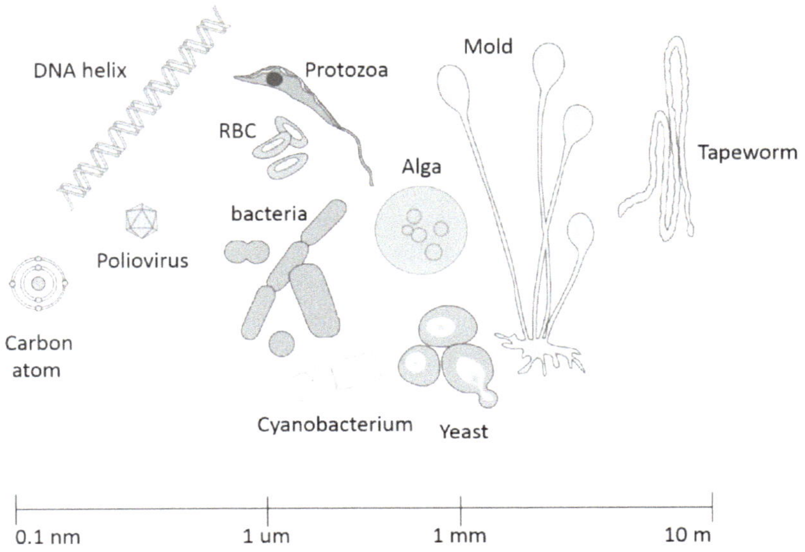

FIGURE 1.3 Size comparison among various atoms, molecules, and microorganisms (not drawn to scale) (From Motifolio Inc.).

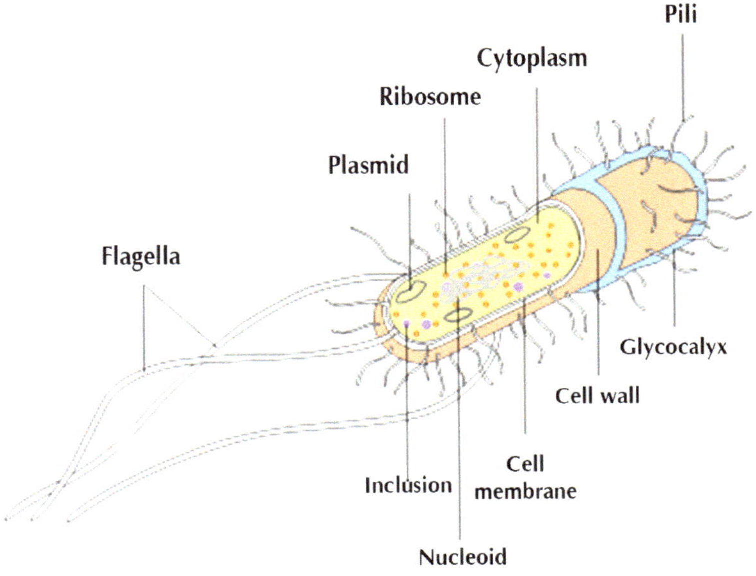

FIGURE 1.4 Diagram of the prokaryotic cell structure (From Motifolio Inc.).

Eukaryotic (Greek *eu*, "true," and *karyon*, "nucleus") cells are more complex and, on average, 10 times larger than prokaryotic cells. They have a nucleus, separated by a double membrane from other cellular structures, which contains DNA. Eukaryotic cells have several types of membrane-enclosed organelles with specific functions: mitochondria, endoplasmic reticulum, Golgi complexes, peroxisomes, and lysosomes (Fig. 1.5). There are also differences within eukaryotic cells. Plant cells, for example, contain a cell wall, vacuoles and chloroplasts that are not present in animal cells (Fig. 1.5).

Most eukaryotic organisms are multicellular. This allows the eukaryotic cells to become specialized. Through a process called differentiation, the cells change from one type to another. Commonly, a less specialized type becomes a more specialized type during cell growth. Differentiation occurs many times during the development of a multicellular organism as it changes from a simple cell to a complex system of tissues and organs. Cell differentiation relates specifically to the formation of functional cell types (e.g., vascular tissue cell types in plants). Differentiated cells contain large amounts of specific proteins associated with cell function.

Eukaryotes may use either asexual or sexual reproduction depending on the organism complexity. Sexual reproduction allows more diversity in offspring by mixing the genes of the parents to form a new combination and hopefully a more favorable adaptation for the environment.

Organisms may also be classified by the source of the energy and carbon they need for synthesizing the cellular material. *Autotrophs* (from the Greek *autos*, meaning "self," and *trophos*, meaning "feeder") are able to make their own energy-rich molecules out of simple inorganic materials. They obtain all needed carbon from CO_2. *Phototrophs* utilize solar energy, whereas *chemotrophs* obtain energy by the oxidation of a chemical fuel. For example, the lithotrophs oxidize inorganic fuels: HS^- to S^0 (elemental sulfur), S^0 to SO_4^{2-}, NO_2^- to NO_3^-, or Fe^{2+} to Fe^{3+}. *Heterotrophs* (from the Greek *heteros*, meaning "other," and *trophos*, meaning "feeder") require organic nutrients. They are dependent on an outside source of organic molecules for their energy. However, nearly all living organisms derive their energy, directly or indirectly, from the radiant energy of sunlight. The light-driven splitting of water during *photosynthesis* releases its electrons for the reduction of CO_2 and the release of O_2 into the atmosphere:

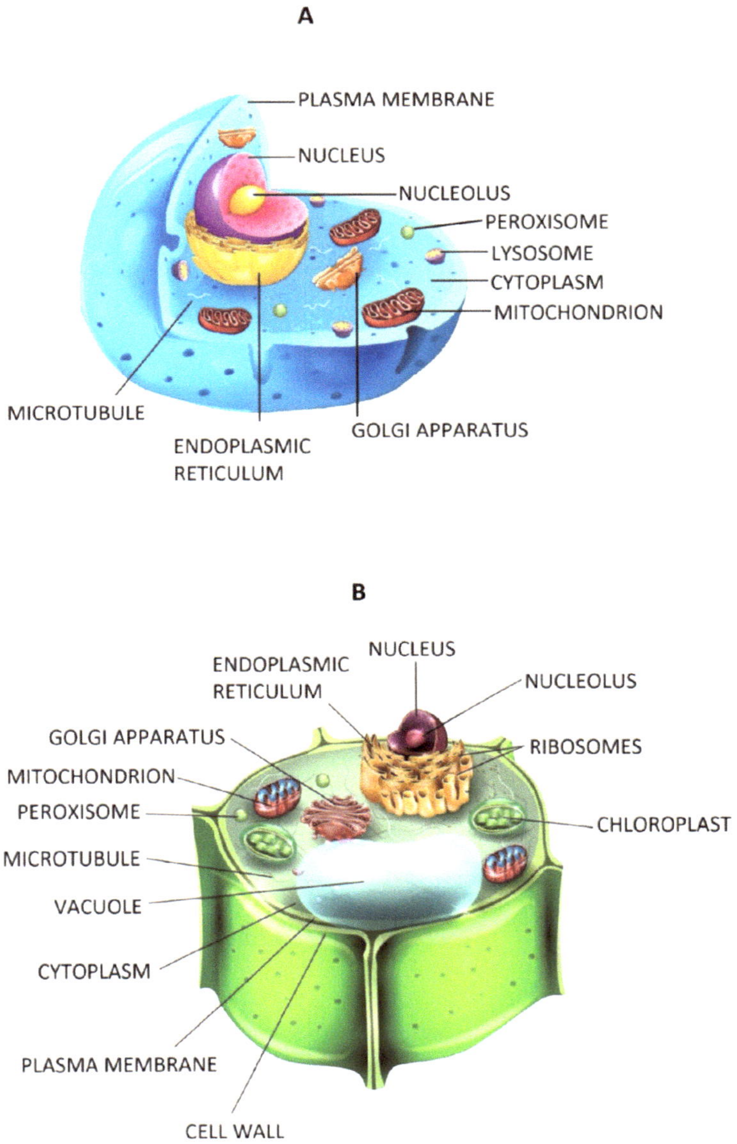

FIGURE 1.5 Diagrams of two major types of eukaryotic cell: (a) a representative animal cell and (b) a representative plant cell. Plant cells are usually 10–100 μm in diameter—larger than animal cells that typically range from 5 to 30 μm. Eukaryotic microorganisms (such as protists and fungi) have structures similar to those in plant and animal cells, but many also contain specialized organelles not illustrated here.

$$\overset{\text{Light}}{\downarrow}$$

$$6CO_2 + 6H_2O \rightarrow C_6H_{12}O_6 + 6O_2$$

(Light-driven reduction of CO_2)

Non-photosynthetic organisms obtain the energy they need by oxidizing the energy-rich products of photosynthesis, then passing the electrons thus acquired to atmospheric O_2 to form water, CO_2, and other products, which are recycled in the environment:

$$C_6H_{12}O_6 + 6O_2 \rightarrow 6CO_2 + 6H_2O + \text{energy}$$

(Energy-yielding oxidation of glucose)

All these reactions are oxidation-reduction reactions: One reactant is oxidized (loses electrons) as another is reduced (gains electrons). Energy is converted into the chemical bonds of adenosine triphosphate.

All living organisms fall into one of three large groups (domains) that define three branches of evolution from a common progenitor: archaea, bacteria, and eukarya. Archaea and bacteria are prokaryotic microorganisms (or prokaryotes). Eukarya (or eukaryotes) include protists, fungi, plants, and animals.

1.2 BIOCHEMISTRY BASICS

Water is the most abundant substance in living systems making to 70% or more of the weight of most organisms. The concentrations of solutes strongly influence the physical properties of aqueous solutions. When two aqueous solutions of different concentrations are separated by a semipermeable membrane (such as the plasma membrane separating a cell from its surroundings), water moves across that membrane to equalize the osmotic pressure in the two compartments. This movement of water across a semipermeable membrane, called *osmosis*, is driven by the *osmotic pressure*. Osmosis is an important factor in the life of most cells. Solutions with osmolarity* equal to that of the cytoplasm are isotonic in relation to that cell. Surrounded by an isotonic solution, a cell neither gains nor loses water. In a hypertonic solution, one with higher osmolarity than that of the

*Osmolarity is the concentration of a solution expressed as the total number of solute particles per liter.

cytoplasm, the cell shrinks as water moves out. In a hypotonic solution, one with a lower osmolarity than the cytoplasm, the cell swells as water enters. In their natural environments, cells generally contain higher concentrations of biomolecules and ions than their surroundings, thus osmotic pressure tends to drive water into the cells. If not somehow counterbalanced, this inbound movement of water would swell the plasma membrane and eventually cause the cell bursting (osmotic lysis). Several mechanisms have evolved to prevent this potentially catastrophic event. In bacteria and plants, the plasma membrane is surrounded by a non dilatable cell wall of sufficient rigidity and strength to resist osmotic pressure and prevent osmotic lysis. Certain freshwater protists living in highly hypotonic media have an organelle (contractile vacuole) that pumps water out of the cell. Cells also actively pump out Na^+ and other ions into the interstitial fluid to keep the osmotic balance with their surroundings.

The pH affects the structure and activity of biological macromolecules; for example, the catalytic activity of enzymes strongly depends on pH. The term pH is defined by the expression:

$$pH = \log 1/[H^+] = -\log [H^+]$$

The symbol p means "negative logarithm of." For a neutral aqueous solution at 25°C, in which the concentration of hydrogen ions (expressed in molar M terms) is 1.0×10^{-7} M, the pH can be calculated as follows:

$$pH = \log 1/1.0 \times 10^{-7} = 7$$

Solutions having a pH greater than 7 are alkaline or basic; the concentration of OH^- is greater than that of H^+. Measured pH values lie in the range 0–14.

The four most abundant elements in living organisms are hydrogen, oxygen, nitrogen, and carbon, which together form more than 99% of the mass of most cells. They are the lightest elements capable of efficiently forming one, two, three, and four chemical bonds, respectively. The trace elements are essential to life, usually for the function of specific proteins, including many enzymes. The oxygen-transporting capacity of the hemoglobin molecule, for example, is dependent on four iron ions that make up only 0.3% of its mass.

Proteins, lipids, polysaccharides, and nucleic acids are the organic compounds that constructed most cellular materials. They are polymers with molecular weights in the range of ~5000 up to several billion, which are assembled from relatively simple monomeric subunits. Each class of molecules has a similar structural hierarchy: Subunits are connected by bonds of limited flexibility to form macromolecules with three-dimensional structures determined by non covalent interactions (hydrogen bonds, ionic, hydrophobic, and van der Waals interactions). These macromolecules then interact to form the supramolecular structures and organelles that allow a cell to carry out its many metabolic functions.

Proteins constitute the largest fraction (besides water) of a cell. They are polymers built from 20 different L-amino acids. The linear sequence of amino acids in a protein, which is encoded in the DNA of the gene for that protein, produces a protein's unique three-dimensional structure—a process also dependent on environmental conditions. Understanding amino acids is therefore central to understanding protein structure. All amino acids have a central carbon atom bound to a hydrogen atom, an amino group, a carboxyl group, and an R group (Fig. 1.6). They differ from each other in their R groups, which vary in structure, size, and electric charge and determine the specific characteristics of each amino acid. Because of the tetrahedral arrangement of the bonding orbitals around the carbon atom, the four different groups can occupy two unique spatial arrangements, and thus amino acids have two possible stereoisomers. Since they are nonsuperimposable mirror images of each other, the two forms represent a class of stereoisomers called enantiomers (Fig. 1.7). Only the L stereoisomers, with a configuration related to the absolute configuration of the reference molecule L-glyceraldehyde, are found in proteins.

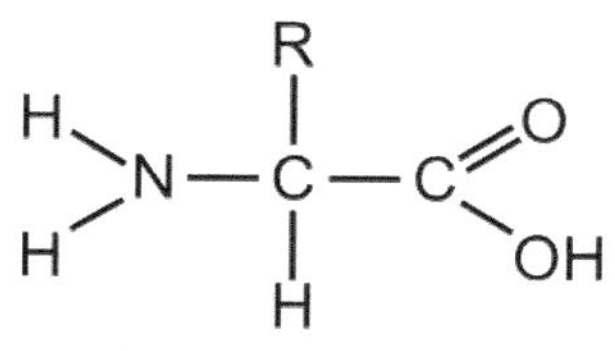

FIGURE 1.6 General structure of an amino acid.

At neutral pH, every amino acid is a zwitterion†: The amino group is protonated, with a positive charge on the nitrogen, and the carboxyl is deprotonated with a negative charge split between the two oxygens.

---

†In chemistry, a zwitterion (from German *zwitter* "hybrid") is a neutral molecule with a positive and a negative electrical charge.

FIGURE 1.7 The two stereoisomers of alanine, L- and D-alanine, are nonsuperimposable mirror images of each other (enantiomers).

The carboxyl group of one amino acid can condense with the amino group of another, eliminating water and forming a covalent C–N bond. The bond is termed a peptide bond and the reaction yields a dipeptide (Fig. 1.8). Each amino acid is subsequently referred to as a residue (what remains after the loss of water). All amino acids and peptides (chains of amino acids) share this property, regardless of identity, length, or composition. Once the dipeptide has formed, additional amino acids can be added through further condensation reactions, forming an extended chain called a polypeptide (protein). Naturally occurring polypeptides range in length from two to many thousands of amino acid residues.

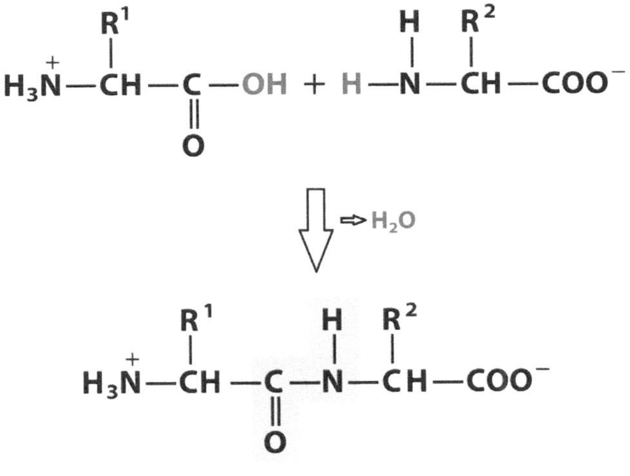

FIGURE 1.8 Formation of a peptide bond by condensation. The amino-group of one amino acid acts as a nucleophile to displace the hydroxyl group of another amino acid forming a peptide bond.

Cells generally contain thousands of different proteins, each with a different biological activity. Four levels of protein structure are commonly defined. The covalent bonds linking amino acid residues in a polypeptide chain is its primary structure. The most important element of a primary structure is the sequence of amino acid residues. A secondary structure refers to particularly stable arrangements of amino acid residues giving rise to recurring structural patterns. A tertiary structure describes all aspects of the three-dimensional folding of a polypeptide. When a protein has two or more polypeptide subunits, their arrangement in space is referred to as quaternary structure. For the intracellular proteins of most organisms, weak interactions are especially important in the folding of polypeptide chains into their secondary and tertiary structures. The association of multiple polypeptides to form quaternary structures also relies on these weak interactions.

Some proteins have catalytic activity and function as enzymes; others serve as structural elements, signal receptors, or transporters that carry specific substances into or out of cells.

Virtually, every chemical reaction in a cell occurs at a significant rate only because of the presence of enzymes. They are biocatalysts that, like all other catalysts, greatly enhance the rate of specific chemical reactions without being consumed in the process. The overall network of enzyme-catalyzed pathways constitutes the cellular metabolism.

Carbohydrates are polyhydroxy aldehydes or ketones, or substances that yield such compounds on hydrolysis. Many, but not all, carbohydrates have the empirical formula $(CH_2O)_n$. Some also contain nitrogen, phosphorus, or sulfur. There are three major size classes of carbohydrates: monosaccharides, oligosaccharides, and polysaccharides (the word "saccharides" derived from the Greek *sakcharon*, meaning "sugar"). Monosaccharides, or simple sugars, consist of a single polyhydroxy aldehyde or ketone unit (Fig. 1.9). The most abundant monosaccharide in nature is the six-carbon sugar glucose sometimes referred to as dextrose. In aqueous solution, all monosaccharides with five or more carbon atoms in the backbone occur predominantly as cyclic (ring) structures in which the carbonyl group forms a covalent bond with the oxygen of a hydroxyl group along the chain.

Oligosaccharides are short chains of monosaccharide units joined by characteristic linkages called glycosidic bonds. The most abundant are the disaccharides, with two monosaccharide units. Typical is sucrose (cane sugar), which is composed of the six-carbon sugars glucose and fructose.

In cells, most oligosaccharides made of three or more units do not occur as free entities but are joined to nonsugar molecules (lipids or proteins) in glycol-conjugates. All common monosaccharides and disaccharides have names ending with the suffix "-ose."

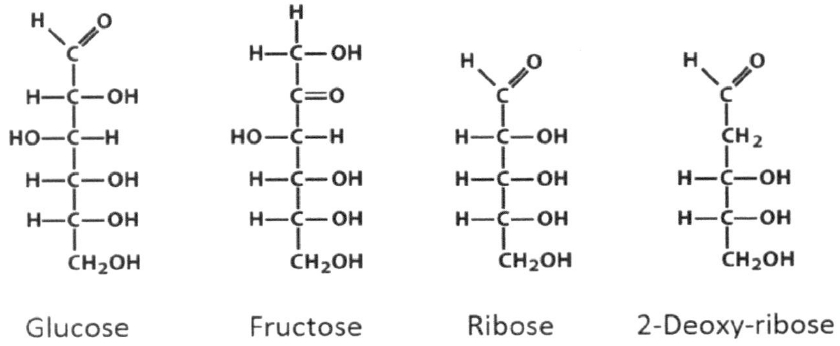

FIGURE 1.9 Representative monosaccharides. Two common hexoses, glucose and fructose. The pentose components of nucleic acids. Ribose is a component of ribonucleic acid (RNA) and 2-deoxy-ribose is a component of DNA.

The polysaccharides (also called glycans) are sugar polymers containing more than 20 monosaccharide units; some have hundreds or thousands of units. Some polysaccharides, such as cellulose, are linear chains; others, like glycogen, are branched. Both glycogen and cellulose consist of recurring units of D-glucose, but they differ in the type of glycosidic linkage and consequently have strikingly different properties and biological roles.

The most important storage polysaccharides are starch in plant cells and glycogen in animal cells. Starch storage is especially abundant in tubers (undergrounds stems), such as potatoes and in seeds. Starch and glycogen occur intracellularly as large clusters or granules. Their molecules are heavily hydrated because they have many exposed hydroxyl groups available to hydrogen bond with water.

Homopolysaccharides contain only a single monomeric unit; heteropolysaccharides contain two or more different kinds. Some homopolysaccharides serve as storage forms of monosaccharides that are used as fuels; starch and glycogen are homopolysaccharides of this type. Other homopolysaccharides (cellulose and chitin, for example) serve as structural elements in plant cell walls and animal exoskeletons.

Heteropolysaccharides provide extracellular support for organisms of all kingdoms. For example, the rigid layer of the bacterial cell wall (the peptidoglycan) is composed in part of a heteropolysaccharide built from two alternating monosaccharide units. In animal tissues, several types of heteropolysaccharides occupy the extracellular space forming a matrix that holds individual cells together and provides protection, shape, and support to cells, tissues, and organs.

Fats and oils (lipids) are the principal stored forms of energy in many organisms. Phospholipids and sterols are major structural elements of biological membranes. Other lipids, although present in relatively small quantities, play key roles as enzyme cofactors, electron carriers, light-absorbing pigments, hydrophobic anchors for proteins, molecules to help membrane proteins fold, emulsifying agents in the digestive tract, hormones, and intracellular messengers. Fats and oils in living organisms are derivatives of fatty acids. Fatty acids are hydrocarbon derivatives, at about the same low oxidation state (i.e., as highly reduced) as the hydrocarbons in fossil fuels. Fatty acids are carboxylic acids with hydrocarbon chains ranging from 4 to 36 carbons long. In some fatty acids, this chain is unbranched and fully saturated (contains no double bonds); in others, the chain contains one or more double bonds. A few contain three-carbon rings, hydroxyl groups, or methyl group branches. The simplest lipids constructed from fatty acids are the triacylglycerols, also referred to as triglycerides, fats, or neutral fats. Triacylglycerols are composed of three fatty acids each in ester linkage with a single glycerol (Fig. 1.10). The cellular oxidation of fatty acids (to CO_2 and H_2O) is highly exergonic.

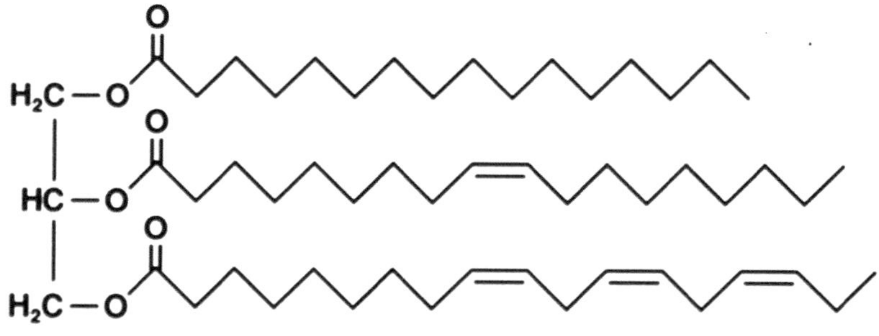

FIGURE 1.10 A triacylglycerol. It has three different fatty acids attached to the glycerol backbone.

The *nucleic acids*, DNA and RNA, are polymers of nucleotides. They store and transmit genetic information, and some RNA molecules have structural and catalytic roles. The amino acid sequence of every protein in a cell, and the nucleotide sequence of every RNA is specified by a nucleotide sequence in the cell's DNA. A segment of a DNA molecule that contains the information required for the synthesis of a functional biological product, whether protein or RNA, is referred to as a gene. A cell typically has many thousands of genes, and DNA molecules tend to be very large. The structure of every protein, and ultimately of every biomolecule and cellular component, is a product of information programmed into the nucleotide sequence of cell's DNA. The ability to store and transmit genetic information from one generation to the next is a fundamental condition for life. The storage and transmission of biological information are the only known functions of DNA.

Despite the near-perfect accuracy of genetic replication, infrequent, unrepaired mistakes in the DNA replication process lead to changes in the nucleotide sequence of DNA, producing a genetic mutation and changing the instructions for a cellular component. Incorrectly repaired damage to one of the DNA strands has the same effect. Mutations in the DNA passed on to offspring—that is, mutations carried in the reproductive cells—may be harmful or even lethal to the new organism or cell; they may, for example, cause the synthesis of a defective enzyme that is not able to catalyze an essential metabolic reaction. Occasionally, however, a mutation better equips an organism or cell to survive in its environment. The mutant enzyme might have acquired a slightly different specificity, for example, so that it is now able to use some compounds that the cell was previously unable to metabolize.

RNA has a broader range of functions than DNA, and several classes are found in cells. Ribosomal RNAs (rRNAs) are components of ribosomes, the structures that carry out the synthesis of proteins. Messenger RNAs (mRNAs) are intermediaries, carrying genetic information from one or a few genes to a ribosome, where the corresponding proteins can be synthesized. Transfer RNAs (tRNAs) are adapter molecules that translate the information contained in mRNAs into a specific sequence of amino acids. In addition to these major classes, there is a wide variety of RNAs with special functions.

Besides their roles as the subunits of nucleic acids, nucleotides have many other functions in every cell as energy carriers, components of enzyme cofactors, and chemical messengers.

1.3 ORGANISMS RESPONSIBLE OF BIODETERIORATION

1.3.1 ARCHAEA

The domain Archaea (single-celled microorganisms) has been recognized as a major domain of life only in the 1980s. The extreme conditions under which many species live have made them difficult to culture, so their taxonomical unicity went unrecognized for a long time.

Like bacteria, archaea have a prokaryotic cell. However, archaeal cell membranes are made of molecules that are distinctly different from those of bacteria. Bacteria and eukaryotes have membranes composed mainly of glycerol–ester lipids, whereas archaea have membranes composed of glycerol–ether lipids (Fig. 1.11). The difference is the type of bond that joins the lipids to the glycerol component. Ether bonds are chemically more resistant and stable than ester bonds. This stability might help archaea survive extreme temperatures and very acidic or alkaline environments.

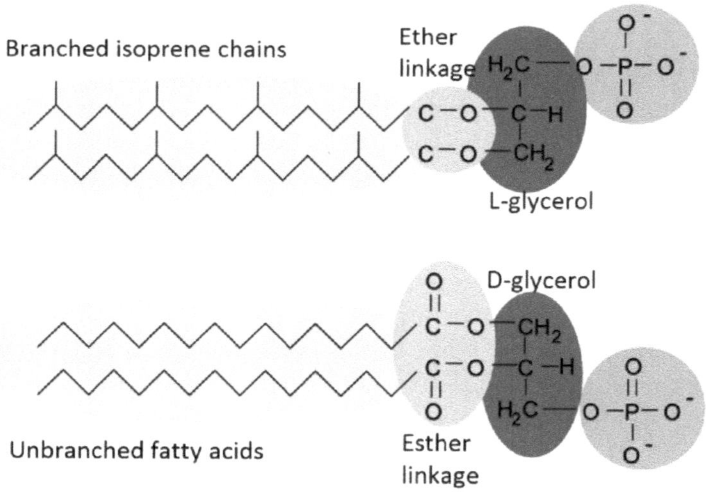

FIGURE 1.11 Like bacteria and eukaryotes, archaea possess glycerol-based phospholipids in the cell membrane. However, archaeal lipids are different. (i) They have an L-glycerol, while bacteria and eukaryotes have a D-glycerol. (ii) Most bacteria and eukaryotes have membranes composed of glycerol–ester lipids, whereas archaea have membranes composed of glycerol–ether lipids. (iii) The hydrophobic side-chains in eukaryotes and bacteria are fatty acids, while the chains in archaea are made up of isoprene that can be branched and joined in a greater number of ways providing much more variability to archaea's membranes. (From Wikimedia Commons).

While the side chains in the phospholipids of bacteria and eukaryotes are fatty acids with chains of usually 16–18 carbon atoms, archaea do not have fatty acids in their membrane phospholipids. Instead, they have side chains of 20 carbon atoms built from isoprene. Isoprene is the simplest member of a class of chemicals called terpenes. A terpene is a molecule built by connecting isoprene molecules together. The terpenes include beta-carotene (a vitamin), natural and synthetic rubbers, plant essential oils (such as spearmint), and steroid hormones (such as estrogen and testosterone).

The isoprene side-branched chains can form rings of five carbon atoms. This property may help prevent archaeal membranes from leaking at high temperatures. Archaea lack peptidoglycan in the cell wall except for one group of methanogens that contains pseudo peptidoglycan.

In most archaea, surface-layer proteins, known as an S layer, and polysaccharides compose the cell wall. In *Halobacterium*, the proteins have a high content of acidic amino acids giving the wall an overall negative charge. The result is an unstable structure that is stabilized by the presence of positive sodium ions that neutralize the charge. Thus, *Halobacterium* grows in environments with high salinity contents. A second type of archaeal cell wall is composed entirely of a thick layer of polysaccharides.

Archaea, unlike other known organisms, are able to produce methane. Methanogenic archaea play a pivotal role in ecosystems where bacteria derive energy from oxidation of methane.

Archaea inhabit some of the extreme environments on the planet. Some live near thermal vents at temperatures well over 100°C (Fig. 1.12). Others live in hot springs, or in extremely alkaline or acid waters. They live in the anoxic muds of marshes and at the bottom of the oceans, and even thrive in petroleum deposits deep underground. Some archaea can survive the desiccating effects of extremely saline waters. They may be extremely abundant in environments that are hostile to all other life forms. However, archaea are not restricted to extreme environments; they are also quite abundant in marine plankton.

1.3.2 BACTERIA

Bacteria have a prokaryotic cell. The plasma membrane is surrounded by a cell wall in all bacteria except one group, the Mollicutes, which includes pathogens such as the mycoplasmas. The cell wall is composed of peptidoglycan (also known as murein, a polymer consisting of sugars and amino

(a)

(b)

FIGURE 1.12 View of two hot springs ((a) Grand Prismatic Spring and (b) Morning Glory) in Yellowstone National Park (USA). Thermophilic ("heat-loving") prokaryotes thrive in such hot springs. The brownish orange color is due to the carotenoid pigments of thickly growing thermophiles.

acids) and gives bacteria structural support. The composition of the cell wall varies among species and is an important character for identifying and classifying bacteria. Heterotrophic bacteria can be either Gram-positive (e.g., *Bacillus*) or Gram-negative (e.g., *Pseudomonas*). Biologist Hans Christian Gram invented a method to differentiate two types of bacteria that have structural differences in the cell walls. Some bacteria, called Gram-positive bacteria, retain the crystal violet dye because of a thick layer of peptidoglycan in their cell wall (Fig. 1.13). In contrast, Gram-negative bacteria do not retain the violet dye and are colored red or pink because of a thinner peptidoglycan layer sandwiched between the inner cytoplasmic cell membrane and the outer membrane (Fig. 1.13). Thus, the cell wall of Gram-positive bacteria consists primarily of the relatively uniform peptidoglycan-based layer. In contrast, the cell wall of Gram-negative bacteria presents as a highly organized outer membrane in which an asymmetrical bilayer of phospholipid and lipopolysaccharide constitutes a permeability barrier. Diffusion pores, formed of aggregates of internal proteins, connect the periplasm of the cell to the external environment.[4] Compared with Gram-positive bacteria, Gram-negative bacteria are more resistant against some chemicals because of their impenetrable cell wall. Most bacteria have the Gram-negative cell wall and only the Firmicutes and Actinobacteria have the Gram-positive arrangement.

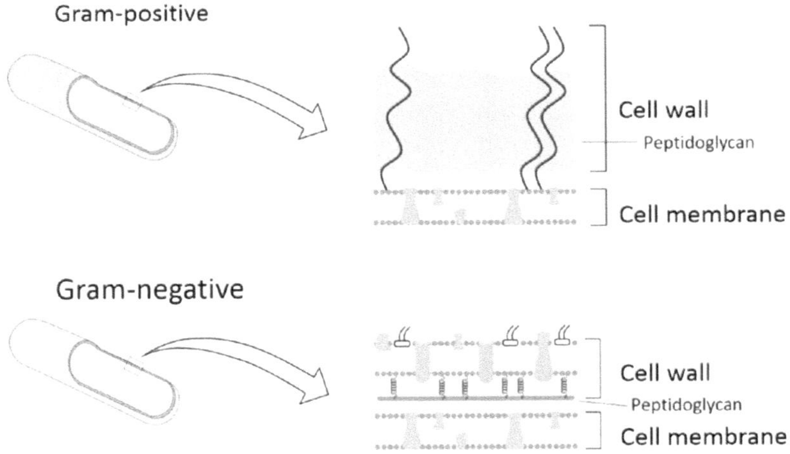

FIGURE 1.13 Comparison of the cell wall of Gram-positive (no outer membrane, thick peptidoglycan layer) and Gram-negative (outer membrane, peptidoglycan layer) bacteria (From Motifolio Inc.).

Bacteria can be aerobic, anaerobic, or facultative anaerobic based on their response to gaseous oxygen. Aerobic bacteria thrive in the presence of oxygen that is required for their growth. Anaerobic bacteria do not tolerate gaseous oxygen. The third group, the facultative anaerobes, can survive in presence or absence of oxygen.

A relevant characteristic of bacteria is their capacity of rapid growth and reproduction. Many species tolerate a wide range of environmental conditions: temperature, pH, salinity, etc. They can also survive adverse conditions by forming endospores.

Some aerobic bacteria became the mitochondria of modern eukaryotes, and some photosynthetic cyanobacteria became the plastids, the likely ancestors of modern plant cells.

Autotrophic bacteria fix inorganic carbon dioxide to produce the organic compounds (carbohydrates, lipids, proteins) they need to grow. The biochemical process uses energy from light (photoautotrophic) or from inorganic chemical reactions such as oxidation of nitrogen, sulfur, or other elements (chemoautotrophic). Photoautotrophs include cyanobacteria (described below) and sulfur bacteria that use hydrogen sulfide as hydrogen donor instead of water like most other photosynthetic organisms.

Bacteria develop in a wide variety of habitats and conditions and play important roles in the global ecosystem. They are involved in the cycling of nutrients such as carbon, nitrogen, and sulfur. When organisms die, the carbon contained in their tissues becomes unavailable for most living organisms. Decomposition, that is the release of nutrients back into the environment, is one of the most important roles of bacteria.

The cycling of nitrogen is another relevant activity of bacteria. Plants rely on nitrogen from the soil for their growth as they cannot utilize atmospheric nitrogen. The primary way in which nitrogen becomes available to them is through nitrogen fixation by bacteria. They convert gaseous nitrogen into nitrates or nitrites as part of their metabolism. Denitrifying bacteria metabolize in the reverse direction, turning nitrates into nitrogen gas or nitrous oxide.

Actinobacteria include a wide range of morphologies from coccoid or fragmenting hyphal forms (e.g., *Nocardia*) to branching filaments (e.g., *Streptomyces*) that resemble the mycelia of fungi (*vide infra*). Most actinobacteria are found in the soil where they play important roles in decomposition and humus formation. *Streptomyces* can develop on surfaces in caves despite the unfavorable conditions of these environments.[5]

Cyanobacteria are a group of bacteria capable of producing energy through photosynthesis (Fig. 1.14). The characteristic blue-green color of their cells is due to the photosynthetic pigment composition: chlorophyll *a* (greenish pigment), carotenoids (yellow-orange pigments), and phycobiliproteins that include phycocyanin, responsible for the blue color.[6] However, not all cyanobacteria are blue; some species are red or pink for the pigment phycoerythrin present in the cells.

(a)

(b) (c)

FIGURE 1.14 A tomb made up of marble in the English Cemetery, Firenze (Italy). It is covered by a black biofilm (a). A sample taken from the biofilm consists of fungi, green algae, cyanobacteria, and extracellular polymeric substances (b). Cyanobacterium *Gloeocapsa* sp. Optical microscope, visible light imaging (c).

In addition to the multilayered cell wall, a gelatinous sheath permits the adhesion of bacteria to the substrate. It is the cement that holds together the cells and form biofilms on the substrates. All cyanobacteria are unicellular, though many grow in colonies or filaments. The cell may be filled with tightly packed folds of its outer membrane to increase the surface area on which photosynthesis takes place. Fossilized cyanobacterial mats composed of layers of microorganisms and trapped sediments are called stromatolites.

Cyanobacteria can be found in almost every terrestrial and aquatic habitat—oceans, fresh water, damp soil, deserts, bare rocks, and even Antarctic rocks. Cyanobacteria often occur in association with algae. They form biofilms on rock surfaces that are deep or bright green under humid conditions and deep black when dry.

They are very important for the health and growth of many plants because they convert inert atmospheric nitrogen into an organic form, such as nitrate or ammonia. Plants need these fixed forms of nitrogen for their growth.

Cyanobacteria played an important role in the ecological changes of Earth's history because they converted the early reducing atmosphere into an oxidizing one, generating the oxygen atmosphere and changing the composition of life forms on Earth.

1.3.3 ALGAE

Algae are included in the Kingdom Protista. They have eukaryotic cells, exhibiting a full range of pigments, highly specific reserve materials and wall components used as discriminating elements for systematic classification. As cyanobacteria, they are autotrophs, obtaining the energy through photosynthesis. Their cell wall contains either polysaccharides (such as cellulose) or a variety of glycoproteins or both. Additional polysaccharides (mannans, alginic acid, sulfonated polysaccharides, etc.) are used as a feature for algal taxonomy.

Algae include unicellular species—one single cell covers all the vital functions—as well as multicellular species. Unicellular species live isolated or form colonies of different shapes and structures. Algae widely differ in dimensions. Unicellular species are microscopic ($\leq$10 μm), while multicellular organisms are mostly macroscopic and may reach a length of several meters in the marine forms.

Most algae that colonize cultural heritage objects belong to the group Green algae. Although most of them are aquatic, they are found in a wide variety of habitats, including in desert microbiotic crusts worldwide. Many green algae are microscopic although some of the marine species are large.

The members of the class Chlorophyceae live mainly in fresh water although a few unicellular species are essential terrestrial living in soil, on rocks, and on tree branches and trunks. Within this class, the genus *Chlorococcum* is worth mentioning because it includes many of the unicellular green algae occurring singly or in a layer on outdoor cultural heritage objects. These algae reproduce only by spores.

Protists of the group known as diatoms have a cell wall made of silica that is called frustule. It consists of two overlapping halves. The biogenic silica is synthesized intracellularly by the polymerization of silicic acid monomers. This material is then secreted and added to the wall. The diatoms are unicellular or colonial organisms that are important components of the phytoplankton.

1.3.4 FUNGI

Fungi are eukaryotic organisms classified as a Kingdom. They are heterotrophs, that is, rely solely on carbon fixed by other organisms. They have a rigid cell wall composed of chitin (acetyl glucosamine polymer), glucans (glucose polymers), and proteins. There is no cellulose in their cell walls. The metabolic versatility of this group of microorganisms enhances their efficiency to colonize very different kinds of substrata (wood, glass, stone, and paper).

They have different structures ranging from unicellular to very complex multicellular organisms. Branched filaments, called hyphae, which originate from the germination of the fungal spore, form most fungi (Fig. 1.15). Hyphae are cylindrical, thread-like structures, 2–10 μm in diameter and up to several centimeters in length. They grow at the tip and divide repeatedly along their length creating long and branching chains. The combination of apical growth and branching/forking of hyphae leads to the development of a mycelium, an interconnected network of hyphae.[7] Fungi reproduce and are dispersed by means of spores that are produced by specialized structures. Spores have a protective coat that shields them from harsh environmental conditions such as drying out and high temperatures. Wind, rain, or animals (including humans) spread spores. Upon reaching a suitable substrate, they germinate and produce new hyphae.

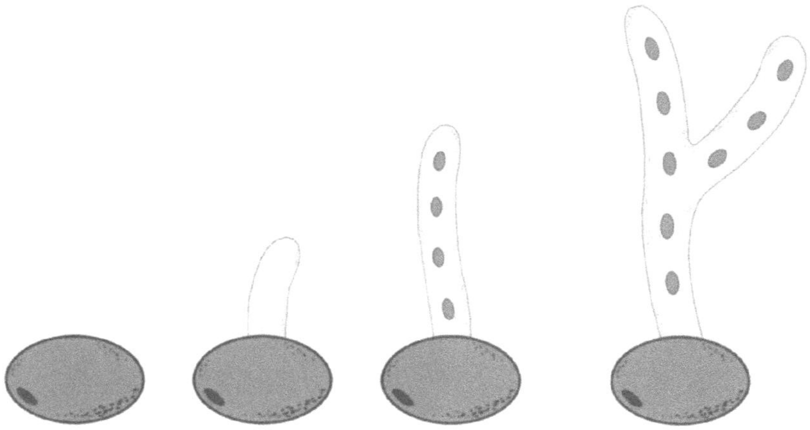

FIGURE 1.15 Growth of a fungal hypha from a spore.

Fungi play an important role in carbon, nitrogen, phosphorus, and other biogeochemical cycles.

Among the most common fungal colonizers of stones, there are the so-called *black meristematic fungi*. They are commonly isolated from the sun-exposed surfaces in Mediterranean countries as well as from dry and cold climates. They represent a wide and heterogeneous group of black pigmented fungi having in common the presence of melanin within the cells. They are also called "rock-inhabiting fungi" to underline the exclusive isolation of many strains from rock surfaces. Many rock-dwelling fungi contain melanin whose implication in fungal protection from metal toxicity has long been established.[8]

Yeasts are unicellular fungi, although some species may also develop multicellular characteristics by forming strings of connected budding cells known as pseudo-hyphae or false hyphae. Yeast sizes vary greatly, depending on species and environment, typically measuring 3–4 μm in diameter. Most yeasts reproduce asexually by the asymmetric division process known as budding. By fermentation, the yeast species *Saccharomyces cerevisiae* (called "ale yeast") converts carbohydrates to carbon dioxide and alcohols. It is also a centrally important model organism in eukaryotic cell biology research.

Fungi produce organic acids that are chelating agents (Greek *chele*, "claw") with high affinity and specificity for metal ions (e.g., Ca^{2+}). Chelating agents contain two or more functional groups that act as electron donors and form several bonds to a single metal ion. Metal ions in solution do not exist in isolation, but in combination with ligands or chelating groups. This combination gives rise to chelates, called also complexes or coordination compounds. Therefore, the chelates contain a central metal ion and a cluster of ions or molecules, called ligands, surrounding it in a cyclic or ring structure. The total number of points of attachment to the central metal ion is termed "coordination number." It can vary from 2 to >12, but is usually 6. Chelates are more stable than nonchelated compounds of comparable composition, and the more extensive the chelation—that is, the larger the number of ring closures to a metal atom—the more stable the compound.

Chelates play important roles in photosynthesis and in oxygen transport in living organisms. Chlorophyll is a chelate that consists of a central magnesium atom joined with four complex chelating agents (pyrrole ring). The molecular structure of the chlorophyll is similar to that of the heme bound to proteins to form hemoglobin, except that the latter contains an iron(II) ion in the center of the porphyrin. Heme is an iron chelate. Furthermore, many biological catalysts (enzymes) are chelates.

One of the principal reasons for fungal success in biogeochemical processes in terrestrial ecosystems is their ability to form mutualistic symbiotic associations with photosynthetic plants, algae, and cyanobacteria (mycorrhizas and lichens), which make them responsible for major transformations and redistribution of inorganic nutrients. A mycorrhiza is a symbiotic association formed by a fungus and the roots of a vascular plant. The hyphae form a closely woven mass around the rootlets or penetrate the cells of the root.

1.3.4.1 LICHENS

A lichen is a mutualistic symbiotic association between green algae or cyanobacteria and fungi. The mutual life form has properties very different from those of its component organisms growing separately. The fungal component of the lichen is called mycobiont while the photosynthesizing organism is called photobiont. Photobionts are green algae and cyanobacteria. Most lichens are associated to green algae. The photobiont produces the nutrients needed by the entire lichen. The nonreproductive tissue, or vegetative body part, is called thallus. Its upper surface is the cortex. Lichens come in many colors (white, black, red, orange, brown, yellow, and green), shapes, and forms. Three major growth forms are recognized: crustose that is flattened and adheres tightly to the substrate (Fig. 1.16); foliose with flat leaf-like lobes that lift from the surface and fruticose that has tiny, multiple leafless branches three-dimensional growing (Fig 1.16). Anchoring hyphae, called rhizines, attach the foliose lichens to the substrate.

FIGURE 1.16 Marialva Castle (Portugal). Crustose and fruticose lichens colonize a granite block. Photo courtesy of José Delgado Rodrigues.

Nearly all lichens have an outer cortex, which is a dense, protective layer of hyphae. Below it, there is the photosynthetic layer (the photobiont). Then, a layer of loose hyphae, the medulla, is present (Fig. 1.17). Foliose lichens have a lower cortex, while crustose lichens attach to the substrate through the medulla.

Mycobionts produce large numbers of secondary metabolites, the lichen acids, which play a role in the biogeochemical weathering of rocks and in soil formation.

As lichens have no means of excreting the elements that they absorb, they are particularly sensitive to toxic compounds. Therefore, they are good indicators of the toxic components of polluted air (e.g., sulfur oxide) and are increasingly being used to monitor atmospheric pollutants. Analysis of lichens can detect the distribution of heavy metals and other pollutants around industrial cities.

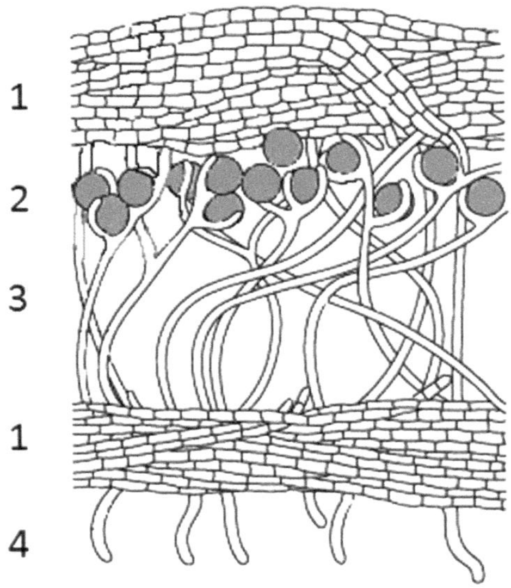

FIGURE 1.17 Schematic cross-section of a foliose lichen. Cortex, the outer layer made of tightly woven hyphae (1). Photobiont layer with photosynthesizing green algae (2). Loosely packed hyphae in the medulla (3). A tightly woven lower cortex (1), with anchoring hyphae called rhizines that attach the thallus to the substrate (4).

Lichens have a low slow rate of growth ranging from 0.1 to 10 mm a year.

Many species are used as sources of dyes. Others have also been used as medicines, components of perfumes, or minor sources of food. Some species are being investigated for their ability to secrete antitumor compounds.

Lichens can reproduce by simple fragmentation producing special powdery propagules known as soredia or small outgrowths called isidia. These fragments, which contain both fungal hyphae and algae or cyanobacteria, act as small dispersal units to establish the lichen in new places.

The mycobiont produces the fruiting bodies, which are spore-producing structures. Nearly all mycobionts are members of the phylum Ascomycota—commonly known as Ascomycetes, though a few are members of

Basidiomycota—commonly called Basidiomycetes. The most common sexual fruiting bodies in Ascomycetes lichens are the apothecium (plural, apothecia) and the perithecium (plural, perithecia) (Fig. 1.18). Apothecia typically are open structures, cup or saucer-shaped, often with a distinct rim around the edge. The hymenium, the tissue containing the asci, forms the disc, the upper surface of the apothecium (Fig. 1.18). Two basic types of apothecium are recognized in lichens, differing in their margins and underside (together named "the exciple") (Fig. 1.18). In lecanorine apothecia, the thallus tissue extends up the outside of the apothecium to form the exciple and the rim. This margin generally retains the color of the thallus and normally contains algal cells. In lecideine apothecia, the exciple is part of the true apothecial tissue and does not contain algal cells (Fig. 1.18).

Discs that become very contorted and appear as cracks in the surface of the thallus are called lyrellae (Fig. 1.19). Lyrellae are apothecia that are long or at least elongated, narrow, often branched, often with hard, black outer margins.

Perithecia are usually flask-shaped fruiting bodies that contain the asci. At maturity, an opening at the top, the ostiole, allows release of the ascospores. Perithecia are usually partially immersed in the thallus or in the substrate and are relatively inconspicuous, rarely more than 1 mm in diameter and commonly much less (Fig. 1.18).

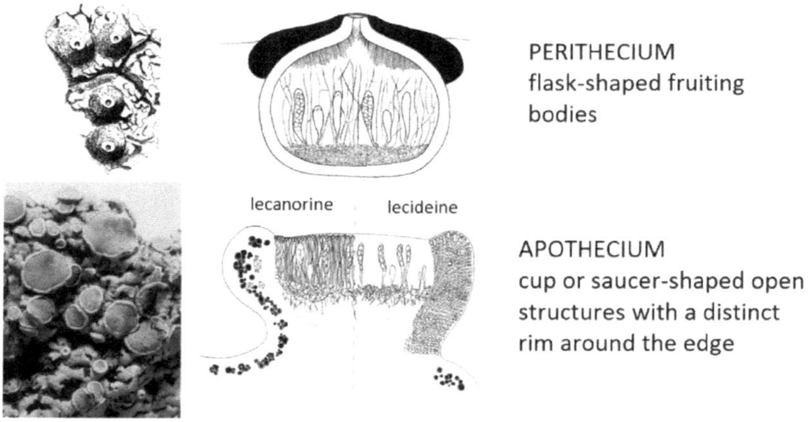

FIGURE 1.18 Diagram of a perithecium and an apothecium, the spore-producing structures of the lichens. (From Caneva, G.; Nugari, M. P.; Salvadori, O. *Plant Biology for Cultural Heritage: Biodeterioration and Conservation.* Getty Conservation Institute: Los Angeles, 2008). Used with permission from Nardini Editore, Firenze.

FIGURE 1.19 Lyrellae of the lichen *Opegrapha atra* (From Nimis & Martellos, 2008[9]). Used with permission.

Lichens grow in some of the extreme environments on Earth—arctic tundra, hot deserts, rocky coasts—and thus they are extremely widespread. They can survive under environmental conditions so severe that exclude any other form of life. This characteristic seems related to their capacity of drying out very rapidly. In this state, photosynthesis ceases. Therefore, lichens can endure even blazing sunlight or great extreme of heat or cold.

Lichens grow on tree barks, leaves, and branches in rain forests, on bare rock and on exposed soil surfaces. The growth form most frequently encountered on rocks is the crustose. The thallus develops upon the surface (epilithic lichens) or inside, in the bulk of the rock (endolithic lichens). Endolithic lichens have only the fruiting part visible growing outside the rock (Fig. 1.20). At times, they are difficult to spot because often they have the same color as the stone. Although endolithic lichens frequently colonize limestone monuments, they received little attention for a long time by the scientific community.

FIGURE 1.20 Natural rock (Altamura, Italy) completely covered by epilithic and endolithic lichens (arrows). The black dots correspond to the fruiting bodies (perithecia of the endolithic lichens) embedded in the stone.

1.3.5 BRYOPHYTES

Bryophytes are a group of nonvascular plants that include mosses (phylum Bryophyta), liverworts (phylum Marchantiophyta) and hornworts (phylum Anthocerotophyta). Bryophytes are small plants that grow closely packed together in mats or cushions on rocks, soil, or as epiphytes on the trunks of trees. Bryophytes are distinguished from vascular plants (tracheophytes[‡]) by several characteristics. Among them, two are relevant. First, in all bryophytes the dominant, photosynthetic phase of the life cycle is the haploid gametophyte[§] generation rather than the diploid sporophyte[¶]; bryophyte sporophytes are very short lived. They are nutritionally dependent on their gametophytes and consist of only an unbranched stalk, or seta, and

[‡]Tracheophyte: Any plants including ferns, conifers, and flowering plants that have vascular tissues (xylem and phloem).
[§]Gametophyte: A plant (or the haploid phase in its life cycle) that produces gametes by mitosis to form a zygote.
[¶]Sporophyte: A plant (or the diploid phase in its life cycle) that produces spores by meiosis to form gametophytes.

a single, terminal sporangium (Fig. 1.21). Second, bryophytes never form vascular tissues (xylem and phloem), the specialized conducting tissues that are found in all vascular plants.

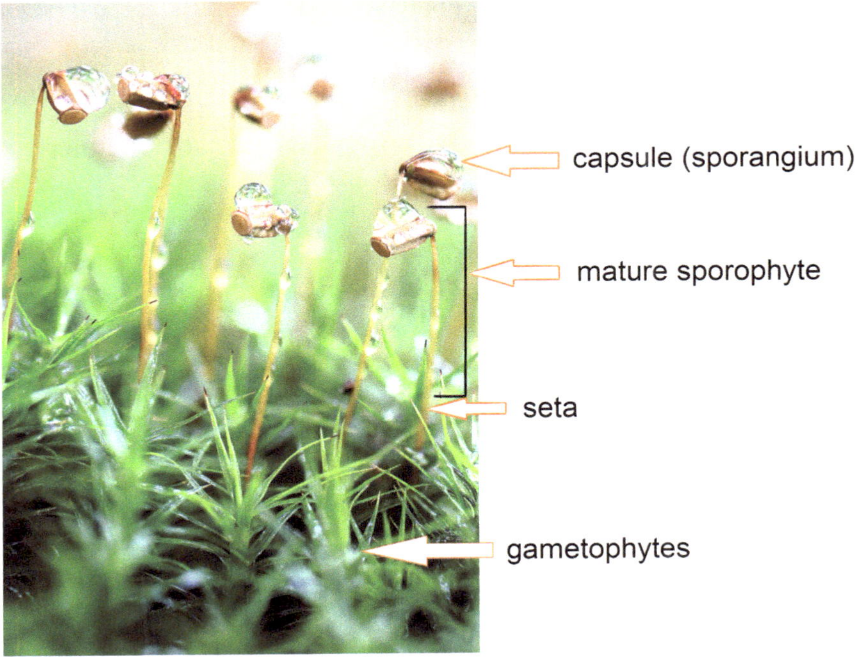

FIGURE 1.21 A common growth form, the cushiony form, found in the gametophytes of different genera of mosses. The gametophytes are erected and have few branches. Sporophytes are composed of a spore capsule atop a long seta rising above the gametophytes.

Mosses are small, soft plants, typically 1–10 cm tall. A moss begins its life cycle when haploid spores are released from the sporophyte capsule where they are produced. Spores landing on a moist substrate begin to germinate. From the one-celled spore, a highly-branched system of filaments, called the protonema, develops. Cell specialization occurs within the protonema to form a horizontal system of reddish-brown, anchoring filaments, called caulonemal filaments and upright, green filaments, called chloronemal filaments. As the protonema grows, some cells specialize to form the adult gametophyte stems. Other cells produce leaves in spiral arrangement on an elongating stem. The leaves have a lamina that is only one-cell layer thick. Near the base of the moss, reddish-brown, multicellular thread-like rhizoids emerge from the stems to anchor the moss to the substrate. Water and mineral nutrients required for the moss to grow are absorbed not by the rhizoids, but by the thin leaves of the plant as rainwater washes through the moss cushion.

Mosses reproduce through spores, not seeds, and have no flowers. Male sex structures, called antheridia, are produced in clusters at the tips of shoots or branches on the male plants. Female sex structures, the archegonia, are produced in a similar way on female plants. The closely packed arrangement of the individual moss plants greatly facilitates fertilization. The sporophyte (i.e., the diploid multicellular generation) stem is called seta (Fig. 1.21). The formation of spore-bearing capsules or sporangia is the final step of sporophyte growth (Fig. 1.21).

Mosses are found throughout the world in a variety of habitats. They flourish particularly well in moist, humid forests. Mosses play important roles in reducing erosion along streams, in capturing and recycling nutrients, and insulating the arctic permafrost.

1.3.6 VASCULAR PLANTS

Vascular plants have vascular tissues that circulate substances through the organism. This feature allows them to evolve to a larger size than nonvascular plants, which lack these specialized conducting tissues and are therefore restricted to relatively small sizes.

The different kinds of cells of the plant are organized into tissues and the tissues are organized into units called tissue systems. Three tissues systems—derma, vascular, and ground—that occur in all organs of the plant—are continuous from organ to organ and reveal the basic unit of the plant body. The dermal tissue system forms the outer, protective covering of the plant. The vascular tissue system is composed of two types of conductive tissues, *xylem* and *phloem* and is embedded in the ground tissue system. Xylem conducts water and minerals upward from the roots of a plant, while phloem transports sugars and other nutrients from the leaves to the other parts of the plant. Tracheary elements, the conducting cells of the xylem, have distinctive lignified wall thickenings, while sieve elements, the conducting cells of the phloem, have soft walls. In most gymnosperms, the water-conducting cells of xylem, called tracheids, are elongated cells with long, tapering ends. Besides channeling water and minerals, they also provide supports for stems. Tracheids are more primitive (i.e., less specialized) than vessel elements that are the principal water-conducting cells in angiosperms.

Vascular plants consist of *roots*, *stems*, and *leaves*. Roots serve in anchorage and absorption of water and minerals from the soil; stems provide support for the principal photosynthetic organs, the leaves, which are well suited to the acquisition of energy from sunlight, of carbon dioxide from the atmosphere, and of water. The stems and leaves together form the shoot system (Fig. 1.22). In younger plants and in annuals (plants with a life span of 1 year), the stem is also a photosynthetic organ. It can change becoming thick, woody, and covered with cork in longer lived plants.

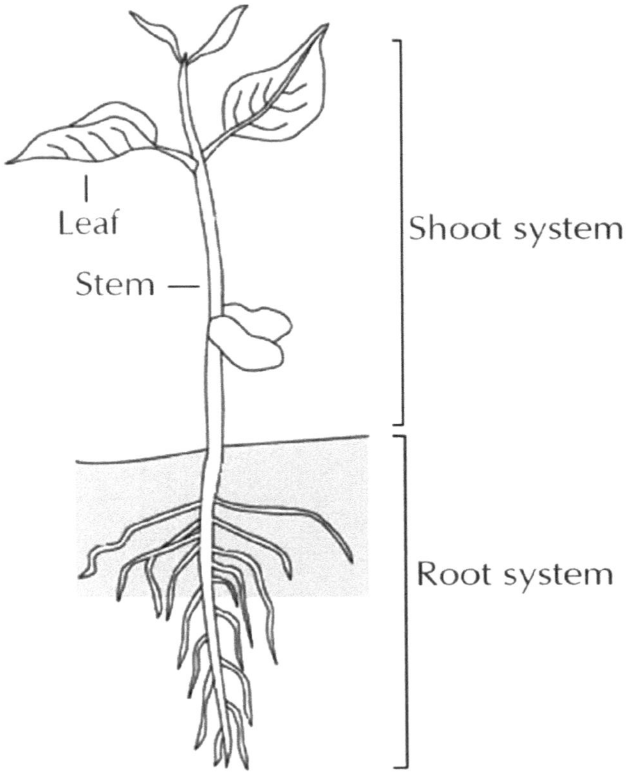

FIGURE 1.22 Diagram of a young plant showing the principal organs—root, stem, and leaf. The roots make up the root system, and the stems and leaves make up the shoot system of the plant.

Plants continue to grow during their lives. They have meristems that are embryonic tissues capable of adding cells indefinitely to the plant body. Meristems are located at the tips of roots and shoots. Therefore, the roots are continuously reaching new sources of water and minerals, and the photosynthetic organs are continuously extending toward the light. This type of growth is called primary growth. There is also the growth that results in the thickening of stems and roots, the so-called secondary growth. It originates from two lateral meristems, the vascular cambium and the cork cambium.

Regarding roots, the taproot system includes a large, central, and dominant root that grows directly downward. It gives rise to branch roots, or lateral roots (Fig. 1.23a). In some plants, like the carrot, sugar beet, and sweet potato, the taproots developed as a storage organ. The taproot system diverges from the fibrous root system because this has many branched roots arising from the stem (Figs. 1.23b and 1.24). No one root is more prominent than the others are. These roots are commonly called adventitious roots.

(a) (b)

FIGURE 1.23 Development of root in a plant. Taproot system (a) and fibrous root system (b).

FIGURE 1.24 Planting Fields Arboretum State Historic Park (Long Island, New York, USA). A plant uprooted by hurricane Sandy. The fibrous root system is visible.

The depth to which the roots penetrate the soil and the distance they spread laterally depend on many factors. Taproot systems generally penetrate deeper into the soil than fibrous root systems. The characteristics of the latter system make the plants having it well suited for the prevention of soil erosion. Some trees (e.g., spruces, beeches, and poplars) rarely produce deep taproots, whereas others (e.g., oaks and many pines) produce relatively deep taproots.

The shoot, which is composed of the stem and the leaves, is structurally more complex than the roots. It has nodes and internodes, with one or more leaves attached at each node. The two principal functions associated with stems are support and conduction. The leaves are supported by the stem. Stems and leaves may undergo modifications performing functions quite different from those commonly associated with them. In some plants, the leaves are modified as spines. Thorns are instead modified branches that arise in the axils[**] of leaves. The so-called thorns on rose stems are prickles, that is, small sharp outgrowths from the cortex and epidermis.

Stems, like roots, have food-storage functions as well. Tubers, for example, may arise at the tips of stolons, which are slender stems growing along the surface of the ground. The tubers may arise also at the end of long, thin rhizomes, which are underground stems. A bulb is a large bud that consists of a small conical stem with numerous modified (scale-like) leaves attached to it. Familiar examples are the onion and the lily. Some plants like succulents have stems and leaves specialized for water storage. Most of them (e.g., the cacti, the euphorbias, the agave) normally grow in arid regions where the ability to store water is essential for their survival. The green, fleshy stems of the cacti are both photosynthetic and storage organs.

The reproductive cells of most vascular plants are enclosed within multicellular protective structures, the *seeds*. They provide the embryonic sporophyte with nutrients and protect it from unfavorable environmental conditions.

The vascular plants that produce only one kind of spore are called homosporous. Homospory is found in almost all ferns, the horsetails (equisetophytes), and some of the lycophytes. Heterospory (the production of two types of spores in two different kinds of sporangia) is found in some of the lycophytes and ferns, and in all seed plants. The reproductive structures of the sporophyte (cones in gymnosperms and flowers in angiosperms), produce two different kinds of haploid spores called microspores and megaspores. They are produced in microsporangia and megasporangia, respectively. The two types of spores differ for their function and not necessarily relative size. Microspores give rise to male gametophytes while megaspores give rise to female gametophytes. The evolution of the gametophyte in vascular plant is characterized by a trend toward reduction in size and complexity, and the gametophytes of flowering plants, the angiosperms, are the most reduced. In angiosperms and in most gymnosperms, male gametophytes, the pollen grains, are carried close to the female gametophytes. The transfer of the pollen is called pollination. Germination of the pollen grains produces special structures called pollen tubes through which the sperm is transferred to the egg to achieve fertilization. In fertilization, the male gamete and the female gamete join to form

[**]The angle between a leaf and the axis from which it arises.

a zygote. The resulting embryo, encased in a seed coating, will eventually become a new sporophyte (the diploid stage).

Seedless vascular plants (ferns, horsetails—equisetophytes, lycophytes), which reproduce and spread through spores, are plants that contain vascular tissue, but do not produce flower or seed. Although they have evolved to spread to all types of habitats, they still depend on water during fertilization, as the sperm needs a layer of moisture to reach the egg. Therefore, they must grow in habitats where water is at least occasionally abundant. The life cycle of seedless vascular plants is an alternation of generations, where the diploid sporophyte alternates with the haploid gametophyte phase. The diploid sporophyte is the dominant phase of the life cycle, while the gametophyte is an inconspicuous, but still-independent, organism.

The spermatophytes (gymnosperms and angiosperms) are instead seed plants.

The *gymnosperms* are a group of seed-bearing plants that includes conifers (pines, cypresses, and relatives), cycads, Gingko, and Gnetales. The term "gymnosperm" comes from the Greek words *gymnos* (naked) + *sperma* (seed), meaning "naked seed," for the unenclosed condition of their seeds (called ovules in their unfertilized state). Gymnosperm seeds either develop on the surface of scale- or leaf-like appendages of cones, or at the end of short stalks (Ginkgo).

Angiosperms are flowering plants with seeds or ovules enclosed in capsules during pollination. They make up the largest segment of plants on Earth today. The term "angiosperm" comes from the Greek words *angeîon* (receptacle) + *sperma* (seed), meaning "enclosed seed," for the enclosed condition of their seeds.

The flowers (Fig. 1.25), the reproductive organs of flowering plants, are the most remarkable feature distinguishing them from other seed plants. The essential structure of the flower is the carpel that contains the ovules. The ovules develop into seeds after fertilization, while the carpel develops into the fruit wall. In most flowers, the carpels are composed of three parts: the ovules; the style through which the pollen tubes grow; and the stigma that receives the pollen. The stamens are the pollen-bearing parts of the flower. They consist of a filament that bears the anther containing the pollen sacs. Usually, other structures are present—the sepals and petals. They protect the fertile parts of the flower (the stamens and carpels) and form an envelope attractive to pollinators. As flowers provided angiosperms with a wide range of adaptability, these plants largely dominate terrestrial ecosystems.

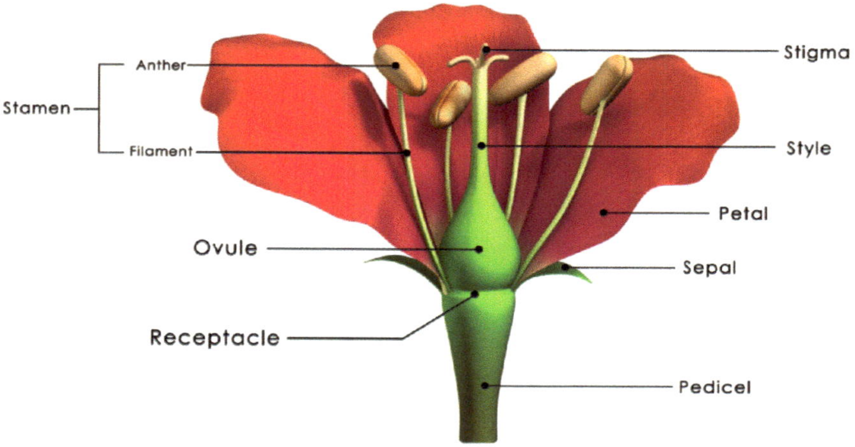

FIGURE 1.25 Diagram of the parts of a flower. The gynoecium is composed of the ovule, style, and stigma. Each stamen consists of a filament and an anther. The sepals, petals, and stamens are attached to the receptacle below the ovule.

An herbaceous plant (herb) is a plant with shoots that dies at the end of the growing season, that is, when the plant produces flowers, fruits, and seeds. It leaves its seeds on the soil; the seeds will produce new plants with good weather. Herbaceous plants do not have a woody stem. They may be annuals, biennials, or perennials. Annual herbaceous plants (therophytes) have a whole life cycle in 1 year and survive harsh seasons only as seeds. Biennial and perennial plants have stems that die at the end of the growing season too, but parts of the plant survive under or close to the ground from season to season (for biennials, until the next growing season, when they flower and die). New growth develops from living tissues remaining on (hemicryptophytes) or under (geophytes) the ground. The underground-living tissues include roots, a caudex (a thickened portion of the stem at ground level), or various types of stems, such as bulbs, corms, stolons, rhizomes, and tubers. Carrots are herbaceous biennials while potato, mint, most ferns, and most grasses are perennials.

Woody plants produce wood as their structural tissue. They have stems with hard lignified tissues above ground that remain alive during the dormant season and grow shoots the next year from the above ground parts. They are usually perennial trees, shrubs, or lianas. New layers of woody tissue reinforce each year their stems and roots, increasing their diameter with new wood deposited on the inner side of a vascular cambium layer located immediately beneath the bark. The growth of the vascular tissues form concentric rings because the secondary growth of the vascular cambium generates a regular periodical increase in thickness with the development of xylem on the inside and phloem on the outside. The soft phloem becomes crushed, but the hard wood persists and forms the bulk of the stem and branches of the woody perennials. Wood is primarily composed of xylem cells with cell walls made of cellulose and lignin.

KEYWORDS

- **prokaryotic cell**
- **eukaryotic cell**
- **photosynthesis**
- **proteins**
- **lipids**
- **polysaccharides**
- **nucleic acids**
- **archaea**
- **bacteria**
- **algae**
- **fungi**
- **lichens**
- **bryophytes**
- **vascular plants**

CHAPTER 2

ECOLOGY

CONTENTS

ABSTRACT

The growth of organisms on stones relates closely to the nature of the substrates as well as to the characteristic of the surrounding environment. The chapter contains an overview of the complex relationships among organisms, materials, and the environment. It emphasizes issues on bioreceptivity of stones and the factors influencing biological growth. Many characteristics make stones favorable or unfavorable to biological growth affecting their susceptibility to hold organisms.

2.1 BIORECEPTIVITY OF NATURAL AND ARTIFICIAL STONES INCLUDING WALL PAINTINGS

Natural stones, which are common components of many culturally important items (statues, tombstones, historic buildings, archaeological sites), encompass marble, limestone, sandstone, granite, limestone, volcanic tuff, etc., while artificial stones include plasters, mortars, stuccoes, bricks, concretes, and ceramics. Wall paintings, composed mainly of inorganic materials, are also included. In the history, they have been made using several techniques. The best known is *fresco* (from the Italian word *affresco* that derives from the adjective *fresco* "fresh"), which uses water-soluble pigments applied on wet plaster or lime mortar (*buon fresco*). The pigments are absorbed by the wet plaster. When drying, the plaster reacts with the air in a process called carbonation in which calcium hydroxide reacts with carbon dioxide and forms insoluble calcium carbonate; this reaction fixes the pigment particles in the plaster. *A secco fresco* uses dry plaster (*secco* means "dry" in Italian); a binding medium, like egg (tempera), glue, or oil, is needed to fix the pigment into the plaster. *Mezzo fresco* is painted on nearly dry plaster so that the pigments only penetrate slightly into the drying substrate. In Classical Greco-Roman times, the encaustic painting technique was in use. Pigments were ground in a molten beeswax binder (or resin binder) and applied to the surface while hot. Today, murals are mostly painted in oils, tempera, or acrylic colors.

The artificial stones vary greatly in composition and can contain high amounts of organic materials. Among artificial stones, it is worth mentioning glass that is a noncrystalline, amorphous inorganic solid. The most familiar types of glass are prepared by melting and rapidly cooling quartz sand, the raw crystalline material, and other ingredients. The high melting-temperature (1723°C) and viscosity of chemically pure silica (silicon dioxide—SiO_2) make it difficult to work with. Therefore, other substances are added to lower the melting temperature and improve the temperature workability. One is sodium carbonate (Na_2CO_3, "soda"), which lowers the glass transition temperature. The soda though makes the glass water soluble, which is usually undesirable, so lime (calcium oxide—CaO), magnesium oxide (MgO), and aluminum oxide (Al_2O_3) are added to provide a better chemical durability. The resulting glass contains about 70–74% silica by weight and is called a soda–lime glass. Soda–lime glass accounts for about 90% of manufactured glass. To color the glass, powdered metals are added to the mixture while the glass is still molten.

The medieval stained-glass windows that decorate many European churches were made using, besides sand, a different ingredient, the so-called potash that is wood ash (K_2O). The stained-glass windows are composed of pieces of colored or clear sheet glass, decorated by grisaille, a vitrified painting used to paint the outlines, the washed tones and the shading, and fitted into H-shaped strips of lead.[10]

Other kind of stone-related man-made objects are mosaics, which have both a natural and an artificial nature, being an assemblage of small pieces (tesserae) of different materials (natural stones, glass, tiles, etc.) inlayed in a mortar bedding.

Both natural and artificial stones differ in surface texture, hardness, porosity*, pH, and, of course chemical composition, characteristics that make them favorable or unfavorable to biological growth. The susceptibility of these materials to hold organisms and to biodeterioration as well is called *bioreceptivity*, a term coined by Guillitte.[11] In other words, it is the aptitude of a stone to be vulnerable to organisms' colonization. Guillitte further defined this concept: The primary bioreceptivity indicates the potential of a healthy material to be colonized; the secondary bioreceptivity is the result of the deterioration caused by nonbiotic and biotic factors; finally, the tertiary bioreceptivity is caused by nutrients contained in the stones (e.g., dead biomass, dust particles, animal feces, water repellents and consolidants, biocides, etc.). The literature reports studies conducted either in situ or under laboratory conditions for the assessment of bioreceptivity.

---

*Porosity is a measure of the void spaces in a material. It is the fraction of the volume of voids over the total volume of the material.

Generally, materials with a high porosity are more susceptible to biocolonization because of their capacity to absorb more water for a longer period.[12,13] Stone surface texture may affect the trapping of moisture and the rate of water run-off as well. Surface roughness concentrates moisture in microfissures where growth is usually more abundant.[14] Moreover, rougher surfaces can be a preferential site for colonization because they provide sheltered microhabitats. The textural properties of ignimbrite from hyper-arid Atacama Desert (Chile) determined differences in the endolithic colonization patterns of cyanobacteria.[15] The patterns related to porosity, pore size distribution, shape and orientation of pores, and color of different areas of the rock. A higher amount of micropores in one area likely allowed a more diffuse and deeper colonization because of a greater water retention capacity and reduced evaporation rate. In addition, the presence of water in micropores, especially those with translucent walls, can enhance light penetration increasing the light available for photosynthesis in the cryptoendolithic habitat.[15]

Although porosity is assumed as the most important aspect to determine bioreceptivity, some studies showed that the chemical composition of stone appears a very relevant factor, in some cases even more important. A study[16] on limestone, granite, and marble samples artificially inoculated with two photosynthetic microorganisms (the cyanobacterium *Gloeocapsa alpina* and the green microalga *Stichococcus bacillaris*) showed that one limestone (Ançã stone) and the marble, which had the highest (>17%) and the lowest (≤1 %) porosity respectively, supported the greatest microorganisms' growth. The results apparently depend mainly on the chemical composition rather than on the physical characteristics of the stones. In fact, the microbial strains developed on carbonate substrates but showed limited growth on granite. In addition, the study pointed out the important role played by the physiological characteristics of the microorganisms on their development. *S. bacillaris* revealed a greater capacity of adaptation to the lithotypes than *G. alpina*, possibly due to its oligotrophy, adhesion capacity, metabolic flexibility, and tolerance to adverse conditions. Similarly, the importance of stone chemical composition emerged from a laboratory experiment[17] that showed higher rates of cyanobacterial growth and substrate dissolution of basalt rocks than rhyolitic rocks. According to the authors, factors accounting for this likely include the higher content of quartz, which has a low rate of weathering and lower concentrations of bioessential elements, such as, Ca, Fe, and Mg.

On the other hand, the issue of stone bioreceptivity deserves further investigation as Figure 2.1 indicates. Limestone blocks with just sporadic spots of biofilms are close to sandstone blocks heavy colonized by biofilms. Here, in contrast with the above-mentioned researches, the limestone is the least colonized stone while the silica-based stone is more prone to

FIGURE 2.1 Detail of a building in the historical center of Delft (the Netherlands). White blocks of limestone surrounded by sandstone blocks heavy colonized by biofilms. Photo courtesy of José Delgado Rodrigues.

biological activity. The different porosity can account but cannot be the main cause of the differential colonization. According to José Delgado Rodrigues (personal communication), the different solubility of calcite and quartz is key factor. As the building is placed in the historical center of Delft, the air pollution plays here a role because it acidifies the rainwater. The acidic runoff produces a progressive and continuous dissolution of the outer layers of limestone while it is quite innocuous for the sandstone. Thus, biofilms have not enough time to settle on limestone surfaces. Furthermore, pores in the sandstone provide excellent sheltered places, and the high translucency of quartz grains allows easy access of sunlight.

In a laboratory study,[18] the meristematic microcolonial fungi *Coniosporium perforans*, *C. uncinatum*, and *Sarcinomyces petricola* showed different penetration patterns on stone specimens. *Coniosporium* isolates exhibited the deepest penetration in marble, limestone, and travertine, which were poorly or not penetrated by *S. petricola*. This fungus instead showed the highest penetration depth within granite (7.4 mm) and slightly more (2.4 mm) in gneiss than the other strains. According to the authors, the rock bulk chemistry can affect the mycelium penetration potential.[18] Moreover, internal discontinuities, weakness planes, macro- and microporosities, and different orientation of schistosity in gneiss contribute to the different orientation and distribution of hyphae in the bulk of the lithotypes.[18]

An interesting review[19] focused on the parameters related to the primary bioreceptivity. The chemical composition, petrography, texture, open porosity of the stones and the physical properties related to movement of water through the rock matrix seem to be the major characteristics that affect colonization.

A study[20] proposed a laboratory test to assess the bioreceptivity of healthy and weathered granites measuring some physical properties. The samples were inoculated with three cyanobacteria belonging to *Nostoc*, *Oscillatoria*, and *Scytonema*. The results showed that open porosity, bulk density, capillary water absorption and abrasion pH were the most important parameters to evaluate the bioreceptivity. Abrasion pH, measured after grinding the rock in distilled water, directly relates to the number of basic cations released by the rock when in contact with aqueous solutions. Bulk density and open porosity are both correlated with void spaces in the rock. Therefore, they refer to the capacity of rock to absorb water. Capillary water absorption provides information about the pattern of the pore network. This parameter is relevant for bioreceptivity because it is connected to the time the rocks remain wet.

Bioreceptivity of ceramics mainly relates to physical characteristics such as porosity and surface roughness.[14] In fact, the smooth and impermeable surface of glazes is more resistant to microbial colonization.[14] Yet when microcracks are present, plants find a favorable substrate to grow (Fig. 2.2).

A study[21] assessed the bioreceptivity of ordinary white cement Portland to three fungal strains (the hyphomycete *Alternaria alternata*, the melanin producer yeast-like fungus *Exophiala* sp., and the meristematic fungus *Coniosporium uncinatum*) that are extensively involved in biodeterioration in natural environments. The samples underwent accelerated weathering with carbonation for 48 h. After carbonation, leaching operations (28 days) were performed. Leached specimens exhibited a higher microbial colonization.

FIGURE 2.2 Queluz National Palace (Portugal). Glazed tiles in the "Canal de Azulejos" showing the growth of plants in the joints and biofilms in the lacunae. Photo courtesy of José Delgado Rodrigues.

Another interesting factor related to bioreceptivity is the perception by observers of the rock color change caused by biological colonization. A study[22] identified seven levels of human perception on granite surfaces and measured the chlorophyll *a* content for each level. The lower appreciable color change is ΔE^* 3.17 corresponding to 0.04 mg chlorophyll a/cm^2. The observers defined the colonization as very intense at 0.56 mg chlorophyll a/cm^2.

The pH of a substrate influences the biological growth because some organisms prefer specific values or tolerate a narrow pH range. Extreme pH values are not favorable to biological growth for the damaging effect of H^+ or OH^- ions. Most organisms tend to live in pH neutral conditions but microorganisms can colonize cement over a wide pH range.[23,24,20, 25,26,19] Many fungi prefer slightly acidic substrates (e.g., granites, some sandstones), while alkaline conditions (e.g., limestone, marble, lime mortars) favor some cyanobacteria.[27] A laboratory test, detecting bioreceptivity of mortars,[28] showed that pH values below or close to 9 allowed the colonization of the fungus *Cladosporium sphaerospermum*, while pH close to, or higher than 10, inhibited such growth. As a fresh concrete has a pH of 12–13, it permits a bacterial growth when pH is lowered by reaction with atmospheric carbon dioxide (carbonation).[25] Thus, the degree of carbonation and pH values play a key role in the susceptibility of mortars to fungal colonization.

A research on the prevention of biological growth in the archaeological Area of Fiesole (Italy)[29] assessed the secondary bioreceptivity of some stones (sandstone, marble, plaster) where any biological growths had been removed. After almost 3 years, sandstone and the original Roman plaster showed very low bioreceptivity with no regrowth detected on their surfaces, while marble showed a high bioreceptivity as a fungal regrowth covered almost entirely the surfaces. As there were no differences in environmental conditions and the porosity was low on all the stones, arguably the differences in bioreceptivity could be due to the chemical composition of the substrates.[29]

Differences in abundance of colonization on stones located in the same site can be due to tertiary bioreceptivity, that is, to different kinds of nutrients contained in the stones.[30] They can derive from existing biological growth on the surface, active or decayed; from bird droppings; from organic compounds used in restoration practice. Moreover, air and rain carry the nutrients in the form of dust and soil particles.[30] Soil fertilization leads to an accumulation (eutrophication) of nitrates and phosphates that are contained in bird droppings too. Some lichens (nitrophilic species) have been adapted to eutrophication.[26] Generally, they have an orange thallus, easily observed on roofs, architectural moldings, and horizontal surfaces of statues if the eutrophication derived from bird droppings (Fig. 2.3), and on monuments in rural areas if air and rain transported the fertilizers. In these situations, nitrophilic vascular plants develop as well.[26]

FIGURE 2.3 Batalha Monastery, Portugal. Biofilms and lichens strongly alter the light color of the stone. Accumulation of nitrates and phosphates from bird droppings affect the growth of orange nitrophilic lichens on the decorations. The architectural design and the exposure influence the colonization. Worth noting are the niches where no growth is present.

Biofilms' growth on glass is correlated to tertiary bioreceptivity. In fact, neither the inorganic composition nor the physical features of glass favor microbial growth, but the organic residues of various origins, such as dust deposits, dead fungi, and bacteria, bird droppings, can be a source of nutrients.[31] The deterioration action of microorganisms on glass is a modest, slow, yet continuous process that can accelerate the weathering of this material.[10] Microorganisms act in synergy with meteorological factors and air pollution.[10] Researches focused mainly on medieval stained-glass windows of European churches. Despite the extensive use of glass as a building material in the 20th–21st centuries, reports on modern flat glass degradation by microorganisms are rare. Nevertheless, biofilms can grow on modern glass affecting its transparency.[32]

Corrosion, orange patinas, pitting, and mineral crusts can occur on external surfaces of medieval stained-glass windows. Microorganisms contribute to all these decay forms.[33–35] Analyses revealed complex bacterial communities consisting of members of the phyla Proteobacteria, Bacteroidetes, Firmicutes, and Actinobacteria.[35] Fungi showed less diversity than bacteria, and species of the genera *Aspergillus*, *Cladosporium*, and *Phoma* were dominant.[33,35] Thus, historical glass windows are a habitat in which both fungi and bacteria form complex microbial consortia of high diversity. The bacteria are phylogenetically related to well-known bacteria that cause mineral precipitation on stones. Regarding the detected fungi, they are ubiquitous airborne species. Therefore, the pitting present on the surfaces of glass windows could relate to other more specialized fungal species that grew in the past and are not detectable now.[35] Glasses made with potash (K-rich composition) are more easily decayed than soda glasses (Na-rich composition). The pitting is associated to microorganisms able to produce mycelia, such as actinobacteria and filamentous fungi. The microbial penetration depends on the chemical composition of the glass. Copper contained in green glasses acts as an inhibitor.[34,35]

A laboratory test that used historically accurate reconstructions of glass windows (15th and 17th centuries, Sintra, Portugal) confirmed some of the results described above. The samples (mixed-alkali colorless glass, purple potash-glass with manganese, brown potash-glass with iron) had two surface morphologies (corroded and non-corroded). They were inoculated with fungi of the genera *Cladosporium* and *Penicillium* previously isolated from the original stained-glass windows. Results showed that fungi produced clear damage on all glass surfaces in form of stains, erosion, pitting, crystals, and leaching. The biodeterioration patterns produced by fungi did not differ between non-corroded and corroded glass surfaces.[36]

A relevant issue connected to bioreceptivity and to climates is the succession and rate growth of organisms on bare rocks. In a glacial recession area in Austria,[37] the first colonizers of a micritic limestone at the rim of the glacier were fungi belonging to the group of Aureobasidiomycetes (black fungi), which actively penetrated the substratum. Then green algae established on stones once the glacier retreated. After about 12 years of exposure, nonlichenized algae, fungi, and lichens were present. The surface became covered mainly by the calcicolous euendolithic lichens *Polyblastia albida*, *Verrucaria hochstetteri*, *Polyblastia dermatodes*, *Sarcogyne regularis*, *Staurothele rupifraga*, *Thelidium decipiens*, *Thelidium minutulum* after 30–100 years of exposure. The pioneer biofilms, followed up by the lichens, grew until they reached a stage of relative high temporal and spatial stability. The lichens photobiont extended from 40 to 250 μm under the rock surface. A further layer, 250 μm–4 mm under the surface, consisted only of fungal filaments. The authors deduced that, on homogeneous carbonate substrata, maintenance of a stable population of lichens for decades corresponds to a "sustainable use" of the rock substratum. In a very different environment (continental climate, Spain), another type of succession occurred on an abandoned dolostone quarry exposed for different periods since Roman times.[38] Endolithic fungi appeared in the most recently exposed surfaces, while epilithic lichens were present on those longer exposed. In the tropical climatic conditions of Angkor, Cambodia, a study[39] identified a clear ecological trend of different biological communities. The first community colonizing the stones of a temple was mainly formed by the green alga *Trentepohlia* sp., widely spread as reddish patinas in relatively xeric and shady conditions. Gray-blackish patinas formed mainly by cyanobacteria (*Scytonema* and *Gloeocapsa*) prevailed in xeric and sunny conditions. With the increasing availability of water and light, different lichen communities (mainly *Lepraria* and *Pyxine*) established, followed by mosses and higher plants.[40]

The bioreceptivity to spontaneous arboreal flora of buildings with different configurations, ages, and stone types in the humid-tropical region of Hong Kong was investigated.[41,42] Most the plants belonged to *Ficus* genus whose natural invasive trait provided ability to grow on walls and buildings. Moreover, building age greatly affected bioreceptivity. The poor maintenance of older buildings resulted in paint loss, open joints, and plaster cracking facilitating moisture ingress and water retention. Exposure to sunlight, debris pockets, surface moisture, and leakage from drainpipes fostered germination and establishment of plants. Volcanic masonry blocks, with more nutrients released upon weathering, were more favorable to tree colonization than granite. On the other hand, according to the authors, interesting if not unique plants enrich urban ecology and deserve to be studied and conserved to maintain the urban biodiversity.

A study of the embankments along the Tiber River in Rome (Italy) investigated the bioreceptivity to plants of travertine, tuff, and cement.[43] The area has a mesoclimate with hot and dry summers, milder winter temperatures, and rainfalls concentrated in autumn and winter. The vascular flora is mainly composed of herbaceous species, both annual (therophytes 44.9%) and perennial (hemicryptophytes 37.4%, geophytes 2.8%). The exposure did not play a relevant role in these environmental conditions.[44] On the contrary, the variation of stones' water content was the most selective factor in terms of quality and quantity of plants. Moreover, the chemical composition of materials affected the colonization, as plants have different preferences for nitrogen content and, at a lesser extent, for pH values. The relationship between species and slopes differed in the lithotypes. As for tuff, slopes did not influence much the plants because its high porosity made it able to retain water. In contrast, the slopes of travertine caused relevant changes in plant composition.[43]

In conclusion, bioreceptivity is a very important characteristic of both natural and artificial stones because it indicates their proneness to biological growth. The primary bioreceptivity, which is the potential of a healthy material to be colonized, was assessed mainly by laboratory tests that included the artificial inoculation of stones with the spores of an organism, the incubation of specimens under optimal environmental conditions, and the quantification of the resulting microbial mass. In some cases, the obtained microbial communities are comparable to natural communities and the laboratory conditions reproduce the colonization simulating competition and/or synergy between microorganisms.[45] Nonetheless, the extrapolation of results toward practice is often difficult because of the large differences of environmental conditions, stones, and organisms in comparison with those observed in situ. Another limitation is the use of different methods for quantifying colonization. Therefore, bioreceptivity assessment should be performed by standardized methods that, unfortunately, are not available. They would allow setting a database on the primary bioreceptivity of lithotypes, leading to bioreceptivity indexes useful in the selection of appropriate stones in new constructions and replacement of materials in existing structures.

The studies agree in defining high porosity, shape and orientation of pores, pore size distribution, and chemical composition as the main factors that determine bioreceptivity. In fact, the smooth and impermeable surface of glazes is resistant to microbial colonization unless fissures are present. Other characteristics of the rocks (surface roughness, internal discontinuities, weakness planes, pH, and different orientation of schistosity) play a role and contribute to the different orientation and distribution of microorganisms in the bulk of the lithotypes. The pH of a substrate, for example, influences the biological growth because some organisms prefer specific values or tolerate a narrow pH range. Extreme pH values are not favorable to biological growth for the damaging effect of H^+ or OH^- ions. In addition, the studies pointed out the importance of the physiological characteristics of the microorganisms on their development. There are microorganisms with a great capacity of adaptation to the lithotypes, possibly due to their oligotrophy, adhesion capacity, metabolic flexibility, and tolerance to adverse conditions.

Differences in abundance of colonization on stones can be due to secondary and tertiary bioreceptivity. Fissures, cracking, and other forms of deterioration facilitate moisture ingress and water retention. These conditions, together with sunlight, debris, and chemical composition of the substrate, foster the development of microorganisms as well as germination and establishment of plants.

The studies on the succession and rate growth of organisms on bare rocks provide relevant information about the relationship of bioreceptivity and climates. In a glacial recession area, the first colonizers were black fungi, which actively penetrated the limestone, and algae. After about 12 years of exposure, a complex community of nonlichenized algae, fungi, and endolithic lichens established. In a continental climate (Madrid), endolithic fungi appeared in the most recently exposed rocks (dolostone), while various epilithic lichens grew on those longer exposed. Finally, in a tropical climate, the green alga *Trentepohlia* sp. colonized the stones in relatively xeric and shady conditions. Gray-blackish patinas formed mainly by cyanobacteria prevailed in xeric and sunny conditions. With the increasing availability of water and light, different lichen communities established, followed by mosses and higher plants.[40]

2.2 FACTORS INFLUENCING BIOLOGICAL GROWTH

Besides substrates' bioreceptivity and capacity to retain water, the environment surrounding the monument and the monument itself act as limiting factors of biological growth. Local microclimate, macroclimate, wind-driven rain, geographical location, pollution, architectural design, and details of monuments or sculptures are remarkable factors influencing organisms' colonization[46,47] (Fig. 2.3). For example, the above-mentioned study on the colonization by pioneer algae and plant communities of travertine embankments along the Tiber River in Rome (Italy)[44] showed that water, particularly rain, represented the main limiting factor for the growth of microalgae. For the macroflora, instead, air humidity appeared more relevant. The plants in fact have greater capacity of making the most of the substrate water thanks to their roots.

A study described the effects of the object's design. As sunny surfaces are more hostile and variable than those shaded, the production of extra cellular polymeric substances (EPS, *vide infra*) tends to be higher to protect the cells from the adverse conditions.[48] In shaded areas, the microbial biomass and the species diversity are usually much higher.[49] On the other hand, many lichens are able to acclimate their photosynthetic metabolism to prevailing temperature and light regimes under different climatic conditions.[50]

Regarding pollution effects on lichens, a study[51] of the urban area of London (United Kingdom) indicated that lichen diversity is today improving following a reduction in sulfur dioxide (SO_2) levels with a further shift in species composition from "acid-loving" to "nitrogen-loving." Moreover, the organic pollutants can be a carbon source for microorganisms.

Temperature and relative humidity are relevant in determining whether fungi can subsist or grow on substrates.[52] Most fungi may be grouped into three categories on a thermal basis. Psychrophiles grow optimally at 0–5°C, mesophiles at temperatures between 20 and 45°C, while thermophiles grow optimally at or above 55°C. Members of the Mucoraceae and Deuteromycetes grow on substrates at relative humidity of 90–100% and not below, while members of the *Aspergillus glaucus* group grow at relative humidity as low as 65%.[52] Wind-driven rain loads can strongly affect mold colonization. The moisture contents in the walls of a brick tower was higher near the edges of the walls than at the center just for wind-driven precipitation, inducing mold growth especially at this position.[53] Moreover, wind-driven rain load caused a significant increase in indoor relative humidity.

Differences in abundance of biological growths on stones located in the same site often occur as on limestone of the Monastery of Cartuja, Granada, Spain.[54] Although two areas were similar in exposure and climatic conditions, one had 56% of lichen cover and an average surface temperature of 20°C, while the other had 3% lichen cover and an average temperature of 23°C. As reported by the authors, a difference of just 3°C in the average annual temperature was enough for lichens to establish. It is likely that the temperature affects the stone water content, a crucial factor for the photosynthetic productivity.[50] In fact, the influence of surface temperature showed a significant positive correlation with green algae biofilms in a survey of four sandstone heritage structures in central Belfast (Ireland) exposed for around 100 years.[55] Areas with lower temperatures were, on average, greener than warmer areas. As reported by the authors, it is possible to model algal greening of sandstones from the scaled-down outputs of regional climate models as it mostly relates to climate and atmospheric particulates.

In some cases, the orientation of a stonework affects biological growth at least in its early development. Different rates of darkening and greening occurred over time on north- and south-facing stone blocks in Northern Ireland.[56] Twenty-one-month exposure trials across nine environmentally variable sites were conducted. The observed differences likely relate to the fact that the north faces receive less direct sunlight in wet climates and northern-latitude positions. Therefore, north-facing blocks, once wet, will remain damp for much longer than other stones. This slow-drying phenomenon is much more hospitable for biological colonization than the hostile environment of rapid wetting and drying cycles experienced on the south faces. In a different climate and latitude (south of Brazil) seemingly happens the opposite, painted surfaces showed higher fungal colonization on south-facing sides[57] that received less solar radiation than north-facing ones. If their surface temperature fell below dew point at night, they remained moist for longer periods after wetting.

The forthcoming expected climate changes will affect the stone decay processes including biological growth. Models predict an increase in temperature and precipitation in northern areas of Europe for the far future (2070–2099), which would lead to a higher accumulation of biomass.[58,59] Projections for the 2020s (2010–2039) indicate that, in the United Kingdom, summer dryness, winter wetness, and precipitation in autumn and spring will increase.[60] In contrast, precipitation will significantly reduce in southern areas of Europe.[58] The increase in wetness and precipitation will make stone structures remain wet longer and possibly the depth of moisture penetration will increase. It appeared that building stone in Northern Ireland has already responded through a bigger incidence of algal "greening."[60] The more moisture retention, caused by algae, the more dissolved salts and moisture penetration to depth of the stone. A possible consequence could be that, while surface decay is apparently quiescent, chemical action is still on-going inside stone (particularly sandstone) when it is wet, leading the way for a rapid future physical damage.[60]

Caves are an "extreme" example of the limiting factors' effects on microbes. High relative humidity, constant annual temperature, and very low photon fluxes characterize these environments. The conservation of Paleolithic paintings placed in caves is extremely important. Cyanobacteria and algae found in Roman catacombs and in caves have adaptive characters that allow them to develop even at the extreme low photon flux densities and the spectral quality provided by the artificial lighting systems.[61–64]

The installation of artificial light and the presence of visitors modify the original conditions of caves promoting the colonization of phototrophic biofilms[61] (Fig. 2.4). Visitors caused daily increases in temperature, water vapor and carbon dioxide concentration in the Roman Catacombs of St. Callistus and Domitilla, Rome (Italy).[65] The microclimatic changes posed risks to their conservation because water condensation could easily occur on walls and ceilings, as rock temperature was lower than air temperature.

The growth of the alga *Chlorella minutissima* with a minor contribution of *Scenedesmus* sp. and *Bracteacoccus* sp. on some surfaces of Moidons Cave (Jura, France) was mainly due to illumination time and, at a lesser extent, to air circulation and visitors.[66]

The identification of species able to grow in caves is a "work-in-progress" because many are still unknown. For example, molecular studies have shown the existence of unknown microbial communities in Altamira Cave (Spain) including anaerobic microorganisms that proliferate on wall paintings because of oxygen low content.[67] They belong to sulfate-reducing bacteria, a significant bacterial group showing a typical metabolism, the reduction of sulfates to sulfides, which can have potential negative effects on painting conservation.

FIGURE 2.4 Lamalunga cave (Altamura, Italy). Green and dark discolorations caused by cyanobacteria (*Aphanocapsa* sp. and *Lyngbia* sp.). The discolorations appeared after the installation of illumination systems inside the cave.

The fountains represent another "extreme" environment, at a certain extent more favorable to biological growth than caves (Fig. 2.5). Microalgae of different genera (*Cladophora*, *Chlorella*, *Chlorococcum*, and many others), diatoms, and mosses grow on the surfaces of fountains. The water microbial contamination may act as a precursor for the formation of detrimental crusts on stones.[68]

(a)

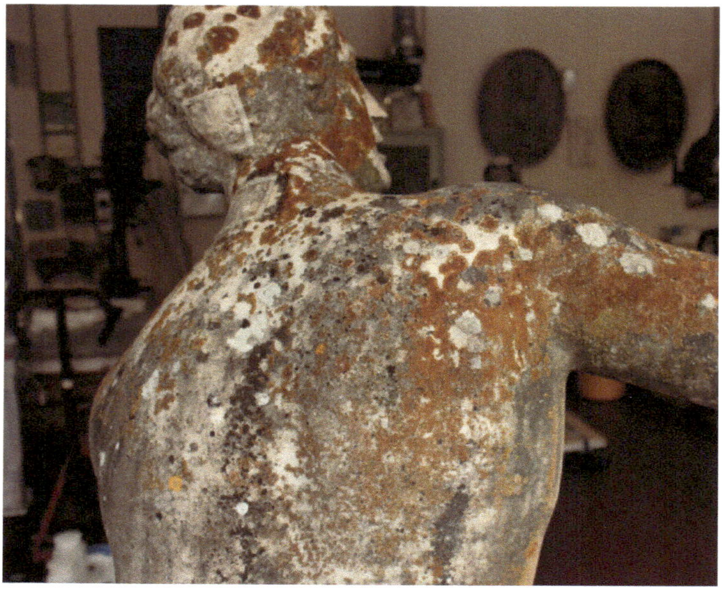

(b)

FIGURE 2.5 *Andromeda* marble statue located on a little isle emerging from the large basin (called Isolotto) of Boboli Gardens, Firenze (Italy). This peculiar location was very favorable to biological growth (a). A detail of the sculpture (b).

As shown in this section, the literature is lavish with excellent publications and books on the factors that affect the growth of biodeteriogens on stone. Nevertheless, the influence of the properties of the mineral surfaces on microbial attachment (in other words, the precise role that surface microstructure plays in directing microbial colonization) requires further investigations. In fact, microbial interactions with specific topographic features, especially submicron features, remain largely unresolved.[69]

As deducted from the multiple intrinsic parameters that influence bioreceptivity of stones, and the wide range of environment factors mentioned as relevant components of the colonization process, the actual proneness of a given substrate to biological growth can hardly be predicted with any isolated subset of parameters. To know how to "read" the actual occurrences is of paramount importance, and the results obtained from the laboratory studies may be of great help to this "reading" process, although they cannot replace the direct interpretation of natural occurrences.

KEYWORDS

- **natural stones**
- **artificial stones**
- **wall paintings**
- **glass**
- **factors that influence biological growth**
- **bioreceptivity of stones**

CHAPTER 3

OUTLINE OF BIODETERIORATION OF STONE OBJECTS

CONTENTS

ABSTRACT

The chapter focuses on the biological alteration of stone objects and refers to the agents of biodeterioration. It outlines the various organisms able to develop on stones and their effect in terms of degradation, including a discussion on an important topic, the bioprotection of stones by biofilms and lichens.

3.1 GENERAL PRINCIPLES

Hueck[70] first defined biodeterioration as "any undesirable change in the properties of a material caused by the vital activity of organisms." This definition is yet in use and applied to any materials, except for living organisms. Another common term in use is biodegradation defined by Allsopp et al.[25] as "the harnessing, by man, of the decay abilities of organisms to render a waste material more useful or acceptable." When we no longer desire materials, constructions, activities, biological cycles are encouraged and biodegradation is positive and useful. Eggins and Oxley[23] detailed and completed the discussion on these terms referring to biodeterioration as a process that decreases the value of an object because implies a damage to it. On the contrary, biodegradation, relating to materials of low or even negative value, increases it. The two processes, caused by organisms, are identical. Their distinction just connects to the consequence for human requirements. In this respect, the authors define biodeterioration as the deterioration of materials, constructions or processes of economic importance. In other words, biodeterioration is the part of the global cyclical processes that we want to be temporarily arrested or removed.[23] Thus, prevention and control of biodeterioration (and any deteriorations) mean to delay as much as possible the natural decay process that normally affects the materials. In this context, the term *biodeteriogen* refers to any organisms that cause damage to materials, although the term is often applied to any organisms discovered on objects, regardless of whether their capacity to damage the substrate has been established or is unknown.

Colonization on natural and artificial stone objects occurs by many systematic groups ranging from bacteria to plants and animals. The microflora of outdoor stones represents a complex ecosystem including bacteria, algae, fungi, lichens. In addition, small animals such as mites may be present.

As explained in the previous chapter, organisms colonize stonework whenever the conditions of moisture, light, temperature, and nutrition are favorable. Variable ecological spatial patterns can occur on monuments because of changes in these environmental factors. Indeed, shortage in light or water conditions can stop organisms' development.[71]

When a biological colonization is evident, the conservator should verify at which extent it damages the materials and know the nonbiogenic agents that take part in the degradation.[72] Many causes have similar effects, act in synergy, or interact in quantitatively variable relations. Thus, the relevance of the biological impact to the entire deterioration process must be evaluated very carefully.[73] Detecting organisms on cultural objects does not automatically imply that they modify the chemical composition or physical properties of the materials. Only in particular conditions and in combination with other factors they can cause and/or enhance deterioration mechanisms. Moreover, the growth of some organisms is very slow and the damage becomes visible only after years or even decades. In some cases, the biological colonization may act as a protective layer shielding the stone from other factors that cause decay, such as wind and rainwater. Therefore, its removal will not necessarily halt the erosion.

The processes of biodeterioration involve the utilization of the substrate as a food source or solely as a support for development. Thus, the growth of biodeteriogens on objects of art can result in visible esthetic and/or physicochemical damages.

The interaction occurring between organisms and inorganic materials is a well-known process that contributes to change the geoenvironments.[73,74] The ability of microorganisms to influence Earth surface and subsurface modifications is increasingly being recognized and understood. Microbial mats were ecosystems that greatly affected the conditions of the biosphere through geological time. These systems, which date back 3500 million years ago, had high metabolic rates and rapid cycling of major elements on very small (μm–mm) scales.[74]

Fungi have important roles in the biosphere as they are involved in biogeochemical transformations of both organic and inorganic substrates at local and global scales.[75,8] In terrestrial habitats, fungi influence biogeochemical processes particularly when considering soil, rock surfaces, and the plant root–soil interface. Of special significance are the mutualistic symbioses, lichens and mycorrhizas.[8] In soil, fungi generally comprise the largest proportion of the biomass (including other microorganisms and invertebrates).[8] Their ability for oligotrophic growth, their filamentous growth habit, and resistance to extreme environmental factors, including metal toxicity, irradiation, and desiccation, make them successful colonizers of rock surfaces and other metal-rich habitats.[8]

In soil-forming processes, an important role is played also by saxicolous lichens and their associated microbial lithobionts (free-living algae, fungi and bacteria), for their capacity of weathering lithic substrata.[76] The relationship between microbial life and geomorphology may be called "microbial geomorphology".[77]

The so-called biomineralization is the process by which biological activity induces the precipitation and accumulation of minerals. It is the result of the metabolism of the organisms, is present in all five kingdoms of life, being widely practiced by plants. Minerals produced by microbes and lichens include iron hydroxides, magnetite, manganese oxides, clays, amorphous silica, carbonates, phosphates, sulfates, sulfides, oxalates.[78–82] Iron and manganese, utilized for the transfer of electrons in oxidation and reduction reactions, are removed from the stone and may be reoxidized at the stone surface, forming patinas or crusts.[83,74] The most visible iron-containing minerals, the ubiquitous subaerial rusts on rocks, composed of ferric oxides or oxyhydroxides, are likely due to microorganisms whose metabolism depends on iron.[84]

3.2 BIOFILMS

The term biofouling (or biosoiling) refers to any alterations caused by the presence of a surface layer of microorganisms and their products on work-stones[72] (Figs. 3.1 and 3.2). The conservative aspects in this case focus mainly on esthetical effects that give the work of art an unpleasant appearance. The presence of microorganisms may be undesirable as they can obscure inscriptions and carvings. They can trap dust particles from the atmosphere through their sticky surfaces, increasing the rate of soiling of the stones and aiding the establishment of higher plants.[85] They can also damage the substrates.

As it is often difficult to recognize a microbe colonization with the naked eye, many alterations caused by microorganisms may be instead ascribed to nonbiological causes. These are the cases of alterations like black patinas, stains, lack of adhesion, pitting, efflorescence, in which the identification of the agent of damage is very important because it will markedly influence the choice of restoring operations, including the possibility of non-removing biological growths.

FIGURE 3.1 Outdoor Buddha statues (Jizo Bosatsu) located in the Buddhist temple Daitoku-ji (Kyoto, Japan). Black biofilms strongly alter the white color of the statues. Photo courtesy of Maria Ruggieri.

FIGURE 3.2 Detail of a marble slab of the *Onze Mil Virgens* chapel (Alcácer do Sal, Portugal). The initial colonization by black biofilms occurs in the intergranular microspaces of the stone where water and airborne particles accumulate. Photo courtesy of José Delgado Rodrigues.

The surface aggregate, sessile community of microorganisms is called *biofilm.*[86–92] Biofilms are ubiquitous in the environment. A first step in the successional development of multispecies biofilms is the coating of noncolonized surfaces with organic substances like polysaccharides and proteins, which improve attachment of initial colonizers. They grow in surface-attached microcolonies, entwined in a matrix of self-produced extracellular polymeric substances (EPS). As the microcolonies develop, additional species, so-called secondary colonizers, are recruited through coaggregation and nonspecific aggregation interactions, increasing the biofilm biomass and species complexity[92] (Fig. 3.3). This process allows species that cannot adhere to the noncolonized surfaces to become part of the biofilm.[90] Organization and cell–cell communication among microorganisms through the production and reception of signal molecules is called quorum sensing. These biochemical signals influence each step in the process of biofilm formation and enhance the persistence of both individual species and the biofilm as a whole.[92,93] Testing has identified quorum-sensing genes in archaea, bacteria, and fungi.[92]

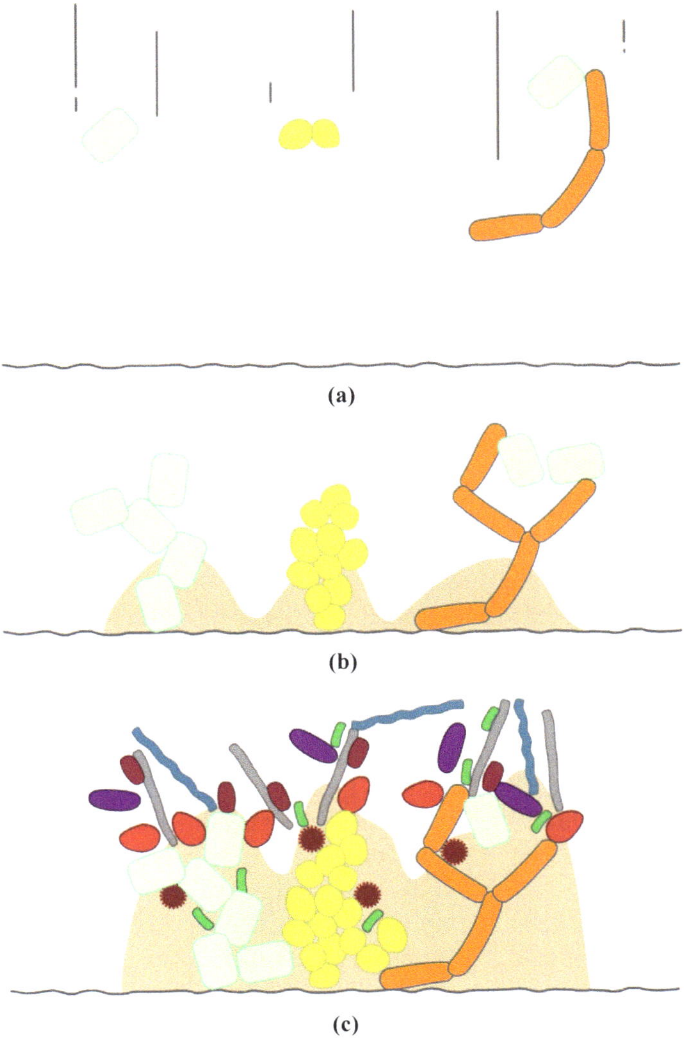

FIGURE 3.3 Scheme showing the sequential development of biofilms on stone. Single cells deposit on the surface (a). Colonization and production of extra polymeric substances (EPS) by primary cells (b). Accumulation and consequent development of other species due partially to the ability of coaggregation by primary cells (c).

From an ecological perspective, the ability of a species to coaggregate likely imparts a selective advantage over non-coaggregating species. While the precise nature and contribution of coaggregation to biofilm development is still being evaluated, it has been demonstrated that the ability to regulate coaggregation has great potential in controlling the rate of biofilm development or altering the composition of biofilms intercellular signal molecules.[94,92,95]

Associations of phototrophic and heterotrophic microorganisms often compose the biofilms that grow on stone heritage objects.[96] Cross feeding and metabolic exchange are the main interactions among species. Photosynthetic microorganisms sustain the growth of heterotrophic bacteria through the production of EPS such as saccharides.[96] The heterotrophs, in turn, can promote cyanobacterial growth by providing key metabolites and other products.

Biofilms are thus accumulations of microorganisms, EPS, multivalent cations, inorganic particles, biogenic materials, colloidal, and dissolved compounds. Polysaccharides are characteristic components of the EPS, but proteins, nucleic acids, lipids, and humic substances have also been identified, sometimes in substantial amount. EPS are involved in the formation of a three-dimensional, gel-like, highly hydrated and locally charged (often-anionic) matrix, in which the microorganisms are placed (Fig. 3.3). In general, the proportion of EPS in biofilms can vary between roughly 50 and 90% of the total organic matter. EPS play significant roles in the attachment of microorganisms to the mineral surfaces and in the protection of bacterial community from toxic compounds.[4]

Despite the number of studies on the damaging action of biofilms on materials (*inter alia*[97,90,26,91,98–102]), the relationship between mineral solubility and the role of microbial surface colonization in weathering reactions is a topic that have yet to be answered in a comprehensive manner.[69,103] Even microorganisms well known for their ability to colonize stone monuments have only rarely been appropriately examined for their actual contribution to stone decay in vivo.[104] There is indeed a difficulty inherent to measuring processes occurring at the microbe-mineral interface. In fact, the effects of microorganisms placed in contact with a substrate have been mainly the subject of simulation experiments under laboratory conditions. They have undoubted advantages over in situ studies as they provide reproducibility and comparison of the substrate before and after exposure. In a few words, the laboratory model systems simplify nature to understand it more easily.[96] On the other hand, due to the large differences between the optimized laboratory conditions and those observed in the field, it is difficult to extrapolate the results, as it has been already underlined. Nonetheless, some laboratory studies are worth mentioning here because they quantified the rates of calcite dissolution induced by bacteria. The heterotrophic bacterial species (*Burkholderia fungorum*) affected calcite dissolution by modifying pH and alkalinity during utilization of ionic N and C species, respectively.[105] The net rate of microbial calcite dissolution in the presence of glucose and NH_4^+ was circa two-fold higher than that observed for abiotic control experiments where calcite dissolved only by reaction with H_2CO_3. In another study,[106] the assessment of deterioration induced by endolithic bacteria was carried out measuring Ca^{2+} released from the stone using the calcium binding fluorochrome Rhod-5N. Calcium ions amount was twice that of non-inoculated controls. Bacteria in the inoculated flask increased exponentially for 5 days and then stabilized, while the Ca^{2+} concentration increased linearly, indicating that endolithic bacteria accelerated deterioration of the limestone. Alginic acid, a polysaccharide produced by *Pseudomonas aeruginosa*, a ubiquitous environmental bacterium, was selected to investigate its effects on calcite dissolution of stone samples.[107] It caused an increase in the dissolution rate of calcite across a range of pH values through surface chelation of calcium.

An interesting study[96] applied an innovative methodology to design and predict biodeterioration/bioprotection processes on lithic surfaces. The method was based on a new *in vitro* model of a fast-growing, phototroph–heterotroph mixed species biofilm on a stone surface. The cyanobacterium *Synechocystis* sp. strain PCC 6803 and the chemo-heterotroph bacterium *Escherichia coli* K12 were used. The system allowed the study of the stone/air interface, the phototroph–heterotroph interactions in an oligotrophic environment, and the survival of microorganisms in harsh environment including desiccation stress and biocide tolerance. The results are very promising because the system was effective at reproducing features typical of biofilms on outdoor stone monuments. Therefore, it helps understand enough aspects to make predictions of field systems.

In fact, the precise role of mineral solubility, and hence dissolution rate, in determining the extent and rate of microbial surface colonization is largely unknown. The initial attachment of a microbial cell to a surface implies nonspecific interactions, such as electrostatic, hydrophobic, and van der Waals forces.[98] Surface hydrophobicity and surface electric charge are of importance in bacterial adhesion.[4] Important as well are environmental factors that influence the production of quorum sensing biochemical signals, known to play a critical role in the establishment and subsequent development of biofilms. All these factors need further studies. One of the most challenging tasks in conservation of cultural heritage will be establishing when biological growth does not pose any problem and is even desirable, when it is merely an esthetic issue, and when it damages the substrate.[30,104] A proteomic and genomic approach[48] would shed light on the physiology and potential deteriorative activity of microorganisms and would help to design new strategies for isolating and successfully culturing new organisms.

3.3 CYANOBACTERIA, BACTERIA, MICROALGAE

Cyanobacteria and microalgae are often the first colonizers of stone surfaces on which they form phototrophic biofilms.

Algae commonly colonize archaeological remains, buildings, mosaics, wall paintings, forming colored (green, gray, black, brown, and orange) powdery patinas, and gelatinous layers. They usually dominate surfaces in wet and rainy areas.

Cyanobacteria typically form dark brown and black patinas but also pink discolorations. They need moisture as well, but survive dry periods and extended irradiation better than algae.[90] They are adapted to resist adverse conditions because of their thick outer envelopes and pigments like scytonemin that contribute to their protection against desiccation and intense solar radiation.[87,108,109] Some cyanobacteria can survive repeated cycles of desiccation and rehydration.[87,90,49] *Synechocystis* sp. strain PCC 6803 recovered in a few minutes when rewetted after desiccation (1 h under a stream of sterile air at 25% RH).[96] Besides the esthetic disturbance caused by the colored patinas, algae and cyanobacteria cause water retention and damage due to freeze–thaw cycles. They also increase ease of establishment for other organisms (for instance fungi and macro organisms).[104]

Extensive cryptoendolithic communities of cyanobacteria (*Chroococcidiopsis* genus) grow inside the ignimbrite rocks of a hyper-arid environment, the Atacama Desert in Chile.[110,111] Xeric conditions were also those of limestone walls of the Church of the Virgin (Martvili, Georgia) where endolithic cyanobacteria (Chroococcales) caused erosion.[112] Cyanobacteria form layers that make temples and monuments in most regions of India look blackish colored despite the adverse conditions during summer when the temperature on these structures exceeds 60°C, and high light intensity and extreme dryness occur.[109]

Some mature biofilms on painted surfaces in tropical and subtropical countries are principally composed of coccoid cyanobacteria.[47] Nevertheless, literature shows that the major biomass on painted surfaces in Latin America is fungal, while in Europe is algal.[47]

Areas colonized by dark biofilms formed by cyanobacteria may absorb more sunlight. Temperature changes increase mechanical stress by expansion and contraction of the biofilms.[91] Temperatures on dark stained areas of stone can differ as much as 8°C from lighter colored areas.

Cyanobacteria showed an active role in the deterioration of surfaces of historic buildings as several studies reported (*vide infra*).

Frescoes and wall paintings are substrates that present often discolorations formed mainly by cyanobacteria, algae, actinobacteria, but also by fungi and lichens (regarding these two colonizers, see next paragraphs). The mural paintings of the crypt of the Original Sin (Matera, Italy) showed discolorations varying in color—brilliant green, dark green, brown, rosy, and black.[113] Cyanobacteria (*Chlorogloea microcystoides*, *Chroococcus lithophilus*, *Gloeocapsa* sp., *Gloeothece rupestris*, *Pseudocapsa dubia*) and green algae (*Apatococcus lobatus*, *Chlorella vulgaris*, *Chlorococcum* sp., *Muriella terrestris*) produced most of these phenomena. Pink-rose patinas had a homogeneous appearance but showed changes in intensity and extension depending on seasons. They were present in the more xeric areas of the paintings. The pink color resulted from the production of a ruberin-type carotenoid pigment produced by photoheterotrophic actinobacteria.[114] Similarly, carotenoids caused pink and yellow discolorations on wall paintings of St. Botolph's Church (Hardham, UK).[115] These pigments provide both photosynthetic and nonphotosynthetic organisms with a protective mechanism against damage caused by photooxidation reactions. They seemingly contribute to the maintenance of the membranes of heterotrophic bacteria.

Rosy-powdered areas on the frescoes in the St. Brizio Chapel (Orvieto Cathedral, Italy) contained instead phycoerythrin, a pigment produced by capsulated coccoid cyanobacteria that grew even in the dark conditions of the chapel being able to use the organic compounds present on the fresco surface.[94] An analogous alteration—pink and yellow discolorations on wall paintings of two churches in Georgia[116]—was due to the bacterium *Micrococcus roseus*.

The green alga *Trentepohlia* formed red powdery spots on a medieval wall painting in Italy.[117] The worldwide famous wall paintings in the Lascaux cave (France) suffered from a green biofilm, called "le maladie verte," formed by green algae belonging to Chlorophyta division[118] (*vide infra*).

The damaging effects of microalgae relate mainly to the promotion of growth of other organisms as the accumulation of their biomass provides an excellent organic nutrient for subsequent heterotrophic microbiota.[87] On the other hand, a study on algae developed on concrete structures showed that they absorbed chemicals such as calcium, silica, and magnesium from the cement paste causing the formation of small cavities and cracks.[105]

The endolithic growth is an adaptive strategy developed by cyanobacteria and, at a lesser extent, algae. The use of the substrate as a shield against external stress proves to be a decisive evolutionary selection advantage.[119] Endolithic communities are those formed by microorganisms that inhabit preexisting cracks and fissures connected to the rock surface (chasmoendoliths) (Fig. 3.4), or grow inside cavities and among crystal grains and are not visible (cryptoendoliths) (Fig. 3.5), or actively produce their own cavities through weathering immediately below the rock surface (euendoliths)[8] (Fig. 3.6). The first form of growth leads to a co-responsibility in the detachment of scales of material due to the pressure exerted by increasing biomass (Figs. 3.4 and 3.5). This process can occur repeatedly, involving areas increasingly in depth.[120] Euendolithic species are able to dissolve the substrate developing within it (Fig. 3.7) and forming microcavities of varying morphologies according to the species. The light that reaches the bulk of the stones limits the growth of these microorganisms. Cyanobacteria and algae colonized the pores in exposed dolomite rocks (Piora Valley, Swiss Alps) appearing as distinct grayish-green bands about 1–8 mm below the rock surface and showing a very active potential of penetration.[121] In the Mediterranean environment of the archaeological site of Baelo Claudia (Spain), cyanobacteria grew inside the mortars up to 2 mm.[122] The mucilaginous sheath of these microbes was a nucleation center of calcium carbonate crystals, which accumulate around cells originating honeycomb-like structures. Successive processes of dissolution–precipitation (e.g., cation exchange, chemical weathering, leaching, diagenesis), eventually produced the loss of external layers of mortars due to the important changes in volume that cells underwent.

(a)

(b)

FIGURE 3.4 Microcolonial meristematically growing fungi on a sample from the façade of Tempio Malatestiano, Rimini (Italy) (a). Internal surface of the same sample with chasmoendolithic algal growth (b). Stereomicroscope, 20×.

FIGURE 3.5 A blister on a limestone slab, Chaalis (France). When a fragment of the blister broke down, the cryptoendolithic algae appeared. They grew in the bulk of the stone, underneath the external surface. It is an example of the severe damage that biofilms can cause in association with other abiotic agents. The overall action of biofilms and associated extracellular polymeric substances is the major cause of the huge plastic deformation of the stone. Photo courtesy of José Delgado Rodrigues.

FIGURE 3.6 Polished cross-section of a marble sample showing algae and fungi in the bulk of the stone. Gravestone, English Cemetery, Firenze (Italy). Visible light imaging.

(a)

(b)

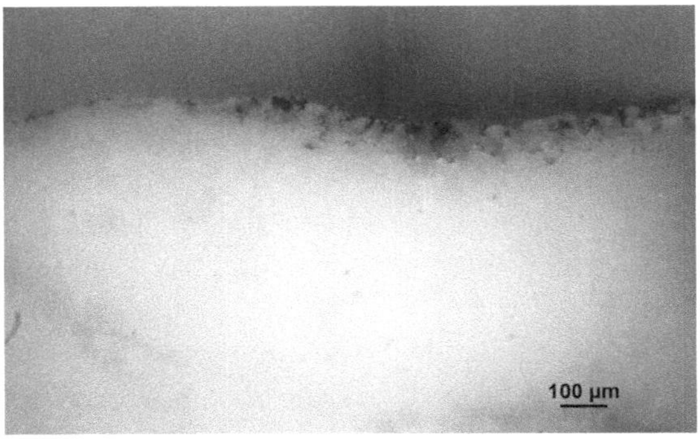

(c)

FIGURE 3.7 Marble well-curb (fortress of Montefiore Conca, Italy) colonized by biofilms (a). Peculiar pattern in stone weathering visible on a sample. Stone parts completely free of biofilms are surrounded by strongly eroded parts where biofilms grew. Stereomicroscope (b). Polished cross-section of the sample showing fungi on the surface and cyanobacteria in the bulk of the stone up to circa 300 μm. Visible light imaging (c).

Euendolithic microorganisms produce on surfaces blind holes that are close together and generally cylindrical. The alteration is called *pitting.*[123] The existing literature on the subject is extremely heterogeneous, with wide differences in the scientific approaches. A review provided a summary of the results.[124] Bio pitting occurs mostly on vertical and sub vertical surfaces, with a preference for southern exposures. The stone frequently associated with this kind of deterioration is marble (53%), followed by carbonate rocks (44%), granite, and concrete (3%). Bio pitting is caused mostly by cyanobacteria along with fungi (11%), lichens (10%), and algae (5%). The ecological conditions favoring this phenomenon are climatic dryness ad scarcity of nutrients.

In subterranean environments like caves, the phototrophic biofilms are a relevant cause of damage.[125] A progressive deepening of their growth into the substrates leads to the mobilization of elements and to enhanced water retention by polysaccharide sheaths. Mineral crystals were observed on the cells of some cyanobacteria (*Scytonema julianum* and *Loriella* sp.), and Streptomyces isolated from caves' walls. *Scytonema julianum* is a typical inhabitant of hypogea environments characterized by high humidity and low irradiance. As reported by the authors, Streptomyces, always associated to *S. julianum*, promote the precipitation of calcium carbonate on the surface of the polysaccharide sheath of cyanobacteria in form of calcite crystals.[88,126,62] Calcium carbonate precipitation and dissolution are major processes in the biotransformation of calcareous substrata in caves. A white crust formation is generally associated to the surface deposition of newly formed crystals and results in a stromatolithic layering on the stone surface. Similar $CaCO_3$ crystals precipitation induced by actinobacteria, Firmicutes, and proteobacteria that grew on surfaces of Roman catacombs showed instead a different morphology being in form of a white fluffy.[82]

In Altamira Cave (Spain), the biomineralization processes formed two main types of $CaCO_3$ deposits: Rhombohedral and spheroid to hemispheroid crystals.[127]

Actinobacteria, especially Streptomyces, and filamentous fungi (*Sporotrichum*, *Aspergillus*, *Cladosporium*, *Penicillium*, *Torrubiella*, etc.), commonly occur together with photosynthetic microbes in the catacombs.[128,113,129] In addition, chemoorganotrophic bacteria grow on caves' walls and their effect is tremendously harmful on decorated surfaces. In Hal Saflieni Hypogeum (Paola, Malta), they extensively formed white alterations on ochre-decorated surfaces dated back to 3300–3000 B.C.[63,82]

Many Streptomyces strains were isolated from mural paintings of Tell Basta and Tanis tombs (Egypt) that showed color changes. According to the authors, the Streptomyces were involved in this alteration by producing a wide range of metabolites such as acids, pigments and hydrogen sulfide.[130]

An alteration, whose origin has been recently object of debate, is worth mentioning here. It is the presence of orange-red stains on outdoor marble monuments and statues.[131–134] They have an inhomogeneous distribution covering large areas of marble or some square centimeters or appearing as small circular spots (Fig. 3.8). The coloration is partially on the surface but often is present inside the marble, about 300 μm in depth from the surface, in the form of very thin films in the inter-granular spaces of calcite crystals. In some cases, the discoloration can be due to microorganisms. A study showed that the red-brown stains of a marble statue (Noguchi's Slide Mantra) were microbial in origin.[135] A microorganism related to the bacterium *Serratia marcescens* was isolated from the surface of the sculpture and cultured in the laboratory. It produced a red-brown pigment identified as prodigiosin. In most cases, however, the red stains contain minium (Pb_3O_4), a lead oxide that can form after lead leaches into the marble from lead support structures and accoutrements used to seal joints and embed iron elements on the monument.[133]

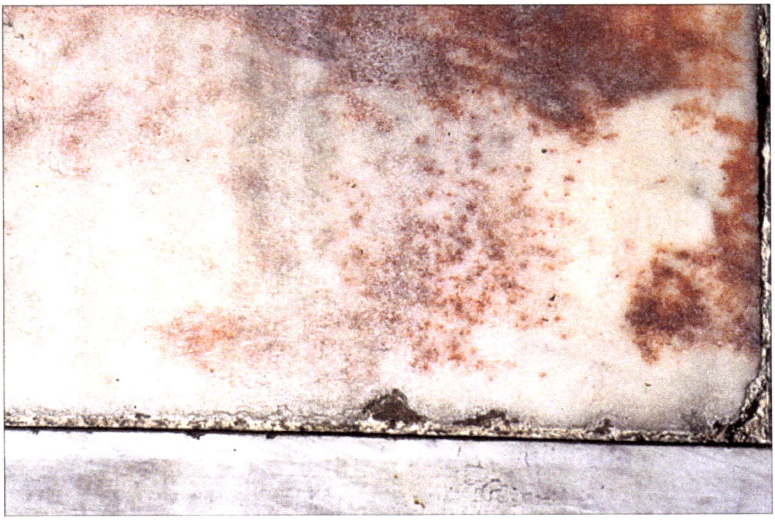

FIGURE 3.8 Red stains on marble surfaces of the tympanum frame of Porta della Mandorla, northern portal of the cathedral, Firenze (Italy), due to lead used in the settling of the marble slabs.

3.4 FUNGI

Many fungi precipitate reduced forms of metals and metalloids from rocks in and around fungal hyphae. Metal mobilization can be achieved by chelation ability of metabolites and siderophores, and methylation. When living in environments of reduced iron content, fungi produce iron(III)-binding ligands, commonly of a hydroxamate nature, termed siderophores.[8]

Most fungi that grow on stone objects belong to Hyphomycetes, a class of asexual or imperfect fungi of the subdivision Deuteromycotina, division Deuteromycota. They lack fruiting bodies, that is, the sexual structures used to classify other fungi. Therefore, the production of conidia (spores) occurs by fragmentation of vegetative hyphae or from specialized hyphae called conidiophores. Hyphomycetes have often been referred to as "fungi imperfecti." They are commonly known as mold.[136] Many Hyphomycetes, notably *Aspergillus*, *Fusarium*, and *Penicillium*, produce toxic metabolites, or mycotoxins. Several Hyphomycetes growing on stone cultural heritage objects are dematiaceous. The term refers to the characteristic dark appearance of these fungi that form dark gray, brown, or black colonies.

Many papers deal with the role that fungi play in weathering of stones and wall paintings[137,138,98,117,139–141] (Fig. 3.9).

FIGURE 3.9 Extensive colonization of many fungal species on a wall painting (Palazzo Milzetti, Faenza, Italy). The growth occurred in a few days and was caused by water flowing inside the wall from the accidental burst of a water pipe.

The growth of fungal hyphae is often accompanied by the production of organic anions and protons, which help to break down weak fragments in solid rock substrates.[142] Many species of Hyphomycetes attack the rock by excretion of organic acids such as acetic, oxalic, citric, formic, fumaric, glyoxylic, gluconic, and to a lesser extent, tartaric acid.[143,144] Particularly, oxalic acid has a high capacity of degrading minerals for its complexing and acid properties. Anion oxalate is a key metabolite playing a significant role in many metal and mineral transformations mediated by fungi.[145] Oxalic acid is a relatively strong acid. Its importance relates though to the ability of the oxalate anion to complex and/or precipitate metals as secondary biominerals that can result in mineral dissolution or metal immobilization.[145] Calcium oxalate precipitation has an important influence on biogeochemical processes in soils, acting as a calcium reservoir.[145] The role of oxalic acid is thoroughly discussed in the next paragraph.

The fungal invasion of prehistoric drawings in Lascaux cave (France) generated worldwide interest because they represent a priceless cultural heritage for all humankind. The fungus *Fusarium solani* formed long white mycelia, with a fluffy appearance. Moreover, dematiaceous fungi produced black stains on the ceiling and passage banks.[118]

Fungi are the principal deteriorating microflora on painted surfaces.[146] The most common genera of fungi that grow on damp wall surfaces are *Aspergillus*, *Penicillium*, *Cladosporium*, *Aureobasidium*, *Alternaria*, *Chaetomium*, *Acremonium*, *Ulocladium*, and *Stachybotrys*.[146,147]

Testing of a mixed fungal inoculum on painted gypsum plasterboards under laboratory conditions (approximately 100% humidity and 20 ± 2°C for 12 weeks) showed that *Ulocladium atrum* mainly developed. *Penicillium purpurogenum* also contributed significantly to the fouling biofilm.[146] The study pointed out antagonism between fungal isolates. *U. atrum* produced an antifungal secondary metabolite, the antibiotic PF 1052 and *P. purpurogenum* was antagonistic to 10 of the other fungi used in the test. While *U. atrum* did not cause apparent changes in the paint surface, *P. purpurogenum* produced mm-sized erosions, which, in some cases, completely penetrated the paint layer, exposing the underlying gypsum.[146]

Fungi belonging to *Cladosporium* genus and bacteria of *Bacillus* genus altered the green pigment malachite of wall paintings in Portugal.[139] As reported by the authors, calcium oxalates caused the discoloration, as these compounds were associated to high levels of microbial contamination in the altered areas.

Fungi have a greater potential than bacteria to dissolve rock phosphates whose solubility depends on the decrease in pH and on chelating reaction of organic acids with minerals containing Fe^{3+}, Ca^{2+}, and Al^{3+}. Hyphomycetes and yeasts are highly effective sorbents for various cations: Fe, Ni, Zn, Ag, Cu, La, Pb, Cr, and Mo. Therefore, fungi can form euendolithic communities with hyphae actively penetrating inside the calcite grains, developing networks of hyphae with dense extracellular matrixes in the bulk of the substrate and leaving tunnels and boreholes on the surface.[141]

Due to their modular and adaptable growth, fungi develop under extremely variable environmental conditions. They utilize the airborne anthropogenic compounds and thus are prevalent in urban conditions.[90] The importance of various weathering mechanisms as a function of fungal species composition, rock type, and external conditions (moisture, temperature, pollution, and climate) is a relevant issue.

Interesting studies[148,149] discuss a very important topic, the quantification of risk by fungal growth on surfaces. In fact, standard methods for the detection, assessment and quantification of biodeterioration are essential to compare results obtained by different researchers.[47] Although the studies did not focus on cultural heritage buildings, they can be relevant all the same. They correlated the amount of ergosterol to the colony-forming unit (CFU*) value of fungi inoculated on a cultural medium and on samples made of concrete, gypsum board, emulsion coat, brick, and plaster. Ergosterol is a basic sterol of cellular membranes in filamentous fungi and yeast. Ergosterol content lower than 2.12 mg/m^2 corresponds to the normal level of spores' contamination without any active growth; values ranging 2.12–3.96 mg/m^2 indicate the activation of mycelium growth, and those over 3.96 mg/m^2 reveal active fungal growth and high contamination. The minimum detection level for ergosterol is 12.5 μg/m^2, well below macroscopic detection levels.[149] Accordingly, it could be possible to estimate the level of fungal contamination of materials, also when the mycelium is inactive and cannot be detected using traditional methods, or before it becomes macroscopically visible. However, some authors point out that the amount of ergosterol depends on many factors, such as the type of material, its moisture content, the microorganism species and age, and the growth conditions, while temperature does not have significant effects.[147]

*In microbiology, the CFU is used to estimate the number of viable bacteria or fungal cells in a sample. Counting with CFU requires culturing the microbes.

Among fungi, microcolonial meristematically growing black fungi were deeply studied recently. Their cells have thick-pigmented (melanin-containing) walls. They form slowly expanding, cauliflower-like colonies that grow by isodiametric enlargement of the cells.[144] In addition to the meristematic growth, many of the black fungi can exhibit a yeast-like growth.[150] They abandoned the hyphal phase adopting the microcolonial or yeast phase characterized by an extremely slow growth, in response to the lack of organic nutrients and to stresses of outdoor substrates (Fig. 3.10). They produce various pigments including carotenoids, mycosporines, and melanins that protect them from UV irradiation. Moreover, melanin provides them with extra-mechanical strength making hyphae able to grow better into fissures. Microorganisms that thrive under extreme conditions have been termed "poikilotrophic," that is, able to deal with varying microclimatic conditions such as light, temperature, salinity, pH, and moisture.[8]

Air-dried mycelia of some meristematic fungi tolerate temperatures as hot as 120°C for at least 0.5 h.[143] As a response to temperature stress, these fungi develop multilayered cell walls and accumulated trehalose. Under NaCl stress, the intracellular amount of glycerol regulates the osmotic potential.[144] A study[151] focused on the effect of temperature stress on black rock inhabiting fungi grown at different temperatures. The protein profiles were analyzed in comparison with each other and with the fungus *Penicillium chrysogenum* as a reference strain since it is well characterized under different growth conditions in nature and at laboratory scale. The black fungi, when exposed to 40°C, a temperature significantly above their growth regime, responded with a reduction of the total number of proteins, a reaction different from that of *P. chrysogenum*. In contrast, they considerably increased the number of proteins at 1°C. As reported by the authors, a special set of proteins, present in black fungi, provide cells' survival to temperature stress. The new, still unknown, proteins do not commonly occur in mesophilic fungi[†].

The microcolonial phase enhances the survival and persistence on these fungi in the biofilms. They are stress-tolerant colonizers involved in biodeterioration.[152,13,98,144,153,154] However, the scarcity of nutrients and the very slow growth limit their contribution to weathering.[142]

[†]A mesophile is a microorganism that grows best in moderate temperature, neither too hot nor too cold, typically between 20 and 45°C. Most fungi are mesophilic.

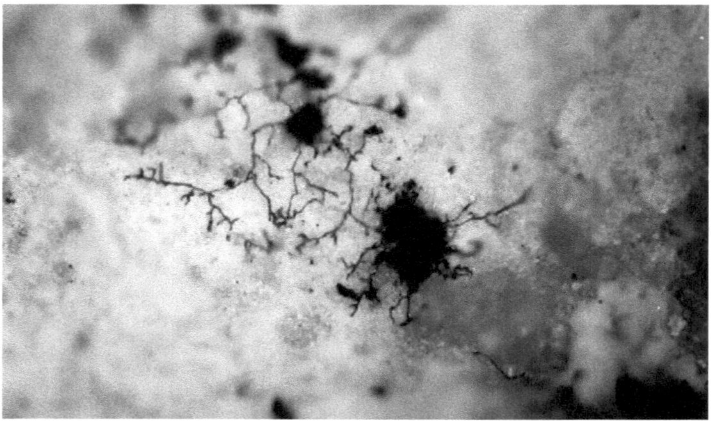

(a)

(b)

FIGURE 3.10 Growth of meristematic black fungi on limestone. Stereomicroscope, 50× (a), SEM image (b). SEM image courtesy of Isetta Tosini.

A laboratory study on meristematic microcolonial fungi growing on carbonate (travertine, limestone, marble) and silicate (granite, gneiss, sandstone) rock slabs added with a culture medium, detected high variability in penetration depth.[18] *Coniosporium perforans* and *C. uncinatum* exhibited a deep penetration within marble (3.5 mm) and limestone (6.6 mm). *Sarcinomyces petricola* instead poorly or not penetrated these stones while showed the maximum penetration (7.4 mm) within granite. The latter fungus was also able to colonize the bulk of gneiss slightly more (2.4 mm) than the other strains. The fungi secreted unidentified siderophore-like iron-chelating compounds, which might be the cause of chemical deterioration of rock slabs.

Fungi develop in hypogea once the environmental conditions are favorable causing great damage as it happened on the wall paintings of Takamatsuzuka Tumulus (Japan) discovered in 1972.[138] The temperature of the tomb was quite stable until 1980, when it began to rise following the outdoor temperature. By 2000, concurrently with the opening to visitors, the temperature reached nearly 20°C creating an ideal microclimate for fungi that widely colonized the wall paintings. The authors reported that the only way to prevent further deterioration would be the detachment of the wall paintings for restoration.

Several black fungal spots were spread on the wall paintings of Tutankhamen's tomb (Valley of the Kings, Luxor, Egypt) when it was discovered in 1922.[155] Fungi caused the spots, mainly *Aspergillus* spp. (*A. nidulans*, *A. niger*, *A. flavus*, *A. terreus*) and to a lesser extent *Cladosporium* spp. and *Aureobasidium* spp. Although located in a desert, over the centuries periodic floods occurred allowing moisture enter the tomb chambers. The perspiration and breathe of visitors was a further source of moisture. According to a report by the Getty Conservation Institute (www.getty.edu/conservation), six people breathing in the tomb chamber for an hour raised the RH level by 3%.

Arthropods are vectors of entomopathogenic fungi (parasite of insects) and can play an important role in caves, catacombs and mural paintings.[156] As large varieties of arthropods live in caves, the investigation of fungi should imply an examination of arthropods, if abundant, to reduce their population, decreasing consequently fungal contamination.[156]

Fungal colonization becomes a relevant problem in the museums when economic crisis or war damages to buildings make it impossible to arrange proper safeguarding of collections. For example, in the National Museum of Belgrade (Serbia) no heating in winter and high temperatures in summer triggered a widespread fungal contamination on many archaeological objects.[157]

3.5 LICHENS

Many studies demonstrate the damaging capability of lichens. Generally, the physical and chemical weathering of rocks by epilithic crustose lichens encompasses the following mechanisms[158]:

1. Penetration of the hyphae through intergranular voids and mineral cleavage planes;
2. Expansion and contraction of the thallus by microclimatic wetting and drying;
3. Freezing and thawing of the thallus;
4. Swelling and deteriorative action of organic and inorganic acids originating from lichen activity;
5. Incorporation of mineral fragments into the thallus.

Effective physical weathering requires chemical weathering since penetration of hyphae is facilitated by the dissolution of minerals along grain boundaries, cleavages, and cracks. The overall effect increases porosity and permeability. Since these weathering processes interact and enhance each other action, it is impossible to separately quantify their role.[159,160] Lichens produce a variety of different organic compounds, generally called lichenic substances, and many organic acids. One of them is oxalic acid, a strong complexing agent. It is an intermediate product of the cellular metabolism of many organisms. In lichens, mycobiont secretes it. In comparison with other organic acids, it is one of the most active agents of chemical alteration of stones because of its complexing and acid properties. In fact, it forms chelating bonds thus producing calcium, magnesium, manganese, and copper oxalate crystals at the rock-lichen interface and in the lichen thalli.[161–165]

Oxalates have been implicated in Fe, Si, Mg, Ca, K, and Al mobilization from sandstone, basalt, granite, calcareous rocks, and silicates.[145] The precipitation of poorly ordered iron oxides and amorphous alumino-silica gels, the neo-formation of crystalline metal oxalates and secondary clay minerals have been frequently identified in a variety of rocks colonized by lichens in nature[158] (Fig. 3.11).

Calcium oxalate occurs in two crystalline forms—the dihydrate (weddellite $CaC_2O_4 \cdot 2H_2O$) and the more stable monohydrate (whewellite, $CaC_2O_4 \cdot H_2O$). Calcium oxalate crystals are commonly associated with lichens and with free-living, pathogenic, and plant-symbiotic fungi. They are formed by precipitation of solubilized calcium as calcium oxalate. Fungal-derived calcium oxalate can exhibit a variety of crystalline forms (tetragonal, bipyramidal, plate-like, rhombohedral, or needles).

FIGURE 3.11 A crustose lichen on a natural outcrop made of a red siltstone (Ericeira, Portugal). The lichen has the same color of the stone because it likely incorporated in the thallus the colored compounds of the rock. Photo courtesy of José Delgado Rodrigues.

Regarding oxalates, it is opportune to make a digression to discuss on calcium oxalate patinas and deposits‡. In fact, this is not a true digression because these patinas/deposits are related to lichens (and biofilms) as it will be explained. Structural, textural, and/or material transformations can cause thin-colored patinas/deposits (20–200 μm) on stone surfaces. *Patina* and *deposit* are all-embracing terms in the terminology of materials' surface changes due to natural ageing by chemical–physical reactions going on near, at and above the original surface of a stone. The interactions of stone surfaces with the environment, the biological activity, and ancient protec-

‡The definition of this alteration is controversial. Some conservation scientists use the term *patina*, while others refer to it as a *deposit*. As this book is not the right context to discuss about terminology, both terms are reported.

tive treatments as well lead to stone changes. These aging processes can form films and deposits over the surface or transform the external part of the stone. The color of patinas and deposits varies from black to gray, from brown to orange and can derive from many substances (fly-ash or other dust, biominerals formed by microbiological activity, Fe content of minerals, organic pigments like melanins, melaninoids, humic substances, rusty decay products of chlorophyll). Calcium oxalate patinas/deposits, frequently found on many monuments, especially on those built in calcareous stones, are composed almost exclusively of calcium oxalates (Fig. 3.12).

FIGURE 3.12 Calcium oxalate layer on Trajan's column, Rome (Italy). Photo courtesy of Giulia Caneva.

Due to their hardness and insolubility, the oxalate layers play a considerable role in the protection of stones from environmental agents and deterioration. Their genesis has been the subject of a far-reaching discussion in the scientific community; the topic is still matter of debate. Lichens and fungi's capacity to produce calcium oxalate has induced some authors to hypothesize that the patinas were originally produced by past biological growth that is no longer present today because of atmospheric pollution. Other authors, on the contrary, affirm that these layers are the remaining traces of intentional surface treatments and contain the decomposition products of organic materials applied for protective and/or esthetic purposes.[166] Over time, these substances reacted with the stone and mineralized into calcium oxalate. It is likely that the action of microorganisms or abiotic agents caused their transformation. Only in a very limited number of cases, which cannot be generalized, the patinas/deposits could be the traces of past lichen growths (see review in Pinna and Salvadori, 2008[167]).

Lichens also develop on wall paintings causing severe damage. The crustose lichen *Dirina massiliensis* f. *sorediata* grew on wall paintings in ventilated areas of a crypt[113] and in a medieval cave exposed to external environmental conditions for the collapse of a portion.[117] This lichen formed gray-green powdery thalli that merged forming a compact and continuous crust including materials derived from the paintings. Old thalli appeared inflated at the center creating blister-like structures and causing the detachment of painting layers.

Many are the studies on the deteriorative effect of crustose and foliose lichens on cultural heritage. The growth of crustose lichens can cause severe alteration of the stones as the thalli detached and incorporated mineral fragments[168,76,13] (Fig. 3.13).

The penetration of *Caloplaca teicholyta* into the dolostone of churches in Segovia (Spain) severely altered the dolomite crystals.[13] Fungal growth first occurred through the inter-crystalline porosity, followed by the intra-crystalline penetration of hyphae and eventually by the complete loss of crystal structure when the thalli surrounded rock fragments. This type of colonization and degradation depends strongly on the dolostone textural properties.

The most eroded areas on marble statues in the gardens of the National Palace of Queluz, Portugal, corresponded to those colonized by algae and lichens.[169] Their colonization rates showed a very high dependence on the local environmental conditions; they grew more rapidly on sculptures located under or very close to large trees. On sandstone sanctuaries in Thailand[168] lichens altered feldspar grains to clay minerals. In arid conditions, lichens may accelerate weathering because they protect the rock against dehydration.[170]

Epilithic lichens (*Aspicilia intermutans*, *Lecanora bolcana*, *Rhizocarpon lecanorinum*, *Tephromela atra*, and *Xanthoparmelia pulla*) caused ruptures of primary minerals (even quartz) of granodiorite in the Sila uplands (Italy), with detachment and progressive incorporation of fragments into the thallus.[171]

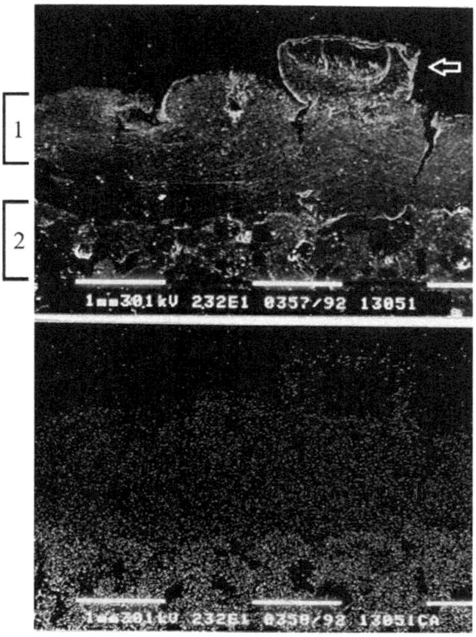

(a)

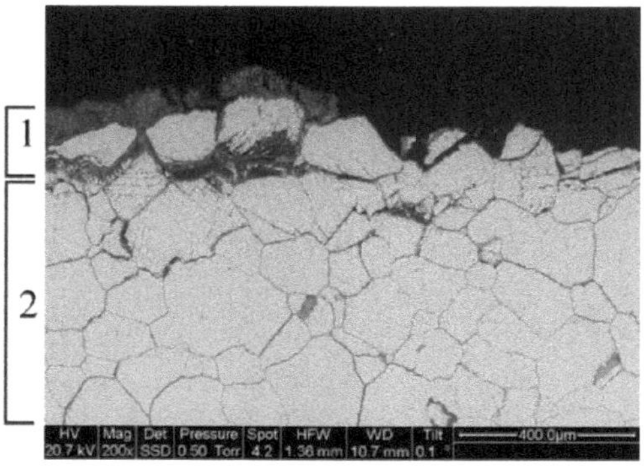

(b)

FIGURE 3.13 Cross-section of epilithic crustose lichens (1) on marble (2). SEM–EDS micrograph and map distribution of calcium in the lichen thallus and in the substrate. An apothecium is visible (white arrow). The apothecium is a spore-bearing structure consisting of a cup-shaped body (a). SEM–BSE micrograph showing that the lichen thallus enters stone microfractures and causes the detachment of stone fragments. The strong backscattering signal from fungal hyphae is due to OsO_4 staining of cytoplasmic lipids (b). Appl. Phys. A, Removal of *Verrucaria nigrescens* from Carrara Marble Artefacts using Nd:YAG Lasers: Comparison Among Different Pulse Durations and Wavelengths, 118(4), 2015, 1517–1526, Osticioli, I., Mascalchi, M., Pinna, D., Siano, S. With permission of Springer.

Several studies have shown that lichen-covered rocks weather an order of magnitude faster than bare-rocks.[170] The chemical weathering rate of outcrops of mica–schist (Hubbard Brook Experimental Forest, USA) by foliose and crustose lichens was measured quantitatively and compared with the weathering rate of a bare rock.[172] The experiment used mini watersheds (<1 m^2). Elemental fluxes of K^+, Na^+, Mg^{2+}, and Si in runoff from rain events of all the lichen mini watersheds were much greater than the elemental fluxes from bare-rock mini watersheds.[172]

The medulla layer of three epilithic crustose lichens (*Sporastatia testudinea*, *Lecidea atrobrunnea*, and *Rhizocarpon geographicum*) that grew on serpentinites (alpine environment, Val d'Aosta, Italy) contained numerous lithic fragments showing the weathering effect of the species.[76] Similar results were obtained studying epilithic lichens that developed on granite.[173,174] The deterioration though varied considerably from species to species independently of environmental conditions and substrate. Some lichens caused an intense damage to the stone, showing actions of dissolution/etching of mineral grains and precipitation of amorphous and crystalline compounds.

A 1-year long laboratory experiment on lichen–rock interactions was conducted using mycobionts and photobionts of the endolithic lichens *Bagliettoa baldensis* and *Bagliettoa marmorea*.[175] They were isolated and then inoculated on marble and limestone samples. The same species growing on limestone outcrops and abandoned marble quarries showed penetration pathways very similar to those reproduced *in vitro*. The study highlighted that lichen-driven erosion processes increased the availability of hyphae passageways only after a long-term colonization. The differences in hyphae colonization and penetration depended on the mineral composition and structure of the lithotypes. Recent researches[176,18] suggested that the chemical deterioration of silicate and carbonate rocks by endolithic lichens depends on the secretion of siderophore-like compounds and of carbonic anhydrase. This enzyme increases the speed of the reaction $CO_2 + H_2O \leftrightarrow HCO_3^- + H^+$, thus accelerates the dissolution of calcium carbonate quickening the hydration of respiratory CO_2. This scenario, which needs further experimental evidence, enlightens the complexity of phenomena involved in carbonate deterioration, which remains an interesting goal for further researches.

There are evidences that different climates affect the endolithic growth of lichens.[119] In the humid Northern Alps (Austrian glacier), the bulk of the calcareous rock, just under the surface, showed three layers. At a depth of 0–150 μm, the substrate is mostly intact, just partly interlocked with the lichen cortex. At 150–300 μm, photobionts (green algae or cyanobacteria) are capable of actively dissolving the substrate, thus creating habitable cavities. Then several mm in depth, there is the mycobiont with hyphae actively solubilizing the substrate and often forming dense networks. Most biomass of endolithics on carbonate rocks in the arid Mediterranean-Maritime Alps (Provence, France) was instead confined just under the surface. The average colonization intensity and penetration depth are markedly deeper in the more humid substrate of the Austrian Alps.

Endolithic lichens are agents of biological weathering particularly in severe environments.[177] The damaging effect of *Lecidea auriculata* on pyroxene–granulite gneiss boulders of the glacier Storbreen (Norway) was proved measuring rock-surface hardness by Schmidt hammer technique[§]. The lichen's action was both rapid and highly effective even during the early stages of colonization. Circa 20 years after the starting of lichen growth, mean values of lichen-covered surfaces were at least 10 units lower than lichen-free surfaces. As reported by the authors, rock weathering by some endolithic lichens can be fast, equivalent to an enhancement of the rate of weathering on lichen-free rocks about 200 times in 50 years.

Although it is now widely accepted that lichens can play a relevant role in both the processes of physical and chemical weathering of rocks and minerals, the overall effect of lichens in this weathering is still matter of debate. The crustose lichen colonization does not always indeed result in a damage to the substrate (Fig. 3.14).

A study of different kinds of sandstone from Wyoming covered with lichens and old petroglyphs[178] demonstrated for example that lichens were not one of the key factors in the deterioration of the petroglyphs, either in a negative (destruction of the outermost layer) or in a positive way (protection from rain, sun, etc.). Lichens filled the gaps between the grains, which were large enough to host them without exercising relevant pressure. Porosity was less toward the outside, since in the outer layer lichens occluded the pores. Counting the lichen thallus, the porosity proved to be the same as in the core of the rock. According to the authors, the deterioration of the sandstone depended mainly on the nature of the sandstone itself,

§Schmidt hammer measures the rebound of a spring-loaded mass from the rock surface. The rebound value can be converted to give the compressive strength of the rock. Rock surface roughness strongly influences the results.

in particular on the dimension of the quartz grains: The larger the grains, the greater the porosity, water absorption, fragility and decohesion of the sandstone. The results may help decide whether to eliminate the lichens from the surface and to account the need for their removal to esthetic or site management reasons rather than to chemical–physical reasons.[179]

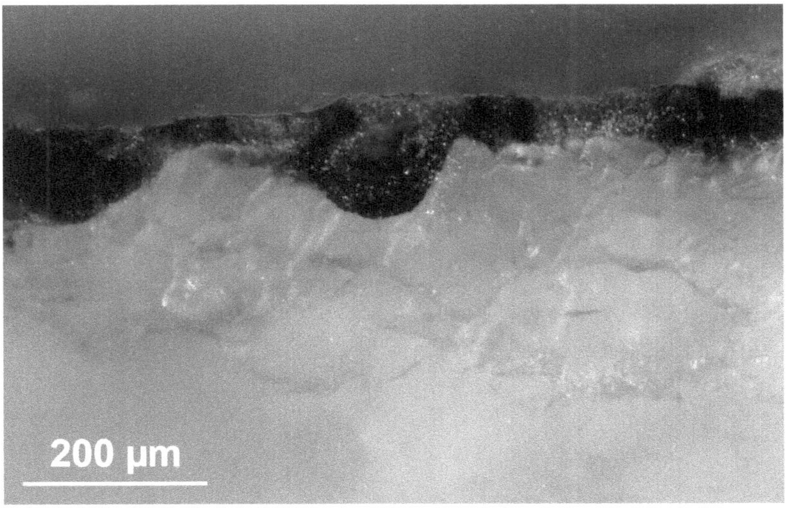

FIGURE 3.14 Polished cross-section of the lichen *Verrucaria nigrescens* showing that it covers the substrate but does not penetrate the bulk of the stone. Visible light imaging. Phys. A, Removal of *Verrucaria nigrescens* from Carrara Marble Artefacts using Nd:YAG Lasers: Comparison Among Different Pulse Durations and Wavelengths, 118(4), 2015, 1517–1526, Osticioli, I., Mascalchi, M., Pinna, D., Siano, S. With permission of Springer.

Epilithic crustose lichens showed different degrees of weathering of sandstone rock carvings in western Norway.[159] A porous gray to beige rind beneath the lichens was the visible effect of weathering. Its thickness varied 3–24 mm. It was circa 16 mm beneath *Ophioparma ventosa*, and circa 8.5, 10, and 9.1 mm beneath *Fuscidea cyathoides*, *Ochrolechia tartarea*, and *Pertusaria corallina*, respectively. The authors explained the differences in weathering with the production of different lichen compounds, which are active in complexing and removing elements from the substrate. In fact, *O. ventosa* and *P. corallina* contain thamnolic acid, and *O. ventosa* contains divaricatic and usnic acids. Moreover, minerals differ in their resistance to dissolution. Quartz, potassium feldspar, and muscovite are the most resistant minerals while calcite, apatite and chlorite are the least stable ones. Plagioclase, in this case, weathers more readily than potassium feldspar because of a greater abundance of mica inclusions.

Beneath lichens, a wide variety of lithobiontic microorganisms develop. A study of the microhabitats beneath foliose and fruticose lichens that grew on granite (Spain) dealt with the role played by the lithobiontic community in the weathering process.[160] The interface lichen-substrate with lithobiontic microorganisms showed the greatest alteration. The microorganisms mainly attacked micas, as already shown by another study.[159]

In the attempt to quantify the lichen activity on stonework and organize information available in the literature, an index of Lichen Potential Biodeteriogenic Activity was suggested.[180] The Index aims to quantify the overall lichen impact on stonework considering the effect of each species, both on the surface and within the substratum. Other parameters—reproduction, physicochemical action, and bioprotection—are reported as well.

The time is ripe to devise few parameters that give an overview of the condition of an object colonized by biological growth. The parameters or indexes allow the evaluation of the impact of a biological growth on an object. Each parameter takes a value within a certain range, from nonimpact to severe impact. Unfortunately, the papers on this subject are very few, as it is described later.

3.6 BIOPROTECTION OF STONES BY BIOFILMS AND LICHENS

Regarding the role that biofilms and lichens play in weathering of natural and artificial stones, an increasing number of researches accounts for a negligible effect and even for a protection. Thus, the axiomatic correlation among biofilms, lichens and stone damage is a matter of controversy.[104] These researches bring a novel perspective in a field where many studies showed that biofilms and lichens do damage stones.

Interesting studies[37,181] distinguish the effects over time of biofilms and lichens on sandstone and on limestone. While these organisms temporarily stabilize loosely to moderately cemented sandstones, they damage limestone actively boring microcavities and gradually weathering the substratum. The authors hypothesized that the avoidance of a rapid decay of sandstone is a necessary condition for the growth of complex biofilms and lichens with moderate metabolic and reproductive rates. If weathering permanently disturbs a surface, they will be not successful competitors with other organisms. They form a tight network of cells and extracellular polymers, enabling their undisturbed growth over several years or even decades. Enwrapping the grains with a biogenic matrix temporarily stabilizes the surface and reduces weathering, which may allow the organisms to persist for years. Only the pioneer colonizers (algae, cyanobacteria, and fungi) can afford living in an unstable environment, surviving at rapid weathering rates. On homogeneous carbonate rocks, lichens and biofilms have a different pattern of growth because they can actively bore cavities. However, a structural weakening of the substratum involving the risk of sudden desquamation and destruction is uncommon. On carbonate substrata, endoliths with their relatively slow growth rates have a chance for a sustainable life for long periods. According to the authors, slow-growing lichens that develop over decades have this particular feature. They describe them as K-strategists, that is adapted to live at a "carrying capacity" of the environment, thus with low growth rate. This stage of colonization is an indication of low weathering rates on limestone because the organisms cannot afford to deteriorate the surface. In contrast, *r*-strategists grow rapidly on easily available substrates but have a relatively low probability of surviving.

Another study[119] specifies the steps of endolithic growth on carbonate substrates. In the initial colonization phases of an endolithic biofilm, a destructive impact on the carbonate rock surface occurs. However, as soon as the endolithic biofilm is established, the erosional activities are reduced, so that a stabilized biofilm may act as protection of the rock surfaces from further damage by other agents. The colonization of subaerial rocks by microbial endolithic biofilms is a slow process. The above-mentioned laboratory experiment[175] on mycobionts and photobionts of the endolithic lichens *Bagliettoa baldensis* and *Bagliettoa marmorea* inoculated on marble and limestone samples, presented similar conclusions. In fact, lichen-driven erosion processes increased the availability of hyphae passageways only after a long colonization. Moreover, the differences in hyphae colonization and penetration depended on the mineral composition and structure of the lithotypes.

An established biofilm is maintained by internal nutrient recycling and regrowth without increase of the biomass. A continuing substrate removal would result in a loss of the residual substrate between microorganisms and the atmosphere, thus depriving the biofilm of its protective cover.

Some recent results from sites in different environments support the above-described theory. After the removal of lichens on marble and sandstone in the archaeological site of Fiesole (Italy), the researchers carried out in situ water absorption measurements.[29] Both the substrates showed a quite low absorption capacity, which was even close to that of healthy substrates. This is very interesting because the only alteration detected on the stones was that of epilithic crustose lichens forming a uniform and continuous covering that could have acted as a protective barrier for the substrates, protecting them from nonbiogenic weathering. A study[182] on the lichen colonization of limestone from four quarries and eight European monuments exposed to various environmental conditions showed that organic matter made waterproof the stone and could act as a barrier to sulfate contamination.

Measurements of the capillary water uptake at different places on a temple of the Angkor Wat site (Cambodia)[99] showed that lichens protected the stone from rapid water uptake, whereas certain algal and blackened cyanobacterial biofilms increased it. Rebound hardness and drill resistance measurements taken on lichens did not show any evidence that they significantly affect the mechanical properties of the stones. As reported by the authors, the lichen layer regulated the humidity, thermal transmission, and water vapor diffusion, reducing thermohydric stresses of the stones.

Lichens have been suggested to provide protection from wind and rain,[183,184] or to limit erosion by reducing the level of water within the rock.[185] A dense lichen cover can form a physical barrier to erosion and bind grains together, buffering the effects of physical and chemical weathering agents.[158] Lichens may provide thermal blanketing, absorb aggressive chemicals, keep surfaces hot and dry more constantly, reduce boundary layer wind speed.[186] The retention of moisture within the thallus reduced thermal stress on a limestone surface.[187] More widespread exfoliation, saline efflorescence, flaking, and honeycombing occurred on noncolonized surfaces in comparison with colonized ones.[188,189] An experimental study showed that endolithic lichens substantially reduced loss of lithic material from a limestone surface.[190] Two types of limestone samples taken from bare natural outcrops and endolithic lichen-covered rocks were exposed to artificial conditions in the lab, simulating rainstorm using distilled water. The analysis of run-off showed that loss of both dissolved and particulate limestone from the bare rock surface was substantially higher than that from the lichen-covered surface, suggesting that a process of bioprotection may well take place.

Some authors propose even a more complex view of the issue suggesting that a single species may act protectively in one environmental context, but can be deteriorative in another.[186] For example, the epilithic crustose lichen *Verrucaria nigrescens* is common in Europe in both wet and dry climates. In a wet context, the lichen may act as a bioprotector, shielding the stone from the direct action of rainfall, acid attack, wind, runoff and dissolution. On the contrary, in a hotter and drier context, with a more extreme temperature regime, the lichen may have a dominantly biodeteriorative effect, amplifying temperature fluctuations at the rock surface and possibly leading to thermal shock and thermal stress.

Even algae seem to play a role in protecting stones. A survey of four sandstone structures in central Belfast exposed for around 100 years and colonized by green algae biofilms[55] showed that algal patches were associated with less weathered surfaces (i.e., harder algal patches were associated with lower coefficient of variation of surface hardness). This might indicate that green algal cover had a bioprotective role. On the other hand, the observed situation may refer to the preference of stable surfaces by algae.

All these results give relevant suggestions to thorough studies of the protective effects of biofilms and lichens in conservation of stone. Moreover, the knowledge of biofilms strategies for colonization would be relevant to understand whether they are capable to stabilize poorly cemented stones. As the literature concerning the degradation of stone by lichens and biofilms is copious, well documented and experimentally proved, any confutation must provide an equal weight of proofs to be valid.[191] For example, the positive effects of the lichen cover in reducing dangerous evaporation processes cannot outweigh the negative effects of their hyphal penetration.[192] Although the protective effects of lichens deserve further research, this aspect cannot be generalized, and each case should be examined on its own merits. The relative importance of various weathering mechanisms as a function of species composition, rock type, and external conditions (moisture, temperature, pollution, and climate) needs further studies. In conclusion, the scientific community involved in this field is mature enough to go beyond the studies that report just unending lists of species. The discussion and assessment of a "common language," that is, standard test methods for the evaluation of the damage by biofilms and lichens, are very important. Standard test procedures are essential to compare the results of different laboratories, and to interpret, understand

and evaluate the research. The application of standard procedures will ultimately result in proposing indexes of risk or danger of biofilms and lichens on different stones.

3.7 MOSSES AND VASCULAR PLANTS

Mosses play a relevant role in trapping debris and building up soil. They can penetrate some types of stone with the rhizoids. Their presence on porous materials can result in frost-related damage and large growth can reduce moisture evaporation from stone surfaces.

Vascular plants colonize stones range from small herbaceous species to woody shrubs and trees[193] (Fig. 3.15). In temperate climates, plants on walls are mainly nonvascular and herbaceous with a limited occurrence of arboreal components.[26,42]

FIGURE 3.15 Belem (Pará, Brazil). A house presumably abandoned, a condition that leads to a rapid and heavy colonization of plants in hot and humid climates. Photo courtesy of José Delgado Rodrigues.

Plants with roots or climbing and adhering parts of leaves and stems cause both esthetic damage and chemical and structural alterations to stones. The roots of trees and larger shrubs may cause damage in various ways including action on walls, foundations, paving, monuments, and below ground drains. The impact of plants is a significant cause of mosaic deterioration. The roots secrete chemical substances that attack building materials.[194] Tree roots can travel outwards from the base of the trunk at least as far as the branches and often much farther.[25,26,195] Other ways of damage are trunk and branch contact with walls and roofs, blockage of rainwater disposal systems by leaf fall, shading of surfaces. Herbaceous plants like *Mercurialis annua*, *Parietaria diffusa*, and *Sonchus tenerrimus* are less destructive than trees and shrubs like *Ailanthus altissima*, *Capparis spinosa*, *Clematis vitalba*, *Ficus carica*, *Hedera helix*, and *Rubus ulmifolius*.[194] Ivy (*Hedera helix*) is commonly found on buildings and monuments. Aerial roots and woody growth can penetrate open joints causing displacement of bricks or stones. Suckers and tendrils can damage the surfaces and leave marks upon removal. The shading effect of extensive growth may reduce moisture evaporation from wall surfaces. Other invasive ruderal plants are *Ailanthus altissima* and *Robinia pseudoacacia*.

The aerial roots of many plants enter the joints to reach the soil behind the walls where they ramify forming a root system to capture nutrients and water and contribute to anchorage. The increase in photosynthetic food production would trigger lignification and further bifurcation, and secondary thickening of the roots (Fig. 3.16).

The roots of plants of *Ficus* genus are exceptionally strong and searching, permitting penetration in the cracks between masonry blocks to explore the soil behind.[42] A peculiar harmful effect is that of roots on underground structures like catacombs, crypts and caves. Plants that grow on the ground placed above the underground structures can develop a strong root system, expanding many meters both laterally and vertically.[195] When the roots become massive, they can detach parts of rocks causing considerable problems of stability and safety. The species' "tendency" to this kind of deterioration depends on the age and vitality of the individuals, the type of soil/rock, and the distance between plants and underground structures. The case of the Jewish catacombs of Villa Torlonia (Rome, Italy) is paradigmatic.[195] The roots penetrated in vaults and walls, even though they are about 10 m beneath the ground surface. Among the various plants that grew in the historical garden above the catacombs, *Ficus carica*, *Quercus ilex*, and *Pinus pinea* developed a strong root system. The roots of *F. carica* grew around 50 m from its location (Fig. 3.17).

FIGURE 3.16 The strangler fig (*Ficus*) extends it long roots to the ground surrounding the temple Ta Phrom at Angkor, Cambodia. Photo courtesy of Barbara Salvadori.

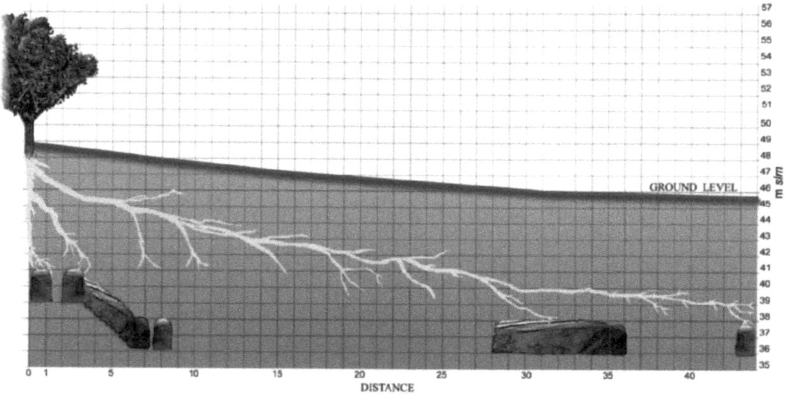

FIGURE 3.17 Graphical representation of the extraordinary development of roots of the common fig tree (*Ficus carica* L.) that damaged the Jewish catacombs of Villa Torlonia (Roma, Italy). (Reproduced from *Caneva, G., Galotta, G., Cancellieri, L., Savo, V. Tree Roots and Damages in the Jewish Catacombs of Villa Torlonia (Roma). J. Cult. Heritage 2009; 10: 53–62*. Copyright © 2009 Elsevier Masson SAS. All rights reserved.).

An interesting work[196] examined the role of native and alien plants in 20 Italian archaeological sites. An alien plant is a plant intentionally or accidentally introduced in a territory because of human activity. The study took into consideration the effect of alien and native weeds on ancient remains based on their frequency, abundance, and danger index (DI)—proposed by Signorini[197]—ranging from 0 to 10, which is a measure of the potential damage that each species can cause to stonework depending on plant and root features¶. Nonnative weeds were potentially harmful because of their generally higher DI, but they scarcely damage the objects because they were mainly restricted to recent habitats. A harmful alien able to grow on walls was the tree-of-heaven *Ailanthus altissima.* This species, native of China, causes serious damage to the stone because it grows quickly and develops vigorous roots. Its dispersal strategies are also particularly well suited to the environment of archaeological remains. It has efficient vegetative regeneration after mowing and its wind-dispersed seeds can reach any habitats. However, the study showed that the most detrimental species, because of generally high DI, abundance, and frequent occurrence on walls and buildings, were native. Among the most harmful were ivy (*Hedera helix*), fig tree (*Ficus carica*), old man's beard (*Clematis vitalba*), and elm (*Ulmus minor*). The most widespread species, not necessarily showing high DI or abundance, were mainly of dry Mediterranean meadow origin such as annual forbs and grasses present on the tops of walls and perennials, which colonize the ruins. Native species dominated all the 20 sites surveyed showing that archaeological areas act as conservation sites for them and are hot spots of floristic richness.

Another study of plants in habitats around 56 castles in Germany[198] showed that alien species that grew near medieval castles today, traced back to historical reasons for their introduction and therefore reflect the history of the castles. A study[199] of plant colonization of Coliseum in Rome (Italy) proposes similar conclusions providing information about the monument's history. Changes over time caused the growth of different floras: A gradual decrease in species typical of mature dynamic stages and a constant increase in alien species, indicators of a strong impact of man on the area.

¶The DI index is the sum of the values assigned to root morphology, plant vigour, and plant life form. Root morphology value varies 0–2, depending on the shape and size of the plant (e.g., considering the root depth and the presence of taproots). Plant vigour value ranges 0–2. Plant life form value varies 0–6 (annual, biennial, perennial herb, subshrub, shrub, vine, and tree).

Although the literature reports numbers of results on the damaging effects of woody plants on stones, recently some researchers focused on the potential bioprotective role of evergreen creeper/climber plants that rooted out of historical walls but grow on their surfaces, as ivy does. Ivy (*Hedera helix*) is a native evergreen creeper/climber common in woodlands throughout Europe. The study wanted to test whether ivy serves as a beneficial (bioprotective) or detrimental (biodestructive) agent on historical walls and buildings.[200] The project examined the temperature and humidity conditions at the walls behind ivy canopies in comparison to uncovered walls. In fact, two major causes of damage to exposed stones are the freeze/thaw and the wet/dry cycles. In the first cycle, water expands as it turns to ice causing damage; in the second cycle, salts within the fabric cause damage by expanding as they come out of solution. According to the authors, anything that moderates either of these cycles is potentially protective. Hourly data showed that ivy reduced extremes of temperature and relative humidity.[201] Differences in the exposure level of studied walls (i.e., whether they are shaded or not by trees or other walls), the thickness of the canopy and the aspect of the walls influenced the degree of microclimatic alteration provided by the ivy.[201] On average, exposed surfaces had a mean daily temperature range that was 3.6 times greater and a humidity range 2.7 times greater than those of the ivy-covered walls did. Therefore, ivy canopy may prevent excessive heating and cooling. The examination of stone blocks to assess any deteriorative effects from rootlet attachment showed no noticeable impact of the surface. On the other hand, aerial rootlets may chemically deteriorate vulnerable minerals such as mortar and brick, and roots may penetrate walls and cause physical breakdown. The authors concluded that ivy might have positive aspects for some walls but less good or even damaging for others.

Another study[202] on ivy was conducted at three sites representing different environmental conditions (Oxford, UK). It showed that ivy trapped particulate matter acting as a "particle sink," especially in high-traffic areas. The authors reported that ivy can retard biodeteriorative processes on historic walls and reduce human exposure to respiratory problems caused by vehicle pollutants through absorbing pollutant particles.

From an ecological perspective, it is opportune to underline that many different surfaces including sediments, bare rocks and soil substrates are subject to bioprotection by many types of higher plants. To date, the protective role of subaerial vegetation has been reported mostly in terms of protection from erosion, preservation and stability of soil by plant root systems.[42] Moreover, in tropical environments, a light forest cover seems advantageous for the conservation of outdoor monuments reducing evaporation rates, but it is advisable in such a context to evaluate and balance the overall effects, either positive or negative, of plant growth.[192]

3.8 BIRDS

Birds, and particularly feral pigeons, pose serious problems for the conservation of monuments. Feral pigeons are well adapted to living on urban buildings and close to humans.[203] Pigeons' droppings contain 2% phosphoric acid (H_2PO_4) that causes deterioration of stones and corrosion of metals.[68] The uric acid in droppings is also very corrosive to building materials.[203] Droppings often leave stains that are very difficult to remove. In the archaeological area of Herculaneum (Italy), pigeons are a major problem in terms of conservation. They nest in the most secluded corners of the site and cause numerous problems. They chemically affect and seriously damage the decorative surfaces in the ancient buildings with the acidity of their excrement. Moreover, they mechanically damage the carbonized wooden artifacts, which are one of the most extraordinary features of this archaeological site, and esthetically compromise with soiling the decorated pavements and frescoed surfaces.[204]

The nests and excrement of the birds cause other types of damage to monuments, such as blocking rain pipes, water conduits, and gutters.[205] Moreover, feral pigeons are a hazard to human health because they carry several bacterial infections that can be transmitted to humans.[206]

Another relevant aspect of pigeons' actions is that their nests attract pests that can enter buildings and museums infesting collections.[203] The debris from pigeons' nests is an ideal food source and nesting material for pest such as moths, dermestids and spider beetles. These insects can then enter the museum and arrive to the collections with devastating results[203] as reported by a conservator[203] at the McManus Galleries (Dundee, Scotland) where nesting pigeons caused a pest infestation.

The limitation of birds and pigeons' populations include preventive measures that discourage them from roosting and nesting on buildings. Therefore, these options are discussed in the chapter dedicated to the prevention.

KEYWORDS

- **biodeterioration**
- **biofilms**
- **cyanobacteria**
- **bacteria**
- **microalgae**
- **fungi**
- **lichens**
- **bioprotection of stones**
- **mosses**
- **vascular plants**
- **birds**

CHAPTER 4

CONTROL METHODS OF BIODETERIORATION

CONTENTS

ABSTRACT

The chapter discusses the methods and the more suitable techniques and products that are useful for the removal of micro and macro organisms that grow on artificial and natural stone works of art, including wall paintings. It provides updated results on mechanical, physical, and chemical methods. Results deriving from studies in other fields—civil buildings and their structures, water systems, plastics, etc.—are presented.

The chapter reports non-conventional methods such as use of laser, microwaves, and nanoparticles. Large space is devoted to the chemical methods using biocide formulations. The chapter provides information to develop a general understanding of the chemistry and mode of action of microbicides and herbicides. The peculiar interactions with the cell of the different groups of chemicals are discussed. Information about major active agents of biocide formulations applied on stone objects are provided.

Topics associated with biocides include methods of application on stones, types of microbicides and herbicides, mechanisms of antimicrobial action, resistance of organisms to biocides, toxicity, recolonization after treatment, novel biocides, and alternative methods for the control of biological growth. The evaluation of biocides' effectiveness, environmental impact, chemical stability, long-term effect, and harmlessness towards substrates are discussed.

The chapter details the potential negative effects of biocides on stone surfaces. The reported results allow evaluating the risks related to the application of the products on stones and the consequences of frequent use and/or over dosage of the biocides.

The chapter accounts for some products that over time disappeared from the market. It contains also up-to-date information on legislation and regulations governing the use of biocides in the European Union.

The control of biodeterioration encompasses the operations undertaken to eliminate biological growth and, possibly, to delay a new colonization. In all fields of restoration practice cleaning of objects including the removal of patinas, crusts and vegetation belongs to the first steps of restoration measures. The current attitude is generally oriented toward a planned removal of spontaneous flora and microflora whenever they cause an objective damage and/or structural impairments to the substrate. Nonetheless, the removal of biological growth can be necessary also for reasons of safety or esthetics. Management of biological colonization must consider that forms of biological growth can have a scarcity or rarity.[193] Where churchyards are concerned, there may also be a preference for conserving biodiversity, including ferns and lichens on monuments.[165] Thus, it is advisable that any conservation project "will additionally involve cognizance of ecological values, both the importance of the vegetation itself and its capacity to serve as a habitat for other species."[2]

The measures against flora and microflora should include the detection, monitoring, removal, and prevention of their formation. The operations for the elimination of biological growth, where not possible by alteration of environmental conditions (such as reduction of humidity and/or incident light), generally encompass three methods—mechanical removal, physical eradication, use of biocides. The different kind of treatments are carried out depending on biodeteriogens, the materials constituting the objects and their state of conservation but the application of chemicals (biocides) has been the most widely used so far. Although the treatments of biodeteriogens are parts of common restoration praxis, problems in the development and establishment of useful and successful treatment techniques are still to be solved. Moreover, this aspect can benefit from a more systematic implementation of a long-term follow-up to treatment procedures.[104]

In addition, it is worth mentioning that no control method offers a panacea, nor it is perfect.[207] In practice, it means that fast biofilm regrowth and proliferation may follow any inadequate control treatment regime. Therefore, before taking any conservation action, it is important to consider a monitoring and maintenance regime.[104]

4.1 PHYSICAL ERADICATION AND INHIBITION

The physical eradication of microorganisms is carried out applying methods such as electromagnetic wavelengths, high temperatures, and laser. Their great advantage lays on lack of the potential risks associated with the irreversible application of microbicides. Physical methods do not introduce any harmful chemicals to humans, to environment, or to the cultural heritage. Within electromagnetic wavelengths, ultraviolet (UV) light has been used to remove algae, cyanobacteria, heterotrophic bacteria and fungi on damp plasters (e.g., interior of churches and caves).[208] The application of UV radiation is simple; its limitation is the very low penetration in substrates[209] and in very thick biofilms.[210] Other drawbacks are the induced photooxidation in organic materials and the interaction with some pigments, effects that limit their use on wall paintings. A 48-h-long application of UV radiation may be effective over 2 months or more.[85] Applications with lamps emitting UV-C radiations for three cycles of 10 h gave good results on algal biofilms.[211] The method appears efficient, easy to carry out and relatively inexpensive. In a Japanese cave (the Kitora Tumulus), ultraviolet rays (254 nm) irradiation killed most biofilms in 30 min. However, a black basidiomycete (*Burgoa* sp.), very resistant to UV irradiation, was sometimes observed in the stone chamber.[208] It is important to underline that many fungi species containing black pigments, for example, melanins that protect them from UV, are stress tolerant and thus resistant to treatments.

A study applied UV-C to remove algal colonization composed mainly of *Chlorella minutissima* on some surfaces of the Moidons Cave (Jura, France).[66] A box containing two lamps (λmax = 254 nm) irradiated the surfaces. Each lamp had on and off intervals of 30 and 15 min corresponding to 8 h of exposure. A black plastic hermetically sealed the box to avoid UV-C dispersion. Each treated biofilm received a dose of 180 kJ/m^2 that is the resulting value obtained by the same authors in laboratory tests to get total chlorophyll bleaching. Greening on some surfaces disappeared 1 month after UV-C treatment. Where the biofilm was very thick this did not happen, indicating that the efficiency of UV-C treatment was primarily influenced by the thickness of the biofilm. Recolonization occurred 16 months after the treatment showing a good effect.

A laboratory testing studied the antimicrobial properties of blue light LED (470 nm) on bacteria and of the same light along with the photosensitizer erythrosine (ERY) on filamentous fungi.[212] Bacteria (*Leuconostoc mesenteroides*, *Bacillus atrophaeus*, and *Pseudomonas aeruginosa*) and fungi (*Penicillium digitatum* and *Fusarium graminearum*) were inoculated in agar plates and then exposed to blue light combined or not with ERY. Blue light alone significantly reduced bacteria and *Fusarium graminearum* viability. When combined with ERY, it considerably affected *Penicillium digitatum* viability. Therefore, blue light is lethal to bacteria and filamentous fungi although its effectiveness depends on light purity, energy levels and microbial species. A study that used blue light to damage and eventually inhibit the growth of cyanobacteria in caves, confirmed these results. The installation of lamps emitting monochromatic blue light (emission peak around 490 nm) in the Roman Catacombs of St. Callistus and Domitilla, Rome (Italy), had good results because cyanobacteria cannot use this spectral emission for photosynthesis and growth.[65,82] After 5 months, no photosynthetic activity by cyanobacteria was detected, and after 10 years there was a drastic reduction in the extent of the phototrophic community. However, bacteria belonging to phyla Proteobacteria and Firmicutes, not affected by the presence of the blue light, increased in number after the treatment.

Another study[213] showed that moderate intensity of light inactivates cyanobacteria causing photoinhibition of photosynthesis, photobleaching of pigments and photodamage to the cells. In that case the blue light was not efficient (150 μmol photons/m^2/s of blue light were not able to damage biofilms), while the same intensity of red, green or white irradiation for 14 days severely damaged the cyanobacteria cells due to radicals' formation. The efficiency of the irradiation depends also on the composition and pigmentation of the cyanobacteria. Red light was the most effective for species rich in phycocyanin and allophycocyanin, as *Leptolyngbya* sp. and *Scytonema julianum*, whereas green light inhibited species rich in phycoerythrin, like *Oculatella subterranea*. White light showed good performance on grayish and black cyanobacteria, such as *Symphyonemopsis* sp. and *Eucapsis* sp. The same authors suggested a treatment of cyanobacteria that involved the interaction of light with photosensitizers.[214] The results are discussed in Section 4.12.

A drawback of the mentioned lighting systems is that they imply the loss of full color vision of the wall paintings located in the caves. Therefore, it is advisable to inform the public about the conservative reasons that led to this choice by means of explanatory systems at the entrance of the caves.[215]

UV light at 260 nm demonstrated good efficacy even in controlling vegetation on a polychrome stone picturing Buddha located outdoor.[216] The conservators did not detect any effect on the pigments (hematite, yellow ochre, kaolin, and Chinese ink).

High-intensity pulsed light from a portable xenon flash-lamp system was employed to remove lichens from 10 large marble statues of Ming dynasty located in the Seattle Art Museum garden.[217] The museum's conservators planned to move them indoors replacing them with concrete replicas. The system emitting ultraviolet and visible radiations successfully removed the lichens without damaging the stone. It irradiated stripes 1 cm wide and 15 cm long at a fluence of 10 J/cm^2.

Different doses of gamma irradiation (5, 10, 15, 20, and 25 kGy) killed Streptomyces present on movable wall paintings of ancient Egyptian tombs.[218] The applied doses did not cause any observable alterations or color changes to pigments and binding media (Arabic gum, animal glue, and egg-yolk) of the paintings.

As mentioned above, recent researches deal with promising physical methods such as high temperatures and laser. Both treatments represent an innovative approach in terms of feasibility, low costs, and eco-compatibility.

Temperature of circa 40°C reduces the activity of many organisms although temperatures above 100°C are necessary for severe reductions in the viability of organisms that produce spores.[25]

Heat shock treatments showed high efficacy when applied to artificially hydrated lichens, mosses and liverworts.[219,220] These are poikilohydric organisms, that is, they lack structures or mechanisms to prevent desiccation and regulate water loss and, hence, the water of the environment determines their water content. They can tolerate dehydration (low cell or tissue water content) and recover from it without physiological damage. They are thermotolerant (up to 65–70°C) when dry, but thermosensitive when wet. Taking advantage of this last characteristic, the researchers wetted the bryophytes before treating them with heat shock. Treatments of hydrated samples at 60°C caused the death of all the bryophytes. The researchers applied also two biocides at 40°C. This temperature was sufficient to increase significantly the effects of the biocides, even at concentrations 10 times lower than those in current use.[220] Another study[219] applied thermal treatments, in parallel to the application of three biocides, on epi- and endolithic lichens. The results showed that a 6-h-long treatment at 55°C is sufficient to kill the lichens if they are fully hydrated. As in the case of bryophytes, the treatment at 40°C with the application of biocides at concentrations 10 times lower than usually were harmful also to the lichens.

Focused on application of high temperatures on lichens and fungi, a study proposed an innovative portable device-producing microwave heating.[221] The availability of portable devices for in situ utilization is very important in conservation of outdoors cultural heritage. The microwaves technique allows treating areas at depths of a few millimeters to minimize the interaction with the substrate. Moreover, the device controls the emitted power, maintaining the surface temperature approximately constant at the desire value. Crustose and foliose lichens treated for 3 min at 50°C showed irreversible damages while black fungi colonies (*Sarcinomyces* sp., *Pithomyces* sp. and *Scolecobasidium* sp.) in agar plate needed a temperature of 65°C for 3 min for the resistance of the fruiting bodies.[222]

A method based on the dual action of high temperatures and water vapor pressure is the so-called steam cleaning. The temperature makes softer the layers, while the pressure provides the mechanical action for their removal.[223] Indeed, it is not an innovative technique as in some European countries such as France and Denmark its application is almost a regular practice. The method can achieve temperatures of 150°C at the nozzle end and the operator can vary the temperature and pressure. In all cases, the operator should carry out trials to ascertain the optimum parameters (such as pressure, temperature, nozzle type, and its distance from the stone) that best suit the condition of substrate and the type of organisms. The technique is not suitable for deteriorated surfaces. In other words, its usage is appropriate only if the stone has a good state of conservation. The way in which a steam cleaner is used can make the difference to its effectiveness. When the nozzle is held too close to the surface, this can result in damage to the surface and uneven cleaning.[224]

The steam/superheated water can remove algae, lichens, mosses.[225,224] The elimination of these organisms is immediate in contrast to biocides that need time to act. The application of the method on some buildings in France ensured efficacy for 2 years when the conservators observed the beginning of a recolonization.[225] Therefore, they suggested the use of a biocide after the steam cleaning as a preventive action to improve the durability of the treatment. Heated water jets (pressure 150–180 bar, temperature 75–85°C) successfully removed the growth of biofilms and lichens.[226]

In the field of cultural heritage, laser cleaning is a well-established technique and a great solution for many conservation projects because it provides fine and selective removal of superficial deposits and encrustations without damaging the substrates.[227] Laser ablation applied to a material may produce thermal, chemical and mechanical damage. Following many types of laser exposure, the material evaporates or decomposes when a critical or threshold temperature is reached. At this point, the affected region becomes gaseous or is broken up into small particles.[228] By varying the laser power and the time of application, it is possible to control the depth of ablation to vaporize and remove only the superficial material without disturbing anything underneath.[228] After ablation, the remaining particulate material can be removed from the stone with a brush or a slightly moistened swab leaving the substrate undamaged.[228] The suitable choice of irradiation parameters through optimization studies carried out on a set of specific conservation problems led to effectiveness and selectivity of the ablation processes. Laser cleaning showed potential for cleaning also deteriorated substrates such as very friable stones that need consolidation. Some laser-cleaning studies focused on the removal of epilithic lichens and fungi from stone. This literature reports the use of the following lasers:

- Free running pulsed Erbium: YAG laser at 2940 nm (fluence 0.382–12.74 J/cm^2, energy range 3–100 mJ). It completely destructed the cell wall of the lichen *Diploschistes scruposus*; the polysaccharide and secondary products of the lichen were much reduced or lost in the material that remained after ablation[228];
- Laser at 532 nm (energy 190 mJ) showed efficacy on biofilms[99];
- Nd:YAG laser at 1064 nm (0.5 ms pulse duration, fluence 5 J/cm^2, 1 Hz) showed efficacy on lichens[217]; inspection of the lichens after long-pulse laser irradiation indicated that thermal energy broke down the cellular structure. The mineral grains transmitted and scattered the laser wavelength without appreciable heating. On the other hand, at more intense laser radiation, the lichens absorbed the radiation and the resulting heating led to thermal decomposition, removing both lichens and calcite grains;
- Nd:YVO4 laser at 355 nm (pulse duration 25 ns, frequency 10 kHz, fluence $\geq$0.5 J/cm^2), used successfully on black biofilms (algae belonging to *Trebouxia* genus and cyanobacteria) that grew on granite.[229,230] Laser caused modifications of biotite crystals;
- Nd:YVO4 laser at 355 nm (pulse duration 25 ns, fluencies 0.14 and 0.21 J/cm^2), applied on the crustose epilithic lichen *Pertusaria amara* that grew on Hercynian granite.[231] The laser successfully removed the lichen but some residues remained on the stone surface. Moreover, laser caused modifications of biotite crystals that appeared slightly molten;
- Q-switched Nd:YAG laser at 1064 nm (pulse duration 5 ns, frequency 10 kHz, fluence 2.0 J/cm^2) removed the crustose lichen *Verrucaria nigrescens* as well as fungi and algae inside the stone[227];

- Q-switched Nd:YAG laser at 532 nm (10 ns pulse duration, fluence 1–1.4 J/cm^2) and short free running Nd:YAG laser at 1064 nm (40–120 ns pulse duration, fluencies 2 J/cm^2 and 4–8 J/cm^2) on the crustose lichen *Verrucaria nigrescens*.[232] The first laser successfully removed the lichen. Regarding the second laser, fluence as high as 2 J/cm^2 was still insufficient for a practicable safe removal. This fluence corresponds to a peak intensity of 200 MW/cm^2, which is an upper limit in the laser application to avoid the plasma-mediated ablation of the substrate. The combined application of this laser with microwaves on lichens showed promising results[233];
- Q-switched Nd:YAG laser at 1064 nm (15 ns pulse duration, fluence 2.5 J/cm^2) and its third harmonic at 355 nm (15 ns pulse duration, fluence 0.5 J/cm^2) on hydrated lichens *Caloplaca* sp. (which was in a unhealthy state) and *Verrucaria nigrescens* that grew on dolomite stones.[234] The thalli thickness was about 500 μm. Infrared laser successfully removed *Caloplaca* sp. Ultraviolet laser had a rather superficial effect and only partially removed *Verrucaria nigrescens*. The researchers then applied the same laser using sequential irradiations, for example, 100 pulses at 1064 nm followed by 100 pulses at 355 nm on *Verrucaria nigrescens*. The dual infrared-ultraviolet sequential irradiation was the optimal condition for the lichen's removal, while ensuring preservation of the stone.

The comparison of these results shows that the application of laser to kill biofilms and lichens is still at an experimental level. Many factors (thickness of biolayers, healthy of lichens, chemical composition and state of conservation of stone, endolithic growth) may contribute to the result. Nonetheless, some indications for future developments of the technique emerge. QS Nd:YAG laser at 532 nm seems promising for the removal of biofilms and lichens at relatively low fluencies, for example, 1–1.4 J/cm^2 (laser fluence is a measure used to describe the energy delivered per unit area). In addition, the application of shorter wavelengths (355 nm) is worth of being studied in depth because it would need much lower fluencies. In fact, high frequencies increase the risk of surface damage due to the heat accumulation, and laser irradiation at a frequency >10 kHz leads to an appreciable darkening of granites.[229] The damage threshold for granites was around 1.5 J/cm^2 because, above this value, morphological/textural changes on the surface occurred. The presence of biotite needed even lower fluencies because this mineral melted even at fluencies considered safe ($\leq$1.5 J/cm^2). Even though the final fluencies were quite low ($\geq$0.5 J/cm^2), they ensured the complete removal of the biofilms.[229]

Stimulating results are those obtained applying dual IR–UV laser irradiation. The method not only efficiently removes an epilithic lichen, but also strongly damages endolithic colonizers.[234] The coupled application of wavelengths can solve situations that would need fluencies higher than the ablation/damage thresholds of stones.

The effects of laser on cells of biofilms and lichens relate to various mechanisms, including thermal damage, photochemical and photomechanical actions associated with microcavitation and propagation of acoustic shock waves.[234] The wetting of lichens favors deeper and more homogeneous ablation because they are thermosensitive when wet. As we saw for the application of high temperatures, this condition is essential to damage these organisms. A factor that can play a positive role in the ablation is the presence of dark melanins because they absorb higher amounts of radiations in comparison with other cell molecules.[232] On the other hand, attention should be paid in the laser cleaning of darky-pigmented lichens and fungal patinas because the pigments can enter the crystal matrix causing black stains, even more difficult to remove.[235]

However, the interaction between biofilms, lichens, and laser treatments needs still further research. A practical limitation of laser cleaning when the objects are large and numerous is its slow rate (the ablation spots are in the range of a few millimeters). In these situations, the only application of laser is too slow and costly to conform to a cleaning program.

4.2 MECHANICAL AND WATER-BASED CONTROL METHODS

A variety of mechanical measures for the treatment of the microbial growth exists such as cleaning using brushing, scalpels, sand blasting, air abrasive, low-pressure washing, vacuuming, and ultrasounds. Water-based cleaning includes steam cleaning, intermittent nebula sprays (which create a fine mist to soften slowly the layers), rinsing, and pressure washing. Water can be used hot or cold, and as a liquid or vapor. Steam technique that uses high temperature has been discussed in the previous section.

Algal and cyanobacterial biofilms should be completely dry before cleaning because dry crusts often readily detach from the materials and can be removed mechanically using brushes, sand blasting, or low-pressure washing.[235] Unfortunately, microorganisms are often located inside the material being able to penetrate pores, fissures, and cracks. Therefore, it is not possible to reach them mechanically and residues in form of single viable cells or whole colonies are a source for rapid reoccurring of colonization.[235] Many crustose lichens show a close interaction with the substrates, and their simple mechanical removal is not a good choice. It could severely damage the stone because the lichen thallus forms an intimate association with the substrate.[13] Moreover, fragments of the lichens can remain on and inside the stone.[231] On the other hand, some studies report a mechanical cleaning of lichens without applying biocides. The treatment of a standing Buddha engraved on a shale wall (South Korea) involved dry cleaning followed by wet cleaning using distilled water.[236]

In indoor environments, a microbiological attack to objects can occur when elevated air contamination and favorable climate conditions are present. Velvety or powdery fungal colonies contain high amounts of spores and therefore are sources of contamination for other objects because air easily transfers the spores. Moreover, the microbes may produce contaminants, that is, aerial particles such as spores, allergens, toxins, and other metabolites that can be serious health hazards to occupants.[147] In such cases, a mechanical cleaning using a vacuum cleaner equipped with high efficiency particulate arrestants (HEPA filters) is a suitable procedure to remove most hyphae and spores.[237,144] However, this method cannot eliminate all microorganisms. Therefore, it should be combined with another antimicrobial strategy, preferably with the alteration of the environmental conditions to prevent further proliferation of the remaining microflora.

The mechanical methods can be effective when facing with mosses and plants growth.[193,225] A management regime of plants may include felling, topping, lopping, and/or selective pruning.[193,195] The major drawback to mechanical control of plants is that many of them respond to cutting by vigorous regrowth.[25] Therefore, cutting of most plants at the bottom and leaving them to die is not the appropriate way to kill them because it will cause most plants to regrow at the cut level. For example, this action makes ivy to root into a wall. On the contrary, in the case of gymnosperms (cypresses, pines, fir trees, etc.), the felling at the base of the trees is sufficient to kill them. Therefore, the killing of gymnosperms does not need the total extirpation of the plant or the application of herbicides to avoid new shoots.[238]

For the periodic control of spontaneous vegetation in monumental and archaeological contexts, the mechanical mowing is effective. Among its advantages: the utilization of toxic chemicals is not necessary; the indiscriminate removal of spontaneous flora that may include elements worthy of preservation is limited; possible effects of soil erosion are impeded.[238] Ignoring the effects of plants removal on the monuments, planting trees paying no attention to distances between walls and plants, neglecting the tree pruning practice, lead to interventions not only useless and expensive, but also dangerous.[195] Moreover, the identification of plants would help avoid wrong choices in managing the plant coverage of archaeological sites or historical gardens.

Mechanical methods are often used in combination with other kind of treatments.[239,208,38,240] In most cases, the partial mechanical removal of biofilms, lichens, and plants is followed by the application of microbicides and herbicides (*vide infra*). A gentle mechanical method, for example, brushing, to remove organic material prior to biocide application, particularly when the growth of biofilms and lichens is extensive, is suitable.

The growth of many plants including ivy and figs is treated by cutting a section out from the main stems and injecting an appropriate chemical.[193,195,200] In the case of trees that grow on the soil above underground structures, it is very important that the trees cut combines with a consolidation of the structures because the death of the roots may cause collapses and structural damages.[195]

Sand blasting is a tool often used to remove the colonization of biofilms and lichens. It relies on the use of compressed air and aggregate; the latter can also include water. The parameters of all these constituents can vary (e.g., air pressure, size, and nature of aggregate). When the procedure is pushed to extremes, the consequences are very negative. It was the case of marble statues in the gardens of the National Palace of Queluz, Portugal, that, after the sand blasting treatment, acquired a brilliant whiteness and an evenly rough surface compared to the uneven and pitted surface left behind by gentler cleaning methods.[169,241]

A commercial product called Hydrogommage—a sand blasting, low-pressure (0.5–1.5 bar) air/water mixture with SiO_2 particles (0.5–0.1 mm)—was used to remove biofilms composed of green algae (*Trebouxia* sp.) and cyanobacteria (*Gloeocapsa* sp. and *Choococcus* sp.) that grew on granite.[230] The average thickness of the biofilms was 40 μm. The method was quite efficient but caused textural changes of granite, that is, an increase in roughness and microfissures.

High-pressure water blasting or heated jets are methods that conservators should avoid as well on porous materials because not only they do not remove all the microorganisms but also push them deep into the material. Consequently, microbial colonization will be faster afterwards and will take place deeper inside the material.[235] Moreover, methods such as brushing or high-pressure vapor are not suitable in caves as they can destroy the fragile crystal structure of speleothems.[210]

Nonetheless, some studies report that washing[242] or the application of 5% solutions of calcium hypochlorite under low pressure[239] successfully removed the growth of biofilms and lichens. The authors chose these methods since the stone was enough healthy. Power water washing is sometimes used to remove organic material prior to biocide application.[85] Such treatment should be used with great caution and on undamaged substrates only.

The mechanical removal of the dead biomass after a treatment, usually biocide-based, is a common practice undertaken by washing with deionized water and then scrubbing,[226,243,169,241] or by using ammonium carbonate.[244,29] In some cases, conservators remove the dead biomass immediately after the treatment, while in other cases they leave the objects untouched for several months, and then lightly brush to remove any detaching remnants reducing the intervention to a minimum,[169,241] with practically no cleaning actions required. A discordant "voice"[245] suggested not scrubbing or removing the treated lichens. The author did not consider that, when the dead biomass stays in place, it provides nutrients for spores and microorganisms to develop.

These considerations show that the distinction between killing and cleaning is relevant. Cleaning means the removal of unwanted biofilms from a surface[246] and is at least equally important as killing.

There are cases where a mechanical method coupled with biocides is the only one permitted. In many archaeological sites, the whole work must be accomplished during a short summer season.[247] Thus, the use of biocides and pressurized water is a technique that can make cleaning on a large scale possible. Many factors must be controlled—the pressure produced by the pump, the flow rate through the nozzle, the shape of the nozzle, the distance from the nozzle to the surface, and the direction of the jet on stone surfaces. As pressurized water can damage the substrate through detachment of loose particles, deep penetration by moisture and accelerated erosion, a careful examination of the substrate must precede its use to determine whether the stone is sufficiently robust to withstand the treatment.[247] When friable areas are present, they may need to be preconsolidated. The technique was applied for example on a sarcophagus in the garden of the Aphrodisias Museum (Turkey). The restorers treated the object by brush with a biocide and wrapped it with wet cloth and plastic for 45 min. Cleaning was undertaken using a small low-pressure water jet (maximum pressure 90 bars). After 2 years of outdoor exposure, there was no regrowth of biofilms and lichens.

A study[244] experimented a treatment that combines mechanical and physical methods. The study tested the applicability of the Ice-Clean system (Sapio Group) for cleaning a black patina, mainly formed by cyanobacteria, on the marble of the Pyramid of Caio Cestio in Rome (Italy). The system used pellets of dry ice (solid carbon dioxide), projected at high pressure (2–6 bars) on the surface. Thus, the action is mainly mechanical, substantially abrasive, with a physical component (low temperature). Ice-Clean system was effective but some cells of chasmolithic and cryptoendolithic microorganisms survived. Areas treated with the dry ice cleaning and a biocide showed the best result. The features of the stone surface after treatment did not undergo substantial changes. In situ observations with a stereomicroscope revealed a rounding of crystals' edges, but this did not produced changes in the surface porosity. Nonetheless, detachments and loss of small fragments were detected in areas with poorly adhering mortars and fillings. The authors concluded that the Ice-Clean system could be used efficiently and safety for removing biological patinas on stones in a good conservation condition and only if followed by a disinfection treatment.

4.3 BIOCIDES

Masonry biocides are an important component in stone conservation[248] to eliminate the macro- or microflora responsible for biodeterioration, although some authors report their application at the end of the conservation treatment to prevent new microbial colonization of restored surfaces.[249]

Nowadays for some conservators, the use of biocides is anathema. Biocides are toxic, and a toxic substance is always hazardous, but the risk it poses depends upon the circumstances of exposure. However, biocides can potentially affect humans or the environment. Major concerns are connected to health and safety issues. For these reasons, biocides are one of the most tightly regulated and controlled type of chemical products.[250] Risks to human and animal health and to the environment can be reduced either by using a biocide with low potential to cause harm or by applications leading to low exposures. Alternatively, the synergistic use of mixtures of biocides can allow significant reduction in concentrations of the individual chemicals.[251] Another approach can be the combination of chemical and physical treatments that potentially have synergistic effects and allow decreasing the concentration of biocides to an acceptable level.[48] The effective use of biocides implies knowledge of their characteristics, for example, chemical and physical properties, mechanisms of action and spectrum of efficacy. A full range covering all species is not always necessary. The removal of macro- and/or microflora with biocides should assess their effectiveness and foreseen how often the treatment should be repeated.[252] Biocides' safe use depends on a great extent on their biodegradability, that is, on biodegradation of these molecules in the environment. The biodegradability of a biocide is therefore an important criterion of choice.[6] Many biocides are short-lived or degradable through abiotic and biotic processes, but some may transform into more toxic or persistent compounds. This review may thus serve as a guide for identification of microbial control strategies to help develop a sustainable path for managing.

As biocide treatments are chemical in nature, small-scale trials are recommended prior to general application to test their efficacy and to confirm that adverse effects such as staining do not occur[253] (Fig. 4.1).

For the application on outdoor stone objects, the ideal biocide is highly effective on the target organisms while causing no hazard and harm to the operator or visitors. It should have a long effective life, and, when leached by rain, it should be sufficiently diluted not to harm the environment. It should not alter the stone color or affect its structure in ways that could lead to long-term damage.[85,249,48,104]

4.3.1 TERMS

Biocide is a generic term that comprises microbicides, herbicides, acaricides, insecticides, etc. A biocide is a chemical agent that inactivates living organisms. Another generic term in use is *pesticide*, a chemical that kills

FIGURE 4.1 Application of biocides by brushing. Wall paintings of the Crypt of the Original Sin (Matera, Italy). The restorer operates on small-scale trials prior to general application to test biocides' efficacy and to assess any negative effects on the object. Photo courtesy of Giulia Caneva.

or otherwise discourages pests. Pests are microorganisms and organisms detrimental to humans or human concerns. Since the mid-1950s, the use of pesticides has grown continuously every year so that the total amount of pesticide active ingredients in use is now around 1200 tons per year. Pesticides, together with fertilizers, play a central role in agriculture and contribute to enhanced food production worldwide.

Microbicide is the generic term for bactericides, fungicides, algaecides, etc. In other words, microbicides are active on microorganisms. A substance is referred as microbicide when it kills microorganisms and as microbistat when it inhibits the multiplication of microorganisms. Whether the action of a substance is microbicide or microbistat depends on the concentration at which it is used.

Disinfectants and antiseptics are biocides primarily used to inhibit or destroy hygienically relevant microorganisms. They are used to prevent infection and transmission of pathogenic or potentially pathogenic microorganisms. Disinfectants are applied to inanimate objects or surfaces while antiseptics are used to inactivate microorganisms in or on living tissues.

Herbicides are used to control the growth of vegetation. They are classified into two main types:

- Contact herbicides destroy only the plant tissue in contact with them;
- Systemic herbicides enter the plant either through the leaves or through the soil and then are transported throughout the entire plant.

  Systemic herbicides are effective on perennial plants, able to regrow from rhizomes, roots or tubers, while contact herbicides are less efficient on these plants.

  The herbicides applied to the soil and taken up by the roots of the plant are classified into three types:
- Preplant incorporated herbicides, applied prior to planting and mechanically incorporated into the soil, preventing dissipation through photodecomposition and/or volatility;
- Preemergent herbicides, applied before the plant emerges preventing germination or early growth of weed seeds;
- Post-emergent herbicides, applied after the plant has emerged.

A further classification, mainly used in the agriculture field, relates to the range of action. Total herbicides control most plant species while selective herbicides kill specific target weeds, leaving the desired crop relatively unharmed. Many modern chemical herbicides for agriculture have low residual activity because they are specifically formulated to decompose within a short period after application. They often do not provide season-long weed control.

The microbicides used for restoration of stone objects were not originally designed for this field of application but for other fields—paint formulations, interior and exterior architectural coatings, water treatments (industrial, domestic water systems, swimming pools, cooling water systems, etc.), hospital practice, and agriculture and food industries. Nowadays, different disinfection and antimicrobial methods and other disinfection alternatives are employed in these fields, for example, chlorination/dechlorination, UV disinfection, physical/biophysical processes (nanofiltration, microfiltration, and ultrafiltration systems), and copper/ silver ions.[254]

Trade products contain the active ingredient (a.i.) or active substance that is the very molecule having biocide activity. They contain also other compounds that generally improve the performance of the a.i. Commercial products are identified by a name followed by numbers and/or letters. For example, Preventol and Kathon are registered trademarks that encompass many products. In other words, they do not correspond to a definite a.i. while Preventol RI80, Preventol A8, and Kathon CG allow for an immediate identification of the product's composition. Unfortunately, many studies often report the incomplete name of trade products or even the generic term "biocides" without any specifications. Therefore, such information is useless because only the specific data of a product can be of help for conservators.

4.3.2 APPLICATION

The chemicals used are in form of liquids, emulsion formulations, and dispersible powders. Biocides may be applied as an aerosol or as a poultice, by brushing, injection, immersion. The mode of application can vary depending on the situation (Fig. 4.2). The treatment of biofilms, for example, can be performed brushing (Fig. 4.1) the biocide solution more times at intervals of 1 week or other periods to permit a deeper penetration into the stone; after disinfection, the restorers can remove the biocide and clean with water.[113]

(a) (b)

(c)

FIGURE 4.2 Archaeological remnant of a Roman monument (Museum of Salt, Cervia, Italy). The image of a Gorgon is carved on the stone. Lichens and algae covered the object (a). Testing of two biocides applied by poultices to remove the biological growth. A negative side effect of a biocide was the lysis of cell membranes and leakage of green pigments (right area with poultice). This could lead to a color alteration of stone. Removal of poultices and dead mass in bottom areas (b). The object after restoration (c). Photo courtesy of Antonella Pomicetti.

One application of the herbicides is the water-based spray using ground equipment. The treatment by direct contact of weeds without affecting desirable plants is also in use. The application of herbicides to control plant growth in monumental and archaeological sites should consider aspects of environmental pollution.[25,255] A misapplication of herbicides is the volatilization or drift of the chemical because it may result in environmental contamination affecting neighboring fields or plants, particularly in windy conditions (Fig. 4.3). The duration of herbicides' application varies depending on the physiology and the biological characteristics of the species as well as on climate and meteorological conditions. In temperate and Mediterranean climates with strong seasonal variations, the treatments should be carried out when the conditions are favorable to vegetative growth. As a rule, treatments should not occur in periods of rainfall or with strong winds.[255]

FIGURE 4.3 Walls of Domus Tiberiana (Rome, Italy). Correct treatment with herbicides using a lifting device. The method allows the selective treatment of shrubs as well as the control of the dose applied avoiding products' drift in the surrounding environment. The operator wears protective clothing, gloves and eyewear. Photo courtesy of Giulia Caneva.

A consideration about the concentration of usage emerges from the results reported in the literature. In some cases, increasing the concentrations of biocides does not give a better effect, meaning that a higher amount of a.i. does not offer a higher efficacy.[256,257]

When planning the application of a microbicide to an external surface, the choice of treating biofilms and lichens hydrated or dried is a quite controversial issue. Some authors prefer hydrated condition because physiologically active biofilms and lichens are more susceptible to biocides than dried or dormant cells.[85,239,258,253,165,247] The application of the biocide on a wetted surface colonized by cyanobacteria and algae proved significantly more effective than application to a dry colonized surface.[259] Moreover, the degree of cleanliness achieved for treated and scrubbed areas and prewetted treated areas without scrubbing appeared similar, confirming the advantage of the prewetting.[259] Regarding lichens, they are poikilohydric organisms (with no mechanisms to regulate water loss, as above mentioned). Their water content relates to the water present in the environment. Consequently, they are stress tolerant when dry and stress sensitive when wet.[219] As they are also able to resist wetting after a long dry spell, it may be necessary to prewet the surface to assist absorption of the biocide. All these are strong points in favor of treating biofilms and lichens hydrated. On the contrary, the prehydration can reduce the biocide absorption, and biocides tend to be washed off more quickly during periods of rainfall. Some authors suggested that algal and cyanobacterial biofilms should be completely dry before cleaning.[235] In some cases, the company that produces the biocide recommends its application on dry surfaces (see Appendix). A biocide applied on dry organisms will have time to be absorbed by the stone and will remain in the stone fabric to become effective when the organisms absorb it in wet conditions.[85] Moreover, the hydration of the organisms after the biocide application might shorten the elapsed time between the start of the treatment and the action of the biocide.[260] This point deserves further scientific examinations since it is very important to give clear instructions to obtain the best results in the shortest time using the lowest quantity of biocide.[260] However, in general, it is important to prevent the premature loss of biocide caused by washing off.

Spores produced by fungi, some bacteria and algae are normally very resistant to biocides due to the low permeability of their thick-walled structure.[85]

The planning of a conservation program should consider the possible long-term negative effects of biocide treatments. In some cases, the removal of epilithic lichens is just a temporary remedy because the freshly exposed rough surface is prone to lichen regrowth as well as to the growth of microorganisms.[13] This happened in Angkor Wat (Cambodia), where the elimination of lichens led in a few months to intensive blackening of the treated gray stone, due to the growth of cyanobacteria.[99] In such a tropical climate, the spread black layers induced thermal and hydric stress in the stone.

Although the commercial microbicides and herbicides must be extensively tested prior to approval for sale and labeling by many countries, concern regarding health effects is diffuse because of the large number of biocides in use. In addition to health effects caused by active ingredients themselves, commercial biocide mixtures often contain other chemicals that can have negative impacts on human health and the environment. Issues related to these effects are discussed in Section 4.8.

The introduction of new, potentially more efficient, biocides is not common due to the very high costs involved in generating technical and safety data, which is necessary to gain legislative approval to market the product. A different and easier way implies actions on existing biocides such as the formulation of new carrier materials that bind to the biocides and delay their release improving their persistence. The advantages are that the amount of biocide and the environmental contamination (transfer by means of wind or rain, cross-contamination of untreated areas and/or animals) can be reduced.[254] Another way uses combination of existing biocides. An example is the product Parmetol MBX by Schülke, an in-can preservative without formaldehyde and adsorbable organic halogens. The product is a blend of biodegradable components—benzisothiazolone, methylisothiazolone, and bis(3-aminopropyl) dodecylamine.

4.3.3 TYPES OF MICROBICIDES AND HERBICIDES

This section presents a review of the most common microbicides and herbicides used for natural and artificial stones in cultural heritage. It lists them in alphabetic order. The characteristics of each biocide—chemical and physical properties, mechanisms of action, effectiveness, and spectrum of efficacy on targeted organisms in cultural heritage field—are discussed. The section is not intended to provide a definitive description of individual agents, but rather to introduce the reader to the general concept of those agents and to general notions about their use. It provides results and information of their application in the conservation of cultural heritage. Tables 4.1, 4.2, 4.3, and 4.4 are connected to this section as they report the following characteristics of biocides used in conservation of cultural heritage:

- Active ingredients and trade products
- Concentrations of use

- Efficacy on the target organism
- Indications about the experimental approach (in situ or under laboratory conditions)
- Side effects
- Bibliographical reference.

Furthermore, the appendix reports information about some active ingredients. It provides data on chemical and physical properties, toxicity, potential health effects, along with names of the manufactures and/or suppliers of trade products containing the active ingredients. The information on stability and incompatibility of the biocides with other compounds can contribute to a correct use of them in restoration practice.

4.3.3.1 ALCOHOLS

Alcohols, usually ethanol or isopropyl alcohol (2-propanol or isopropanol), are used as microbicides (Table 4.1). They have limited residual activity due to evaporation, resulting in a short time contact unless the surface is submerged, and have a limited activity in the presence of organic material. It is well established that, for the disinfection of solid materials, alcohols are most effective when combined with purified water. Ethanol 70% solution has higher antimicrobial activity than absolute ethanol. The phenomenon is likely due to the facilitation imparted by water to the diffusion of alcohols through the cell membrane while 100% alcohol typically denatures only external membrane proteins.[48] A solution of 70% ethanol or isopropanol diluted in water is effective on a wide spectrum of bacteria, though higher concentrations are often needed to disinfect wet surfaces. Alcohols are not effective on fungal and bacterial spores.

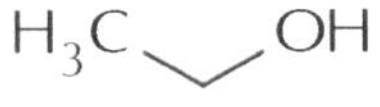

Structural formula of ethanol

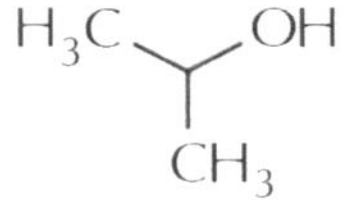

Structural formula of isopropyl alcohol

TABLE 4.1 List of Alcohols, Aldehydes, Azoles, Carbamates, Chlorine-containing Compounds, Hydrogen Peroxide, and Other Biocides. The List Includes Concentrations of Use, Efficacy on the Target Organisms, Indications about the Experimental Approach (In Situ or Under Laboratory Conditions), Side Effects and Notes, Bibliographical References.

Commercial product/Active ingredient	Concentration	Efficacy	Organisms/Substrate	Notes/Side effects	References
Ethanol	Blotting with 70% ethanol until soaked	The treatment was successful and after 5 years, no recolonization occurred	Removal of lichens from a 3000-year-old deer stone, Mongolia		De Priest & Beaubien, 2011
Ethanol	96% ethanol, brushing	Not effective	Granite (Galicia, Spain) colonized by filamentous green algae (*Trebouxia* sp.) and cyanobacteria (*Gloeocapsa* sp. and *Choococcus* sp.). The average thickness of the biofilm was 40 µm		Pozo et al., 2013
Ethanol and biocides Ronkosal (benzyl alcohol, glycerin, benzoic acid, sorbic acid) Diesin (lactic acid, benzyl-C12-16-alkyldimethyl ammonium chlorides) Sagrotan (ethanol, propanol, glyoxal)	Aqueous solutions of Rokonsal 3%, Diesin 3%, Sagrotan 1.5%	The high contamination rate after 2 months required further treatment of several objects with ethanol	Objects (glass, metal, textiles) contaminated by biofilms. Germanisches National Museum, Nürnberg, Germany		Drewello et al., 2004
Formaldehyde	Concentration not known	Effective for a year on fungal biofilm	Cave Kitora Tumulus, Japan.	After a year, recolonization of the walls	Kigawa et al., 2010
Econazole nitrate	0.2% in water or alcohol	Laboratory tests: very efficient. In situ tests: not efficient on all the fungi	Laboratory tests. Agar diffusion tests: fungi *Cladosporium*, *Ulocladium* and *Humicola* spp. In situ tests: wall paintings		Orial & Bousta, 2005
VANCIDE® 51 (blend of sodium dimethyldithiocarbamate and sodium mercaptobenzothiazole)	1% in isopropanol, brushing three times	Effective	Laboratory tests. Marble samples inoculated with bacteria and fungi		Diakumaku et al., 1997

TABLE 4.1 *(Continued)*

Commercial product/Active ingredient	Concentration	Efficacy	Organisms/Substrate	Notes/Side effects	References
Mirecide-TF/580.ECO (aqueous dispersion of 3-iodine-2-propylbutylcarbamate and 2-*N*-octyl-4-isothiazolin-3-one)	2.5% w/w, brushing until the complete imbibition of the lichens 10% w/w poultices for 48 h	Not efficient at 2.5%. Efficient at 10% on almost all the lichens	Crustose and foliose lichens on marble, sandstone and plaster. Archaeological area of Fiesole, Firenze (Italy)		Pinna et al., 2012
Biotin R (mixture of iodo-propynyl butylcarbamate, *n*-octyl-isothiazolone and 2-2′-oxydiethanol)	5% in ethanol	Effective on all the tested microorganisms	Laboratory tests. Agar diffusion tests: fungi, algae, bacteria, actinobacteria. The species were among those growing on monuments		Borgioli et al., 2006
Biotin R (mixture of iodo-propynyl butylcarbamate, 2-*N*-ottil-2*H*-isotiazol-3-one and 2-2′-oxydiethanol)	2% in ethanol, brushing	Efficient	Archaeological site at Ostia Antica (Rome). Extensive growth of cyanobacteria, green algae and bryophytes		Bartolini et al., 2007
Biotin R (mixture of iodo-propynyl butylcarbamate, 2-*N*-ottil-2*H*-isotiazol-3-one and 2-2′-oxydiethanol)	5% in white spirit, brushing till saturation of the stone	Effective on crustose lichens, cyanobacteria and endolithic fungal hyphae	Segovia cathedral cloister (Spain), dolostones, and granite. Crustose epilithic lichens *Aspicilia contorta*, *Lecidella stigmatea*, *Verrucaria nigrescens*		de los Ríos et al., 2012
Biotin R (mixture of iodo-propynyl butylcarbamate, 2-*N*-ottil-2*H*-isotiazol-3-one and 2-2′-oxydiethanol)	4% v/v, ethanol, brushing	Not effective. The lichens partially or completely recovered the normal photochemical efficiency after few days	Crustose epilithic and endolithic lichens *Acrocordia conoidea*, *Aspicilia contorta*, *Bagliettoa marmorea*, *Candelariella vitellina*, *Protoparmeliopsis muralis*, *Verrucaria nigrescens*. Limestone, Karst plateau, Italy		Tretiach et al., 2012
ACTICIDE® IOG (a.i. 2-*n*-octyl-3(2*H*)-isothiazolone+ 3-iodopropynyl butylcarbamate)	0.5% in water-repellent resin, brushing	Effective in reducing around 80% of the lichens	Abandoned dolostone quarry (Redueña, Madrid, Spain)		Camara et al., 2011

TABLE 4.1 *(Continued)*

Commercial product/Active ingredient	Concentration	Efficacy	Organisms/Substrate	Notes/Side effects	References
Mergal S97 (blend of 5-chloro-2-methylisothiazolinone, 2-methylisothiazolinone, methyl-benzimidazol-2-ylcarbamate, and 2-*n*-octylisothiazolinone)	1 and 2% in ethanol	Very good efficacy on fungi and bacteria; very scarce growth on limestone slabs inoculated with *Ulocladium* sp. after prolonged incubation	Laboratory tests. Agar diffusion tests: fungi, bacteria, and yeasts isolated from limestone and granite Rock slabs tests: limestone slabs inoculated with the fungus *Ulocladium* sp.	The product crystal-lizes in stone pores. It could make diffi-cult the penetration of the solutions	Blazquez et al., 2000
Sodium hypochlorite	3.5% in water	Effective when applied with hot water	Epilithic and endolithic cyanobacteria, lichens, algae. Fossiliferous limestone with high porosity. Palace of Saints George and Michael, Corfù, Greece		Pantazidou & Theoulakis, 1997
Calcium hypochlorite	5% in water	Effective when applied with hot water	Epilithic and endolithic cyanobacteria, lichens, algae. Fossiliferous limestone with high porosity. Palace of Saints George and Michael, Corfù, Greece		Pantazidou & Theoulakis, 1997
Calcium hypochlorite	5% in water The restorers applied preheated water (45–65 °C) which helped to soften the lichens, followed by calcium hypochlorite under low pressure (17.5 kg/cm^2)	Effective	Cyanobacteria and lichens on sandstone, the Washington Legislative Building (USA)	The northerly sides of the structure showed signs of recolonization after 12 months	Twilley & Leavengood, 2000
Hydrogen peroxide	200 mg/l, 130 volume	Efficient on biofilm mainly formed by green algae	Cango Caves, South Africa, karst formations		Grobbelaar, 2000

TABLE 4.1 *(Continued)*

Commercial product/Active ingredient	Concentration	Efficacy	Organisms/Substrate	Notes/Side effects	References
Hydrogen peroxide	5% in water	Efficient. It provided the best results at the lowest concentration respect to calcium hypochlorite, sodium hypochlorite, benzyl-dimethyl-alkyl-ammonium chloride (Dimanin), and glutaraldehyde	Laboratory tests. Algae, fungi, bacteria, and cyanobacteria isolated from the external facade of the Colombian National Museum, Bogotà, Colombia		Martinez et al., 2003
Hydrogen peroxide	10% in water, brushing	Noneffective on lichens, cyanobacteria, algae	Historic cemetery of Drapano, Kefalonia, Greece		Spathis et al., 2012
Hydrogen peroxide	25% in water, poultice for 5 h	Effective on lichens, cyanobacteria, algae	Historic cemetery of Drapano, Kefalonia, Greece		Spathis et al., 2012
Hydrogen peroxide	15% (v/v) in water and carbonate/bicarbonate buffer, spraying	40 ml of solution were sprayed on 1 m^2. Effective on algae	Surfaces of Slovenian caves	Two applications of solution per year prevented algal regrowth	Mulec, 2014
Preventol A9D (a.i. *N*-dichlorofluor-methyltio-N0,N0-dimethyl-*N*-*p*-tolil sulphamide)	1, 2, and 3% in acetone	Very good efficacy on fungi and bacteria; very scarce efficacy on limestone slabs inoculated with *Ulocladium* sp. after prolonged incubation	Laboratory tests. Agar diffusion tests: fungi, bacteria, and yeasts isolated from limestone and granite crusts Rock slabs tests: limestone slabs inoculated with the fungus *Ulocladium* sp.	The product crystallizes in stone pores that could make difficult the penetration of the solutions	Blazquez et al., 2000
Zinc chloride solution	1,5–3% w/v	Effective on a gray biofilm after 9 months from the application.	Marble statues, National Palace of Queluz, Portugal	Not effective on lichens	Charola et al., 2007

TABLE 4.1 *(Continued)*

Commercial product/Active ingredient	Concentration	Efficacy	Organisms/Substrate	Notes/Side effects	References
Natamycin	0.25%	Laboratory tests: very efficient In situ tests: not efficient on all the fungi	Laboratory tests. Agar diffusion tests: fungi *Cladosporium*, *Ulocladium*, and *Humicola* spp. In situ tests: wall paintings		Orial & Bousta, 2005
Cypermethrin	1%, 3%	Laboratory tests: very efficient In situ tests: not efficient	Laboratory tests. Agar diffusion tests: fungi *Cladosporium*, *Ulocladium*, and *Humicola* spp. In situ tests: wall paintings		Orial & Bousta, 2005
Polybor (mixture of cyclic borates)	4% in water	Not effective on the microflora	Maya site at Xunantunich, Belize. Biofilms, lichens, and mosses on limestone		Ginell et al., 1995
Koretrel (resin alkyleneoxide 97.6%, alkylaminotriazine 0.98%, *N*0-(3,4-dichlorophenyl)-*N*,*N*-dimethyl urea 0.98%, and denaturated alkyl trihydroxybenzene polyoxide 0.48%)	No dilution	Very efficient	21 saxicolous lichens, Trieste Karst, Italy	The biocide considerably decreases capillary water absorption, and alters the surface color of Aurisina limestone, Muggia sandstone, and a gray granite	Tretiach et al., 2007

Regarding cultural heritage, among various chemical agents tested in the laboratory (Mergal K14, Parmetol DF12, Troysan S97, Preventol R50, hydrogen peroxide, and ethanol) on microbial biofilms on stone samples, ethanol (70%) was the most effective.[48] Possible explanations may be that the small molecule of ethanol penetrated better through the channels of the biofilm matrix to reach the microbial cells, and functional groups of the EPS may deactivate some biocides as well.[48] However, ethanol is highly flammable and thus its use in large quantities is inappropriate.

A recent research on a 3000-year-old deer stone covered with lichens and placed in a natural environment[245] showed that the lichens died by blotting the object with 70% ethanol until soaked. The result needs further tests to assess this effect because it is the only positive result reported on lichens.

A contrasting result was reported by some researchers[230] that applied ethanol (96%) to remove biofilms composed of green algae (*Trebouxia* sp.) and cyanobacteria (*Gloeocapsa* sp. and *Choococcus* sp.) that grew on granite (Spain). The average thickness of the biofilms was 40 μm. The application was not effective.

Formaldehyde fumigation and ethanol were used to control and remove fungal colonies in a Japanese cave with wall paintings, the Kitora Tumulus.[208] After a year, a viscous gel appeared on the walls. As biofilms of fungi and bacteria formed the gel, a new treatment with isothiazolones (Kathon CG) was applied.

When microorganisms heavily contaminate objects in a museum, ethanol proved to be effective, as in the case of a German museum where the treatments of objects (glass, metal, textiles) with biocides containing ethanol, propanol, and quaternary ammonium chlorides gave good results.[237]

4.3.3.2 ALDEHYDES

Aldehydes act by cross-linking of cell wall molecules and by general coagulation of cytoplasmic constituents. They are alkylating agents that react with amino, carboxyl, thiol, and hydroxyl groups on proteins causing the irreversible modification of their structure.[261] Aldehydes have a broad-spectrum activity on microorganisms including a good sporicidal action. Organic matter partly inactivates them. Their use has been restricted because of their sensitizing and carcinogenic properties. They are still in use as biocides for paints and adhesives, for example, in combination with isothiazolinones.

Formaldehyde (also called formalin) is a very reactive substance and the most effective monoaldehyde. It is a strong irritant agent for the skin and mucous membranes. Several European countries restrict the use of formaldehyde, including the import of formaldehyde-treated products. Starting September 2007, the European Union (EU) banned the use of formaldehyde due to its carcinogenic properties. It was banned from use in certain applications (preservatives for liquid-cooling and processing systems, metalworking-fluid preservatives, and antifouling products). The European Regulation 689/2008 (export/import of dangerous chemicals) did not ban formaldehyde from manufacturing into the EU.

$H_2C = O$

Structural formula of formaldehyde

Structural formula of glutaraldehyde

The treatments applied in 1963 for defeating the so-called "maladie verte" of the famous 17,000-year-old paintings inside Lascaux cave (France) included a combined spray application of streptomycin and penicillin for bacteria and a subsequent treatment with formaldehyde for algae.[118] The growth of a green biofilm caused by lighting (the first of various biological colonization suffered over time by the cave) and the damages produced by visitors' breath, led to its closure in 1963. The treatments were effective until 1969 when a new colonization developed. The conservators planned periodic maintenance and cleaning excluding the application of formaldehyde.[118]

Even though formaldehyde has high toxicity and carcinogenic properties, it was used even recently. It was effective on fungi, bacteria, algae, mosses, and lichens of an Austrian fountain.[235] On the contrary, the treatment with formaldehyde fumigation and ethanol to control fungal colonies on wall paintings in a Japanese cave, the Kitora Tumulus, was ineffective because after a year a viscous gel, formed by fungi and bacteria, appeared on the walls[208] (Table 4.1).

In a museum (the National Museum, Belgrade) where fungal growth colonized many archaeological objects because of very unfavorable climate conditions (no heating, water infiltration, etc.), conservators applied formalin (3%), heating it until evaporation.[157] The treatment was successful.

Glutaraldehyde is the most effective dialdehyde as it contains two aldehyde groups. It has a broad activity spectrum and high sporicidal efficacy (2–8 times more sporicidal than formaldehyde). As it works more rapidly with increasing pH, pH is the most important factor in regulating its activity under typical use conditions. The degree of polymerization is negligible in acidic solutions but high at basic pH. It has higher activity and rapidity of action, greater ability to penetrate the cell wall, lower irritation potential than formaldehyde. It reacts with ammonia and substances containing amines, which can cause its exothermic polymerization.[262] Some bacteria have developed resistance to glutaraldehyde. Moreover, as glutaraldehyde can cause asthma and other health hazards, ortho-phthalaldehyde is replacing it. When associated to ethylenediamine-disuccinate (EDDS), glutaraldehyde can prevent the bacterial growth on carbon steel coupons while the compounds alone did not prevent biofilm establishment.[263]

4.3.3.3 AZOLES

Azoles are used in clinical treatment of systemic and superficial fungal infections. They are antifungal agents acting as chelate formers and preventing the synthesis of ergosterol, a major component of fungal cell membranes, thus affecting membrane permeability. They represent a class of microbicides with favorable toxicological properties and low ecotoxicity. The azoles contain either two or three nitrogen atoms in the azole ring and are thereby classified as imidazoles (e.g., ketoconazole, miconazole, clotrimazole, econazole) or triazoles (e.g., itraconazole, tebuconazole, fluconazole), respectively.

Azoles had a limited application on stone cultural heritage objects just because they have a selective spectrum of action (fungi) (Table 4.1). The trade product Preventol A8, containing a triazole as a.i., was very efficient on fungi.[264,235]

Structural formulas of:

2-imidazolin-4(5)-one imidazole sodium imidazolide

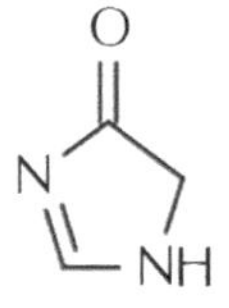

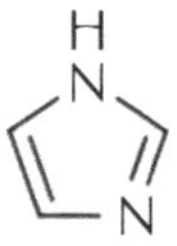

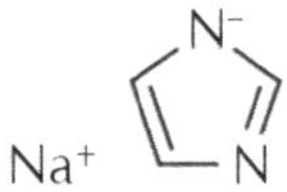

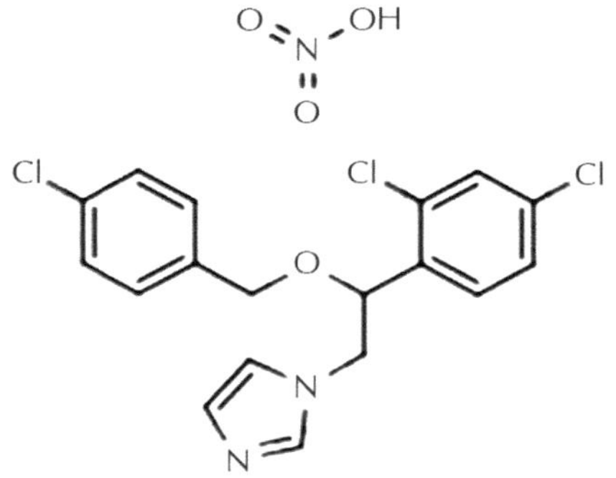

Structural formula of econazole nitrate

4.3.3.4 CARBAMATES

Carbamates are insecticides, fungicides, herbicides, nematocides. They are used for industrial or other applications and in household products. Carbamates encompass the esters and the salts of carbamic and dithiocarbamic acids.

Structural formulas of

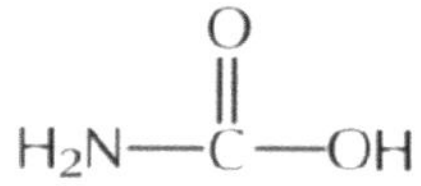

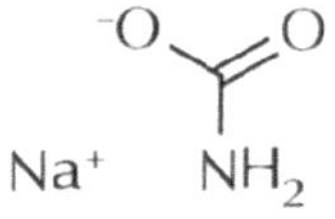

carbamic acid dithiocarbamic acid carbamate sodium salt

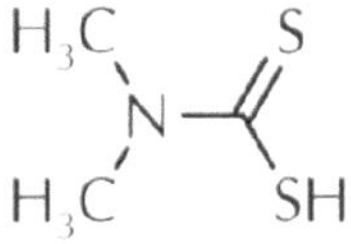

N,N-dimethyldithiocarbamic acid

Carbamic and dithiocarbamic acids lack chemical stability while their esters are more stable. Dithiocarbamates are fungicides with a carbamate structure where sulfur atoms replace both oxygens in the amide functional group.

The acute toxicity of the different carbamates ranges from highly toxic (mostly insecticides) to only slightly toxic or practically nontoxic. In general, the toxicity of carbamates for wildlife is low, but exceptions exist. For example, birds are not very sensitive to carbamates, while bees are extremely sensitive.

Several factors influence the biodegradation of carbamates in soil, such as volatility, soil type, soil moisture, adsorption, pH, temperature, and photodecomposition. As carbamates have different chemical structures, one may be easily decomposed, while another may be strongly adsorbed on soil. Some leach out easily and may reach groundwater.

As herbicides, they act as shoot growth inhibitors. This kind of herbicides are soil-applied herbicides and control weeds that have not emerged from the soil surface. They generally control grass weeds and small-seeded broadleaf weeds.

Microbicides of this class of substances act as chelate formers and widely differ in terms of efficacy. The carbamate fungicides include carbendazim, benomyl, and thiophanates.

The dithiocarbamates (trade product Vancide 51 by Vanderbilt Company, blend of sodium dimethyldithiocarbamate 27.6% and sodium mercaptobenzothiazole) were used in cultural heritage field because they have a broad spectrum of action.[265,235] As they are stable though only at pH 7–10, the application on highly alkaline materials such as lime-slurries, plaster, or mortar can be problematic.[235]

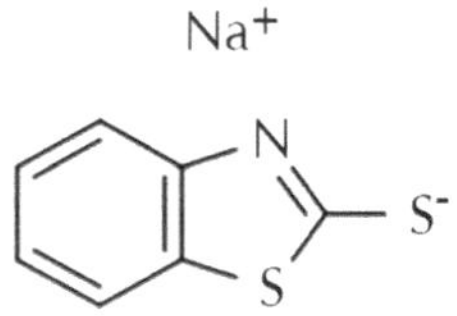

Structural formula of sodium salt of 2-mercaptobenzothiazole

Some studies[266,267,257,268,29] report the application of products containing blends of carbamates with other biocides (Mirecide-TF/580.ECO, Biotin R, Mergal S97) (Table 4.1). The results of in situ treatments of cyanobacteria, algae, endolithic fungi, lichens, and mosses were very good. The concentration needed to kill phototrophic microorganisms and mosses was quite low (2%), while it was higher (5–10%) to kill crustose lichens. Similarly, the concentration necessary to eliminate fungi, algae, and actinobacteria in laboratory tests was 5%.

4.3.3.5 CHLORINE-CONTAINING COMPOUNDS

The chlorine-containing compounds in aqueous solutions consist of a mixture of Cl_2, OCl^-, HOCl, and others. They are among the most used microbicides for the control of biofouling in cooling water systems. Chlorine is an excellent algicidal and bactericidal agent even at very low concentrations.[262,269,270]

Cl – OH

Structural formula of hypochlorous acid

Hypochlorous acid is responsible for the microbicidal action of these compounds.[261] The pH and temperature have a strong influence on their efficacy. In water at pH values between 4 and 7, they predominantly exist as hypochlorous acid in equilibrium with the hypochlorite ion (OCl^-), which predominates above pH 9 and lowers the efficiency of these chemicals. Proteins and ammonia-containing compounds inactivated chlorine-containing compounds reacting with them.[25] An increment in temperature increases their microbicidal activity. High reactivity and low persistence can leave some microorganisms unharmed, especially those that proliferate in biofilms. As an alternative, chlorine dioxide was used because it offers the advantages of selectivity, effectiveness over a wide range of pH, speed of kill, and formation of fewer byproducts.

Calcium hypochlorite is widely used for water treatment especially as a swimming pool additive. It is also a bleaching agent. A weak point is that it adds calcium to the system. Sodium hypochlorite solution, commonly known as bleach, is frequently used as a disinfectant or a bleaching agent in household, public buildings, food industry, hospitals, and water treatment. It is effective on a wide spectrum of microorganisms. The weakness of this compound is its limited chemical stability.

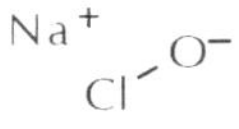

Structural formula of sodium hypochlorite

Although sodium hypochlorite was very effective at killing biofilms, concern over the production of sodium salts in the treated stone restricts its use for most building materials.[85] The recent literature on biocides applied on monuments is very scarce in reporting the application of sodium and calcium hypochlorites that have been instead in the past among the most applied substances for the control of algal, bacterial and fungal growth. In fact, a study that provides a framework for cleaning projects and a methodology for masonry cleaning,[253] suggests the application of sodium and calcium hypochlorite solutions. The authors recommend the application of the solution to a prewet surface that is then scrubbed and thoroughly rinsed with low-pressure water. Hypochlorite solutions were applied on Thai sanctuaries[168] and on the Legislative building in Washington (USA)[239] (Table 4.1). Cyanobacteria and lichens grew extensively on sandstone of the latter building. The restorers applied preheated water (45–65°C) which helped to soften the lichens, followed by 5% solution of calcium hypochlorite applied under low pressure.

4.3.3.6 COMPOUNDS WITH ACTIVATED HALOGEN ATOMS

The addition of a halogen atom to a carbon atom of an electronegative group causes a strong electron deficiency at the carbon atom because the electron affinity of halogen atoms is much higher than that of carbon atoms. The halogens or halogen elements are a group in the periodic table that consists of five chemically related elements: fluorine (F), chlorine (Cl), bromine (Br), iodine (I), and astatine (At). The name "halogen" means "salt-producing." When halogens react with metals they produce a wide range of salts, including calcium fluoride, sodium chloride (common salt), silver bromide, and potassium iodide. Halogens are highly reactive, and thus they can be harmful or lethal to organisms. This reactivity is due to their high electronegativity. Because the halogens have seven valence electrons in their outermost energy level, they can gain an electron by reacting with atoms of other elements to satisfy the octet rule. Fluorine is one of the most reactive elements, attacking otherwise inert materials such as glass, and forming compounds with the usually inert noble gases. It is a corrosive and highly toxic gas. The reactivity of fluorine is such that, if used or stored in laboratory glassware, it can react with glass in the presence of small amounts of water to form silicon tetrafluoride (SiF_4). The high reactivity of fluorine leads to the formation of some of the strongest bonds possible, especially to carbon.

Bronopol, a biocide with activated halogen atoms, reacts with nucleophilic groups and organic compounds of the cell but has a slow microbicidal effect. Under aerobic conditions, bronopol catalytically oxidizes thiol-containing materials, such as cysteine, with atmospheric oxygen as the final oxidant. By-products of this reaction are active oxygen species such as superoxide and peroxide, which are directly responsible for the bactericidal activity of the compound.

Bronopol has a broad spectrum of antimicrobial activity, particularly on bacteria, but a limited antifungal activity. To overcome this drawback, it is combined with other agents such as isothiazolones. Thanks to its characteristics—colorless, odorless, and easily soluble in water and alcohol, low toxicity, good skin compatibility, broad effective spectrum—it is used on a large scale as a preservative of cosmetics and pharmaceuticals. Bronopol has the higher stability in acidic solutions.

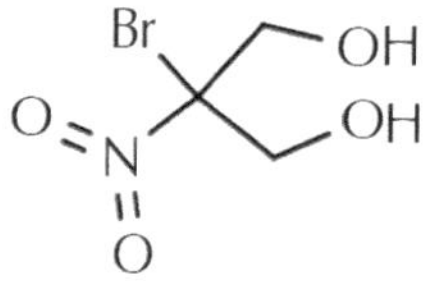

Structural formula of bronopol

4.3.3.7 GLYPHOSATE

Glyphosate (*N*-(phosphonomethyl) glycine) is an aminophosphonic analogue of the natural amino acid glycine, and its name is a contraction of glycine, phospho-, and -ate. It is a broad-spectrum systemic herbicide used to kill weeds, especially perennials. It has low toxicity and low persistence in the soil. Monsanto Company introduced it in 1974 for nonselective weed control. The typical uses are spray and absorption through the leaves, injection into the trunk, or application to the stump of a tree from where it moves to growing areas.

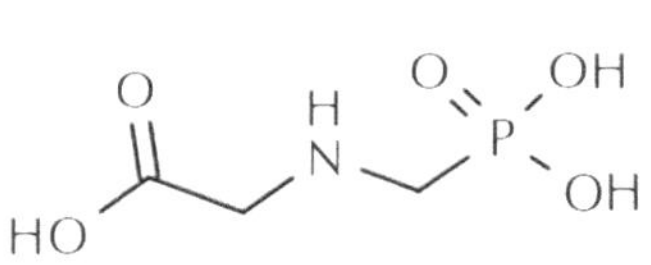

Structural formula of glyphosate

The extraordinary success of the simple and small molecule of glyphosate is due to its high specificity for a plant enzyme. It inhibits an enzyme involved in the synthesis of the amino acids tyrosine, tryptophan and phenylalanine, which are building blocks of peptides. Because of this mode of action, it is only effective on actively growing plants.

Monsanto also produces patented seeds that grow into plants genetically engineered to be resistant to glyphosate. Such crops allow farmers to use glyphosate as a postemergence herbicide against most broadleaf and cereal weeds. Soy was the first of these crops.

In March 2015, the International Agency for Research on Cancer (IARC, within the World Health Organization) established that glyphosate belongs in a 2A category as probably carcinogenic to humans. Group 2A agents are a risk mostly for occupational exposure, that is, for people who work with (or around) the chemical on a regular basis over a long period.

The brand name Roundup is a water-based solution containing glyphosate (as the main a.i.) and the surfactant POEA (polyethoxylated tallow amine). The surfactant, known for its toxicity in wildlife, increases herbicide penetration in plant cells.

Many applications of glyphosate in cultural heritage conservation showed good performance on weeds, shrubs, and small trees[271–275] (Table 4.4). Used on mosses, it showed effective results only after 4 months.[256]

4.3.3.8 HYDROGEN PEROXIDE

Hydrogen peroxide reacts with thiol and sulphydryl groups, and has a biocide-induced autocidal activity through the accumulation of free radicals.[261] This action increases with increasing temperature. The results of its efficiency are controversial. Some authors affirm that it can inhibit the microbial growth even at low concentrations (200, 600, and 1000 ppm) causing a very fast and efficient removal of attached biofilms.[262,276] On the contrary, as other authors reported,[269] hydrogen peroxide at 150 mg/l concentration has no significant microbicidal effect on bacteria and yeasts in laboratory tests. Moreover, it does not have sporicidal efficacy.[277] Hydrogen peroxide has been introduced frequently into practice as a biocide because it decomposed in water and oxygen thus leaving no long-term residues. On the other hand, it is hazardous, and its solutions are a primary irritant. Hydrogen peroxide is used in hospitals to disinfect surfaces in solution alone or in combination with other chemicals. It can cause corrosion of structural metals and possible deterioration of rubber, plastics, wood.[262]

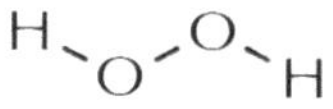

Structural formula of hydrogen peroxide

The demand for microbicides by professionals involved in conservation of cultural heritage led at first to the use of oxidizing agents as hydrogen peroxide, which, though very effective, is unstable. Because of this, it is stabilized in commercial products adding acids that cause a dangerous drop in pH leading to possible corrosion effects on minerals such as calcite.[277] Moreover, it can oxidize metal ions (e.g., iron) causing red, rusty or black stains.[277] The application of hydrogen peroxide is still performed even if it can negatively affect marble[125,240] (Table 4.1). For example, a study[240] reported the repeated application of a poultice containing hydrogen peroxide at 25% for 5 h. The treatment provided very satisfactory results on lichens, cyanobacteria, and algae colonizing marble statues. The choice of such a high concentration was because a solution of hydrogen peroxide at 10% showed no efficacy. The same study showed the effectiveness on lichens, cyanobacteria, and algae on marble statues of a mixture of different compounds (hydrogen peroxide, EDTA, NH_4HCO_3 solution, and organic solvent). Unfortunately, this study, as many others, did not report the concentration of the substances.

Hydrogen peroxide was applied on the surfaces of Slovenian caves colonized by algae.[278] The addition in the aqueous solution of a carbonate/bicarbonate buffer maintains pH neutrality. The compound was used at a quite high concentration (15% v/v); 40 ml of solution were sprayed on 1 m^2. In a cave, the conservators applied the solution even three times. In addition, the study reports that two applications of the hydrogen peroxide solution per year were effective at preventing algal regrowth.[278] As a general consideration, this kind of applications (repeated with the same compound) are not recommended because they can cause the resistance of microorganisms to that biocide and many others (see Section 4.7).

4.3.3.9 IMIDAZOLINONE HERBICIDES

Imidazolinone herbicides act inhibiting the enzyme acetohydroxy-acid synthase (AHAS), also called acetolactate synthase (ALS), which is involved in the biosynthesis of branched-chain amino acids (valine, isoleucine, leucine) in plants. They represent a powerful tool for weed management. Roots and shoots readily absorb imidazolinones. They move to the plant meristem where they exert the toxic action. In general, imidazolinone herbicides are more active against dicotyledons than against monocotyledons. They include the active ingredients imazapyr, imazapic, imazethapyr, imazamox, imazamethabenz, and imazaquin. The chemical structures of these chemicals are closely similar one another. They share some characteristic such as low toxicity on animals, birds, fishes, and invertebrates, but they differ in crop selectivity.

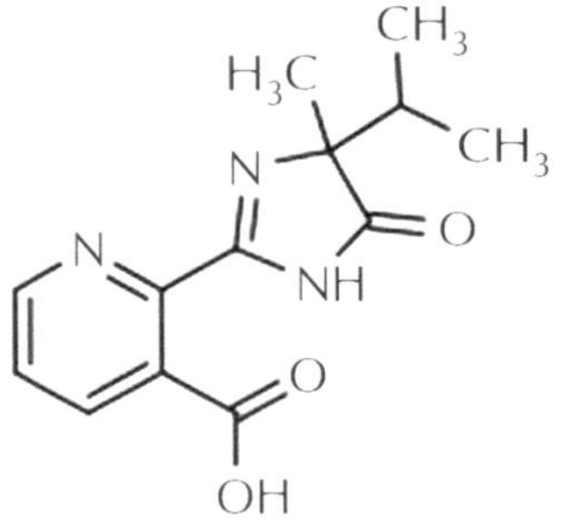

Structural formula of imazapyr

Imazapyr, sold under the trade name Arsenal, is a nonselective herbicide used for the control of a broad range of weeds, including terrestrial annual and perennial grasses and broadleaf herbs, woody species. It was applied in archaeological areas in Rome and Pompei (Italy) and in tropical areas.[255] This a.i. showed high efficacy even on very resistant species like *Ailanthus altissima* and trees of the genus *Ficus*.[255]

4.3.3.10 ISOTHIAZOLONES

They are electrophilic active microbicides whose antimicrobial mechanism of action has been intensively studied over the last decades. The isothiazolone group, which contains N and S atoms in pseudo-aromatic ring systems, is characteristic of these substances and exerts the antimicrobial activity. These chemicals interact with thiol groups of amino acids, thus degrading various proteins with multiple consequences for cell metabolism such as inhibition of enzyme reactions and respiration.[279]

Structural formulas of

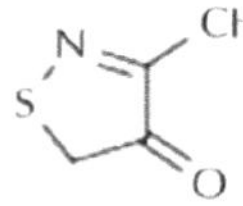

methyl isothiazolone

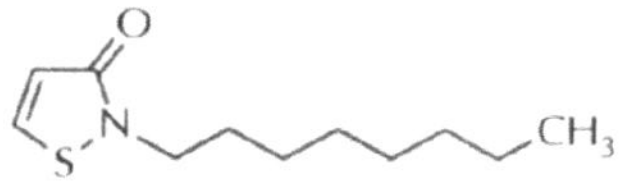

2-octyl-3(2*H*)-isothiazolone

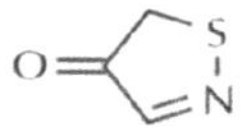

4-isothiazolone

They also have a biocide-induced autocidal activity through the accumulation of free radicals. Isothiazolones have gained widespread acceptance because of their effectiveness at low concentrations on bacteria, fungi, and algae in a wide variety of materials. Their environmental characteristics are also attractive because the degradation can happen via several pathways including hydrolysis, biodegradability and photolytic breakdown. Moreover, they can be used over a broad pH range with no decrease in activity.[262] Among their drawbacks, there are a relatively slow kill and safety concerns arising from the fact of their sensitizing effects. The biggest application is in paint industry especially as marine antifouling agent.[280] They are added also in plaster, adhesives, water systems, cosmetics, household goods, as pulp and wood impregnating agents as well as in leather, fur, and polymer processing.

Isothiazolones containing one or two chlorine atoms (e.g., 5-chloro-2-methyl-isothiazolin-3-one and 4,5-dichloro-2-(*n*-octyl)-isothiazolin-3-one) are more reactive and efficient in their antimicrobial activity because of their second toxophoric structural element. Their efficacy as compounds with activated halogen atoms strongly exceeds that of nonhalogenated isothiazolones.

Regarding cultural heritage (Table 4.2), recent treatments with isothiazolones and with blends of these agents with other active ingredients gave good results. In laboratory tests, for example inhibition halo in agar diffusion tests and stone slabs inoculated with microorganisms, octyl-isothiazolinone was very effective even at 0.05% on many fungi that frequently grow on stones. The only fungus showing resistance belonged to the genus *Helminthosporium*.[281] The product Mergal S97 (blend of isothiazolones and a carbamate) at 1 and 2% in ethanol was effective on fungi, bacteria, and yeasts isolated from limestone and granite.[266] The product ProClin TM 950 (2-methyl-4-isothiazoline-3-one, 9.5% solution in water) at a concentration of 500 μmol/l had a severe impact on membrane integrity and physiological performance of algal cells (*Stichococcus* sp.) but killed only 20–40% of them.[280] On the other hand, laboratory tests on bacteria and algae that grow on materials of industrial water systems showed that isothiazolones at concentrations in the order of ppm kill the microbes but were not able to remove the dead biofilms from the surfaces.[262,282,251]

Isothiazolones and blends of these agents with other active ingredients gave positive results in all the applications in situ (Table 4.2). The commercial products Rocima103, Parmetol DF12, Acticide LV706, Acticide MBL5515, Mirecide-TF/580.ECO, Biotin T and mixtures of isothiazolones and quaternary ammonium salts showed very good efficacy on foliose and crustose lichens, cyanobacteria, algae, fungi in a variety of environments[257,225,283,38,99,29,268,219,284,146] (Table 4.2). The product Parmetol DF12 was even effective at inhibiting biofilms regrowth after 4 years in the extremely favorable environment of Angkor Wat monuments.[99] Regarding lichens, the concentrations of use vary 2.5–10% and the solutions were applied by brushing or poultice (the latter for 48 h), depending on the species and the substrates. Rocima103 was more efficient on crustose lichens than Acticide LV706 and Mirecide-TF/580.ECO. A mechanical treatment of lichens preceded the application of Acticide LV706 while the other products were applied directly on lichens.

TABLE 4.2 List of Isothiazolones Including Active Ingredients and Trade Products, Concentrations of Use, Efficacy on the Target Organisms, Indications about the Experimental Approach (In Situ or under Laboratory Conditions), Side Effects and Notes, Bibliographical References.

Commercial product/ Active ingredient	Concentration	Efficacy	Organisms/Substrate	Notes/Side effects	References
Isothiazolones	0.9–10 ppm	They are fast-acting biocides inhibiting growth, metabolism and biofilm development of both algae and bacteria	Materials of industrial water systems with biofilms formed by bacteria and algae	They are deactivated by H_2S	Videla, 2002
2-Octyl-4-isothiazolin-3-one	0.05%	Very efficient. It prevented the growth of many fungi except *Helminthosporium* sp. This was the most resistant fungus showing growth on all biocide-containing specimens	Biocide-containing phosphogypsum specimens inoculated with the fungi: *Aspergillus niger*, *Cladosporium* sp., *Aspergillus* sp., *Helminthosporium* sp., *Cordana* sp., *Phialophora* sp., *Penicillium* sp., *Trichoderma* sp.	It has a higher leaching rate than other biocides (including different isothiazolones), and thus, it lives shorter in the material	Shirakawa et al., 2002
n-Octyl-isothiazolone	5% in ethanol	Effective on all the tested microorganisms	Laboratory tests. Agar diffusion tests: fungi, algae, bacteria, actinobacteria. The species were among those growing on monuments		Borgioli et al., 2006
Mixture of 5-chloro-2-methyl-3-isothiazolone/2-methyl-3-isothiazolone (3:1)	4 ppm	It was not able to remove biofilms from the surfaces. It can be used as an additional disinfectant after mechanical/chemical cleaning surfaces. Biofilms showed a decrease in their response to the biocide on subsequent exposures 24–72 h later	Materials of industrial water systems with biofilms formed by the bacterium *Sphaerotilus natans*		Ludensky, 2003

TABLE 4.2 *(Continued)*

Commercial product/ Active ingredient	Concentration	Efficacy	Organisms/Substrate	Notes/Side effects	References
Mixture of dichloro-*N*-octylisothiazolone and *n*-octyl-isothiazolone	5% in ethanol	Effective on all the tested microorganisms. Very good performance on fungi and bacteria	Laboratory tests. Agar diffusion tests: fungi, algae, bacteria, actinobacteria. The species were among those growing on monuments		Borgioli et al., 2006
Mergal S97 (blend of 5-chloro-2-methyl isothiazolinone, 2-methyl isothiazolinone, methyl-benzimidazol-2-ylcarbamate and 2-*n*-octylisothiazolinone)	1 and 2% in ethanol	Very good efficacy on fungi and bacteria; very scarce growth on limestone slabs inoculated with *Ulocladium* sp. after prolonged incubation	Laboratory tests. Agar diffusion tests: fungi, bacteria and yeasts isolated from limestone and granite. Rock slabs tests: limestone slabs inoculated with the fungus *Ulocladium* sp.	The product crystallizes in stone pores. It could make difficult the penetration of the solutions	Blazquez et al., 2000
Rocima™103 (blend of 2-octyl-2*H*-isothiazolin-3-one and didecyl dimethyl ammonium chloride in propanol and formic acid)	2% in water, brushing	Efficient	Archaeological site at Ostia Antica (Rome). Extensive growth of cyanobacteria, green algae, and bryophytes		Bartolini et al., 2007
Rocima™103 (concentrated blend of 2-octyl-2*H*-isothiazolin-3-one and didecyl dimethyl ammonium chloride in propane-2-ole and formic acid)	5% v/v, brushing until the complete imbibition of the lichens 10% v/v poultices for 48 h	Not efficient at 5%. Efficient at 10% on the lichens	Crustose and foliose lichens on marble, sandstone, and plaster. Archaeological area of Fiesole, Firenze (Italy).		Pinna et al., 2012

TABLE 4.2 *(Continued)*

Commercial product/ Active ingredient	Concentration	Efficacy	Organisms/Substrate	Notes/Side effects	References
Biotin T (mixture of di-*n*-decyl-dimethyl ammonium chloride, 2-*N*-ottil-2*H*-isotiazol-3-one, isopropanol, and formic acid)	2% v/v, brushing	Efficient in removing almost all colonizations after 2 weeks	Mortar slabs inoculated with cyanobacteria and green algae *Gloeocapsa dermochroa*, *Stichococcus bacillaris*, *Chlorella ellipsoidea*, and exposed outdoor. Walls on buildings in Sintra, Portugal. Lichens and green algae		Fonseca et al., 2010
Biotin T (mixture of di-*n*-decyl-dimethyl ammonium chloride, 2-*N*-ottil-2*H*-isotiazol-3-one, isopropanol, and formic acid)	2% v/v, brushing	Effective. Just a partial recovery of some thalli of *P. muralis* after 18 days	Crustose epilithic and endolithic lichens *Acrocordia conoidea, Aspicilia contorta, Bagliettoa marmorea, Candelariella vitellina, Protoparmeliopsis muralis, Verrucaria nigrescens*. Limestone, Karst plateau, Italy		Tretiach et al., 2012
Biotin T (mixture of di-*n*-decyl-dimethyl ammonium chloride, 2-*N*-ottil-2*H*-isotiazol-3-one, isopropanol, and formic acid)	2% v/v	Not effective, partial recovering of the photosynthetic activity	Mosses and liverworts, Karst plateau, Italy		Bertuzzi et al., 2013
Biotin T (mixture of di-*n*-decyl-dimethyl ammonium chloride, 2-*N*-ottil-2*H*-isotiazol-3-one, isopropanol and formic acid)	2.5% in water, brushing till saturation of the stone	The treatment applied to areas extensively colonized by crustose lichens, resulted in a clean stone surface free of lichen remains. Mechanical cleaning followed it	Segovia cathedral cloister (Spain), dolostones and granite. Crustose lichens (*Aspicilia contorta, Lecidella stigmatea, Verrucaria nigrescens*). Foliose lichen: *Phaeophyscia orbicularis* Cyanobacteria, mosses	Living fungal hyphae were detected in fissures of the stone 2 months after treatment	de los Ríos et al., 2012

TABLE 4.2 *(Continued)*

Commercial product/ Active ingredient	Concentration	Efficacy	Organisms/Substrate	Notes/Side effects	References
Algophase (*n*-butil-1,2 benzoisotiazolin-3-one)	3% v/v in water	Efficient on bacteria (Firmicutes and Actinobacteria)	Marble monuments, Acropolis, Athens Greece	After 14 months, there was no regrowth	Savvides 2014
ACTICIDE® LV706 (a.i. 2-*n*-octyl-3(2*H*)-isothiazolone + benzalkonium chloride)	10% in distilled water, brushing	Effective at reducing around 80% of the lichens	Abandoned dolostone quarry (Redueña, Madrid, Spain)		Camara et al., 2011
ACTICIDE® CF (2-*n*-octyl-3(2*H*)-isothiazolone + terbutryn	1.5% in water-repellent resin, brushing	Effective in reducing around 80% of the lichens	Abandoned dolostone quarry (Redueña, Madrid, Spain)		Camara et al., 2011
ACTICIDE® IOG (a.i. 2-*n*-octyl-3(2*H*)-isothiazolone + 3-iodopropynyl butylcarbamate)	0.5% in water-repellent resin, brushing	Effective in reducing around 80% of the lichens	Abandoned dolostone quarry (Redueña, Madrid, Spain)		Camara et al., 2011
Parmetol DF 12 (a blend of an imidazole and three isothiazolinones)	3% v/v in water	Effective after four years. Ethanol was applied first to destabilize the biofilms	Biofilms on Angkor Wat (Cambodia) monuments		Warscheid & Leisen, 2009
ProClin TM 950 (a.i. 2-methyl-4-isothiazoline-3-one). 9.5% solution in water	500 µmol/l	It severely affected all performance and structural parameters of the algae	Laboratory tests. Aero-terrestrial algal strain of the genus *Stichococcus*	Even though the impact on membrane integrity was severe, the active agent killed only 20–40% of the cells	Gladis et al., 2010

Although Biotin T was effective on crustose lichens, still active fungal hyphae were detected in fissures of the stone 2 months after treatment. This was likely due either to the chosen concentration or the method of application (2.5%, brushing). As the product contains a much higher amount of di-*n*-decyl-dimethylammonium chloride (40–60%) than of isothiazolone (7–10%), it needs a higher concentration and a different application (e.g., poultice) to kill endolithic fungi.

The product Biotin R, a blend of a carbamate and an isothiazolone, is not included in this section because the amount of carbamate is much higher than that of isothiazolone. Therefore, the corresponding results are reported in the section dedicated to carbamates.

4.3.3.11 PICLORAM

Picloram is a pyridine herbicide, mainly used to control unwanted trees in pastures and edges of fields. It is a synthetic auxin, a plant hormone-type herbicide, and belongs to the group of plant growth regulators. The naturally occurring auxin (indole-3-acetic acid, IAA) is a plant hormone that regulates plant growth by modulating gene expression leading to changes in division, elongation, and differentiation of cells.[285] There are evidences of strong carcinogenic activity of picloram in studies on rats. In some countries, it is a restricted use herbicide.

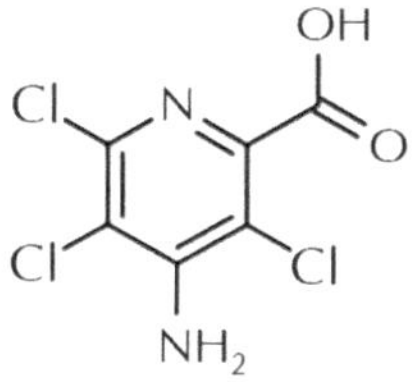

Structural formula of picloram

Tordon 101 (a commercial product containing picloram) was successful in killing plants of an archaeological area by absorption through the leaves[274] (Table 4.4).

4.3.3.12 QUATERNARY AMMONIUM COMPOUNDS (QACS OR QUATS)

Quaternary ammonium compounds (QACs or Quats) are surface-active microbicides whose molecule has two different structural elements, a hydrocarbon water-repellent group, and a water attractant polar group. They are also known as surfactants, a word that is a contraction of the three words "Surface Active Agents." Depending on the charge of the hydrophilic polar group, they are classified as anionic, cationic, amphoteric, and nonionic compounds. Surfactants are materials that reduce the surface tension (or interfacial tension) between two liquids or between a liquid and a solid (Fig. 4.4) acting as detergents, wetting agents, soaps, emulsifiers. Using surfactants, hydrophobic compounds can be suspended in an aqueous solution by forming micelles (Fig. 4.5). A micelle has a hydrophobic core that consists of the hydrophobic compound and the hydrophobic "tails" of the surfactant. The hydrophilic "heads" of the surfactant cover the surface of the micelle. A suspension of micelles is called an emulsion. The more hydrophilic the head group of the surfactant, the greater is its capacity to emulsify hydrophobic materials. When using soap to remove grease from dirty dishes, the soap forms an emulsion with the grease that is easily removed by water through interaction with the hydrophilic head of the soap molecules.

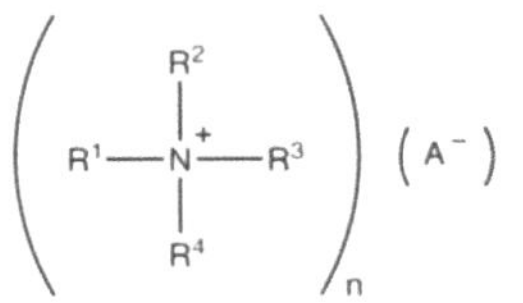

General structural formula of QACs (surfactants).

QACs are a large group of homolog compounds. Their activity results from the cationic surfactant nature of the hydrophobic portion of the molecule. They have a broad-spectrum activity, acting as algicides, bactericides, and fungicides. They are effective on bacteria, but not on some species of *Pseudomonas* or bacterial spores. QACs serve as additives in large-scale industrial water systems to minimize undesired biological growth.

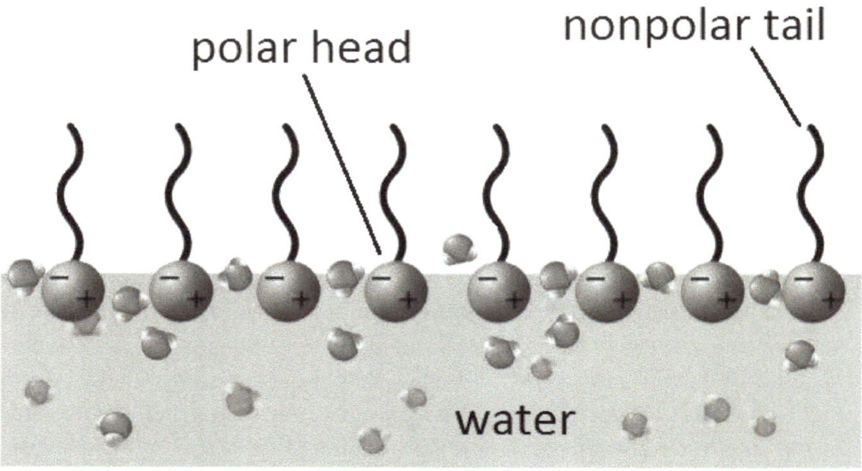

FIGURE 4.4 Surfactants reduce the surface tension of water by adsorbing at the liquid–air interface. (From Wikimedia Commons).

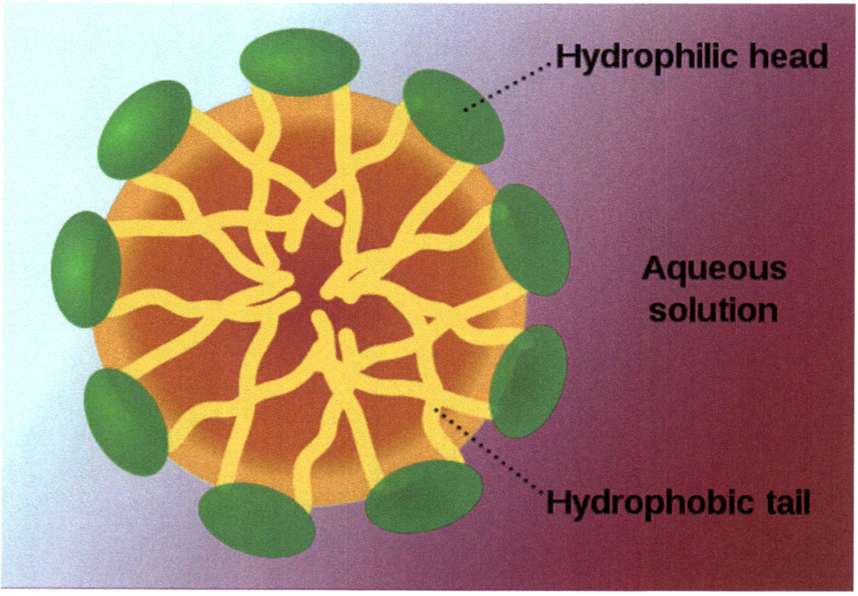

FIGURE 4.5 Diagram of a micelle of oil in an aqueous suspension. The hydrophobic tails form the core of the aggregate and the hydrophilic heads are in contact with water. (From https://commons.wikimedia.org/wiki/File:Micelle_scheme-en.svg)

The antimicrobial activity depends on the length of the alkyl side-chain, optimal efficacy being in the range C8–C16. Moreover, the efficacy increases with temperature and pH. Alkaline media are most favorable while at pH 3 QACs are ineffective as the negatively charged surface of the cell is protected from interaction by protonation (i.e., the addition of a proton H^+ to an atom or molecule). High amounts of proteins and salts severely reduce their efficacy. QACs are incompatible with strong oxidizing agents such as chlorine, peroxides, chromates and permanganates. Most QACs are biodegradable and do not require chemical deactivation following their use.[262]

Biocides containing QACs as active ingredients can contain a variety of additives, including potassium hydroxide, alcohol, water, etc. The diluents are alcohol and water. Alcohol has certain biocidal properties and good penetrating ability that can help QACs' action.

Benzalkonium chloride (short for alkyl dimethyl benzyl ammonium chloride, see Appendix for detailed information) has a broad spectrum of activity covering bacteria, algae, yeasts, and fungi. It is particularly effective at pH 6–8. The number of carbon atoms in the alkyl group may differ between *n*-C8 and *n*-C18, and benzalkonium chloride may contain several corresponding alkyl groups. As a surface-active compound, it greatly reduces the surface tension of water. BC is used as an a.i. in disinfectants. It is applied in food industries, in pharmaceutical, and personal care industries, in the treatment of cooling and swimming pool water, textile materials, leather, wood. Its usefulness in practice is limited by the fact that it can produce foam.

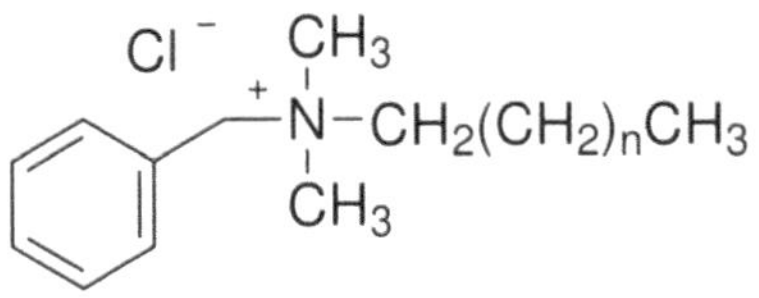

Structural formula of benzalkonium chloride (n = 8, 10, 12, 14, 16, 18)

Di-n-decyl-dimethylammonium chloride (DDAC) exhibits microbicidal activity on a wide range of bacteria, fungi, and yeasts. Compared to other QACs, it maintains an unusually high level of activity in presence of organic matter and hard water. Because of its favorable performance characteristics, it is used in many applications: disinfectants, sanitizers, and cleaners for use in hospitals, homes, institutions, farm and industrial areas, protection of textile material, and timber.

QACs are active ingredients extensively applied in the past and still widely present in the literature on cultural heritage preservation (regarding not only objects made of natural and artificial stones). They bear relatively small hazard to the substrate, have a broad spectrum of action and a middle-to-low toxicity[258,168,243,277,225] (Table 4.3). Benzalkonium chloride, the most widely employed QAC, was applied as a pure agent or as component of a trade product. Although the numerous studies that report its application, the results obtained are somehow controversial. The literature provides similar data in showing that QACs (especially the commercial product Preventol R80, now Preventol RI80, see Table 4.3 and Appendix) are efficient on mosses at 2% and 3% concentrations.[271,256,257] The drawback in the tropical climate of Maya monuments was the regrowth of mosses after 1 year.[286] There is agreement also about the efficacy of benzalkonium chloride in killing cyanobacteria and algae. Both laboratory tests and in situ trials employed products containing this a.i. at concentrations ranging 1–3%.[287,288,273,169,113,140,244] In contrast, Neo Des (known also as New Des or Neo Desogen) was efficient in laboratory tests at a very high concentration, 30%.[289] This was because it contains a low amount of benzalkonium chloride as a.i. (10%). Therefore, a very high concentration is needed which, on the other hand, corresponds to a final 3% of the a.i. For the same reason, Neo Des at 2% was not able to kill all the cyanobacteria that grew on wall paintings (Orvieto Cathedral, Italy).[94]

The combination of isopropyl alcohol and tridecyl ceteth alcohol with benzalkonium chloride produces the molecule isopropyl–tridecyl–dimethyl ammonium chloride, a.i. of the product Umonium 38. It showed good efficacy on green algae but resulted ineffective on cyanobacteria, for example the genus *Gloeocapsa* that grew on wall paintings of a crypt in Italy.[113] The authors attributed the resistance of cyanobacteria to the gelatinous sheath functioning as a barrier to biocide's penetration. Another QAC (didecyl-*n*-methyl-poly (oxyethyl) ammonium propionate), as a.i. of the product Anios DDSH, was not effective on algae and cyanobacteria[283] (Table 4.3).

QACs (principally benzalkonium chloride) applied on bacteria, showed good efficacy only at 2 and 3% concentrations.[266,290] A discordant "voice" reported that Preventol RI80 was not efficient on bacteria (Firmicutes and actinobacteria) that grew on monuments in Acropolis, Athens (Greece)[284]

TABLE 4.3 List of quaternary ammonium salts including active ingredients and trade products, concentrations of use, efficacy on the target organisms, indications about the experimental approach (in situ or under laboratory conditions), side effects and notes, bibliographical references.

Commercial product/ active ingredient	Concentration	Efficacy	Organisms/Substrate	Notes/Side effects	References
Preventol R80 (now Preventol RI80)	3% in water, sprayed twice with a week interval	Efficient in killing all the taxa. No recolonization after a year.	17 taxa of mosses growing on mosaics (Baths of Caracalla, Rome, Italy).		Altieri et al., 1996
Preventol R80 (now Preventol RI80)	0.5%,1%, 10% in water	0.5-1% concentrations showed efficacy against heterotrophic bacteria, actinomycetes, and autotrophic microorganisms. Only high concentrations (10%) managed to eliminate fungi.	Laboratory tests. Fresco samples inoculated with microorganisms.	Color change of cobalt blue fresco samples after 14 days (darkening effect). No changes in white samples.	Monte et al., 1999
Preventol R80 (now Preventol RI80)	2% w/v in water	High efficiency against endolithic cyanobacteria. Low efficiency against epilithic crustose lichens *Thyrea*, *Aspicilia*, *Verrucaria* and *Caloplaca* spp.	"Lioz" limestone, Jeronimos Monastery (Lisbon, Portugal).		Ascaso et al., 2002
Preventol R80 (now Preventol RI80)	2% in water, brushing	Efficient against mosses	Mosaics, archaeological area of Ostia Antica, Rome, Italy.	After 30 days, green staining on the substrate	Bartolini et al., 2006
Preventol RI80	3% in water	Efficient	Algae and lichens on a Roman floor mosaic, mausoleum on the Appian Way, Rome, Italy.		Alberti et al., 2006
Preventol R80 (now Preventol RI80)	2% in water	Effective against mosses only after 15 days (*Tortula muralis*, *Tortula* sp., *Bryum capillare*, *B. caespiticium*)	Mosaics, archaeological area of Ostia Antica, Italy.		Bartolini et al., 2007

TABLE 4.3 *(Continued)*

Commercial product/ active ingredient	Concentration	Efficacy	Organisms/Substrate	Notes/Side effects	References
Preventol R80 (now Preventol RI80)	1,5-3% v/v in water	Effective against lichens and a greyish biofilm	Marble statues, National Palace of Queluz, Portugal.		Charola et al., 2007
Preventol R80 (now Preventol RI80)	2% in water, brushing	Efficient against cyanobacteria and green algae even though some cyanobacteria (*Gloeocapsa* and *Myxosarcina* spp.) were resistant. The biocide solution was applied three times at one-week intervals to permit a deeper penetration into the stone; after disinfection, biocide removal and cleaning by water were performed.	Wall paintings of the Crypt of the Original Sin (Matera, Italy).		Nugari et al., 2009
Preventol RI80	2% in water, two treatments with an interval of seven days, followed by ammonium carbonate cleaning.	Efficient against epilithic and endolithic microorganisms	Black patina (mainly cyanobacteria) on marble of the Pyramid of Caio Cestio, Rome, Italy.		Giovagnoli et al., 2011
Preventol R80 (now Preventol RI80)	4% w/v in absolute ethanol	Limited biocidal action against bacteria (Firmicutes and Actinobacteria)	Marble monuments, Acropolis, Athens, Greece.		Savvides 2014
New Des (10% benzalkonium chloride) and Preventol R80	New Des 2.5 and 5%, Preventol R80 1 and 2% in water	No high efficiency against fungi by New Des. Quite good performance of Preventol R80.	Laboratory tests. Agar diffusion tests: fungi, bacteria and yeasts isolated from limestone and granite crusts. Rock slabs tests: limestone slabs inoculated with the fungus *Ulocladium* sp. In situ tests: limestone and granite blocks.	No chromatic changes of stones.	Blazquez et al., 2000

TABLE 4.3 *(Continued)*

Commercial product/ active ingredient	Concentration	Efficacy	Organisms/Substrate	Notes/Side effects	References
Neo Desogen (10% benzalkonium chloride)	1% in water, dropped on the surface	Efficient	Artificial and natural biofilms on marble, formed by cyano-bacteria and algae.		Tomaselli et al., 2002
New Des	5%, 15%, 30% in water	Efficient only at the highest concentration. Inefficient against some fungi	Laboratory tests. Algae, fungi, and actinobacteria.		Borgioli et al., 2003
Neo Desogen (10% benzalkonium chloride)	2% in water, sprayed on the surface	It did not kill all the cells	Biofilms of cyanobacteria on frescoes, St. Brizio Chapel (Orvieto Cathedral, Italy).		Cappitelli et al., 2009
New Des 50 (a.i. didecyl-dimethyl ammonium chloride)	5% v/v in water, brushing	Effective on both lichens	Crustose epilithic lichens *Protoparmeliopsis muralis* and *Verrucaria nigrescens*. Limestone, Karst plateau, Italy.		Tretiach et al., 2012
New Des 50 (a.i. didecyl-dimethyl ammonium chloride)	5% v/v	Effective	Mosses and liverworts, Karst plateau, Italy.		Bertuzzi et al., 2013
Rhodaquat RP50 (50% benzalkonium chloride)		Effective against lichens. It removes the thalli and is more effective on wet lichens. It inhibited lichen re-growth for 3 years.	Portland limestone monuments. Applied bi-annually by spraying.		Sheppard, 2007
Benzalkonium chloride	30 ppm	It was not able to kill and remove biofilms from the surfaces. It can be used as an additional disinfectant after mechanical/chemical cleaning of the surfaces.	Materials of industrial water systems with biofilms formed by the bacterium *Sphaerotilus natans*.	Biofilms treated with non-oxidizing biocides demonstrated significant regrowth, usually within several hours.	Ludensky, 2003

TABLE 4.3 *(Continued)*

Commercial product/ active ingredient	Concentration	Efficacy	Organisms/Substrate	Notes/Side effects	References
Benzalkonium chloride	3% brushing, 10% poultice and spraying	3% concentration efficient in killing algae, cyanobacteria, fungi. 10 % concentration applied by poultice for 72 hours efficient against lichens. After cleaning, spraying of 10 % concentration solution on the surfaces.	Red sandstone. Extensive colonization of algae, cyanobacteria, fungi, lichens.		Magadán et al., 2007
Benzalkonium chloride	1.5 % v/v in water	*Acrocordia conoidea* was apparently more resistant than *Bagliettoa marmorea.*	Calcareous rocks, mainly limestone. Treatment of two endolithic lichens.		Tretiach et al., 2010
Benzalkonium chloride	3% in distilled water, poultices for 3 h	Effective against bacteria and fungi.	Marble statues, La Recoleta Cemetery, Buenos Aires, Argentina.		Guiamet et al., 2013
Benzalkonium chloride	3% v/v in distilled water, brushing	Not effective	Granite (Galicia, Spain) colonized by filamentous green algae (*Trebouxia* sp.) and cyanobacteria (*Gloeocapsa* and *Choococcus* sp.). The average thickness of the biofilm was 40 μm.		Pozo et al., 2013
Benzalkonium chloride	10 % v/v in water	Inhibition of the growth of *Bipolaris spicifera, Epicoccum nigrum, Aspergillus niger, Aspergillus ochraceus, Penicillium* sp., *Trichoderma viride*. MIC values 0.1 to 4.0 $\mu L.mL^{-1}$.	Laboratory tests on fungi.		Stupar et al., 2014

TABLE 4.3 *(Continued)*

Commercial product/ active ingredient	Concentration	Efficacy	Organisms/Substrate	Notes/Side effects	References
Cetyltrimethylammonium chloride	1% in water	Efficient	Maya site at Xunantunich, Belize. Biofilms, lichens and mosses on limestone.	After one year, it was not effective in controlling regrowth of mosses.	Ginell et al., 1995
Cetyldimethylbenzy-lammonium chloride	1% in water	Efficient	Maya site at Xunantunich, Belize. Biofilms, lichens and mosses on limestone.	After one year, it was not effective in controlling regrowth of mosses.	Ginell et al., 1995
D/2 Architectural Biocide solution, a proprietary combination of octyl decyl dimethyl ammonium chloride (0.30%), dioctyl dimethyl ammonium chloride (0.12%), didecyl dimethyl ammonium chloride (0.18), and alkyl dimethyl benzyl ammonium chloride (0.40%)		Effective against fungi, algae and mosses	Indiana limestone blocks.		Allanbrook, 2007
D/2 Architectural Biocide solution	No dilution	Efficient against bacteria (Firmicutes and Actinobacteria)	Marble monuments, Acropolis, Athens Greece.	After 14 months there was no regrowth	Savvides et al., 2014

TABLE 4.3 *(Continued)*

Commercial product/ active ingredient	Concentration	Efficacy	Organisms/Substrate	Notes/Side effects	References
D/2 Architectural Biocide solution	No dilution, spraying with three modalities: 1) scrubbing after application followed by rinsing; 2) pre-wetting the stone before the application; 3) no scrubbing and rinsing.	Efficient against blackening by cyanobacteria and green algae	Blocks of dolomitic limestone.	After 11 months, there was no regrowth. The degree of cleanliness achieved for treated and scrubbed areas and pre-wetted treated areas without scrubbing appeared similar.	Charola et al., 2012
Kimistone Biocida (a ready-to-use blend of didecyldimethylammonium chloride and benzalconium chloride in water).	Ready to use, 5%. Brushing until the complete imbibition of the lichens.	Not completely efficient	Lichens *Diploschistes actinostomus*, *Parmelia conspersa*, *Parmelia loxodes* on sandstone. Archaeological area of Fiesole, Firenze (Italy).		Pinna et al., 2012
Anios DDSH, mixture of n,n-didecyl-n-methyl-poly(oxyethyl) ammonium propionate with alkyl-propylene-diamine guanidium acetate.	No dilution, spraying	Not efficient	Mortar slabs inoculated with cyanobacteria and green algae *Gloeocapsa dermochroa*, *Stichococcus bacillaris*, *Chlorella ellipsoidea*, and exposed outdoor. Walls on buildings in Sintra, Portugal. Lichens and green algae.		Fonseca et al., 2010
Quaternary ammonium compounds	1000 ppm	Not efficient	Bacteria biofilms on stainless steel.		Meyer, 2003

TABLE 4.3 *(Continued)*

Commercial product/ active ingredient	**Concentration**	**Efficacy**	**Organisms/Substrate**	**Notes/Side effects**	**References**
Quaternary ammonium compounds	8-35 ppm	Efficient	Materials of industrial water systems with biofilms formed by bacteria and algae.	They are incompatible with strong oxidizing agents.	Videla, 2002
Quaternary ammonium compounds		Effective against fungi, bacteria, algae, mosses and lichens	Fountain of Undine in Baden, Austria.		Sterflinger & Sert, 2006
SINOCTAN-10 (10% quaternary ammonium compounds)	20% aqueous solution, repeatedly applied brushing until complete impregnation.	Effective against fungi, bacteria, algae.	Wall paintings, Italian sanctuary.	After more than 3 years, no further biodegrading phenomenon was observed.	Giustetto et al., 2015
Benzalkonium chloride with the addition of the permeabilizers A) Polyethylenimine B) Ethylenediaminetetraacetic acid C) Meso-2,3-dimercaptosuccinic acid.	0.001%, 0.01% with A) 10 μg ml^{-1} B) 0.1, 1 mM C) 1 mM	A, B and C are efficient permeabilizers of the membranes of *Pseudomonas* and *Stenotrophomonas* genera. *Pseudomonas* cells treated with A or B revealed damage in the outer membrane structure. A especially increased the surface area and bulges of the cells. The effect of B was visualized as large areas with varying hydrophilicity on cell surfaces. In liquid culture tests, A and B supplementation enhanced the activity of BC toward the target strains.	Liquid cultures of Gram-negative bacteria: *Sinorhizobium morelens, Pseudomonas* sp., *Stenotrophomonas nitritireducens.*		Alakomi et al., 2006

TABLE 4.3 *(Continued)*

Commercial product/ active ingredient	Concentration	Efficacy	Organisms/Substrate	Notes/Side effects	References
Benzalkonium chloride combined with permeabilisers (A polyethyleneimine, B ethylenediaminetetraacetic acid), an exopolysaccharide inhibitor (bismuth dimercaprol)) and photodynamic agents (nuclear fast red and methylene blue).	BAC 0.1% PEI 10 μg ml^{-1} EDTA 0.1, 1 mM	A and B enhanced the activity of the biocides (A better than B). The two photodynamic agents investigated here (NFR and MB) have the potential to destroy cyanobacteria on stone samples and, since the PDAs are broken down under visible light, no harmful or colored residue remains in the substrate. NFR was more effective and reliable than MB.	Sandstone samples inoculated with *Pseudomonas aeruginosa, Cladosporium* sp., *Penicillium* sp., *Phoma* sp., *Synechoccus leopoliensis, Anabaena cylindrical.*		Young et al., 2008

(Table 4.3). Studies of bacterial biofilms on stainless steel and on materials of industrial water systems treated with very low amounts of QACs ranging from 30 to 1000 ppm showed that they were not able to kill and remove bacterial biofilms from the surfaces.[270,282] On the other hand, other studies, although limited to lab tests, showed that benzalkonium chloride could be effective on bacteria also at very low concentrations (0.001, 0.01%) when mixed with permeabilizers such as polyethylenimine and ethylenediaminetetraacetic acid (EDTA).[89,291] These compounds help the biocide action because they act enhancing the damage to the membrane structure.

Contrasting results are those connected to QACs performance on fungi and lichens. Some studies affirm that QACs at 3% are enough to kill fungi[235,274,290] (Table 4.3). A study[266] reports that Preventol RI80 is instead efficient at 1 and 2%, whereas another[292] shows that it killed fungi only at 10%. These controversial and confusing results are amplified for lichens. Two QACs were effective at a very low concentration, 1%.[286] The product RhodaquatRP50 that contains 50% of benzalkonium chloride was even able to inhibit lichen regrowth for 3 years.[165] Benzalkonium chloride at 1.5% v/v showed different results on two endolithic lichens, one being more resistant than the other.[293] While Preventol RI80 showed good performance on lichens at 1.5–3%,[273,169] benzalkonium chloride killed lichens when applied at 10% by poultice for 72 h.[274] Another study reports that Preventol RI80 at 2% was not effective at killing some epilithic lichens.[288] The product Anios DDSH, with didecyl-*n*-methyl-poly(oxyethyl) ammonium propionate as a.i., was not effective on lichens.[283]

However, when blends of different QACs are used, such as D/2 Architectural Biocide, a combination of four QACs, or Kimistone Biocida, a blend of two QACs, they demonstrated good performance on bacteria,[284] some lichens,[29] cyanobacteria, algae, fungi, and mosses[242,259] (Table 4.3). Moreover, QACs proved to be effective at killing lichens when applied as blends with isotiazolones (see discussion about results of commercial products Rocima103, Biotin T and Acticide LV706 in the section dedicated to isothiazolones).

All these contrasting results lead to believe that it is impossible to draw definitive conclusions about the efficacy of QACs on lichens and fungi. Diverse could be the causes, such as differences related to lichens and fungi species, extension of lichens and fungi growth, thickness of lichen thalli, differences in substrate structure, and composition or even different environmental conditions of the treatments. The microbial biomass and its activity are usually much higher in protected shaded outdoor areas, while EPS production tends to be higher at a sunny site to protect cells from the adverse conditions, including biocides.[48] Moreover, benzalkonium chloride is not a single chemical but a class of quaternary ammonium chloride salts in which the nitrogen is substituted by a benzyl group, two methyl groups and an even-numbered alkyl chain. The differences in alkyl chains may also have a role in the different behavior of the chemical.

To make things still more difficult, it is worth mentioning the way researchers evaluate the biocides' performance. Many studies report just the inhibition of microbes' growth without any specification about membrane integrity. The complete inhibition of microbes' growth and the lack of physiological (e.g., photosynthetic) performance do not automatically imply cell death because recovery is possible. The loose of membrane integrity is instead equivalent to mortality.[280] Algae, for example, can become dormant, resistant or can repair damages. Moreover, they are also able to use alternative substrates and energy sources.

In conclusion, the mentioned results lead to assume QACs act well as bactericides and algaecides. Fungi and lichens should be treated with other compounds like isothiazolones that showed good results, unanimously recognized by the literature. Furthermore, researches and field evidences discourage the widespread use of nitrogen containing biocides (as benzalkonium chloride), which can be nutrients for micro- and macroorganisms and thus favor rapid recolonizations.[48]

A further drawback of QACs application is due to the consequences of repeated treatments with the same product. This practice places heavy selection pressure on a microbe population and may eventually select for resistant individuals that, over time, will multiply. An outstanding case showing such a bad effect is that of the wall paintings of Lascaux cave (France).[294] In 2001, a *Fusarium solani* species complex invaded the cave. The conservators treated the walls repeatedly with benzalkonium chloride from 2001 to 2004, and again in 2008 because in some galleries, white and gray-black fungi regrew. In 2009, a study[294] detected the presence of bacteria belonging to *Ralstonia* and *Pseudomonas* genera. The genus *Ralstonia* encompasses bacterial species that are opportunistic human pathogens able to survive in oligotrophic environments. *Pseudomonas* bacteria species can metabolize chemical pollutants and survive in a variety of environments for their ability to form biofilms. The authors concluded that, because of years of benzalkonium chloride treatments, microbial populations selected by repeated biocide application have replaced the indigenous microbial community.

4.3.3.13 TRIAZINES

The triazine class of herbicides, which includes atrazine, simazine, and hexazinone, were introduced in the 1950s. Triazines target very specifically the D1 key protein in the reaction center of photosystem II (PSII), thereby blocking photosynthetic electron transport. They are biocides commonly added to paints and plasters.[280] There is a concern about atrazine regarding groundwater contamination. In fact, atrazine molecule does not break down within a few weeks after soil application. Thus, when carried into the water table by rainfall, it causes contamination. In some countries, atrazine is a restricted use herbicide.

Structural formulas of

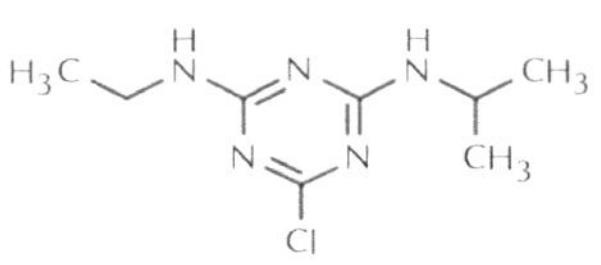

atrazine

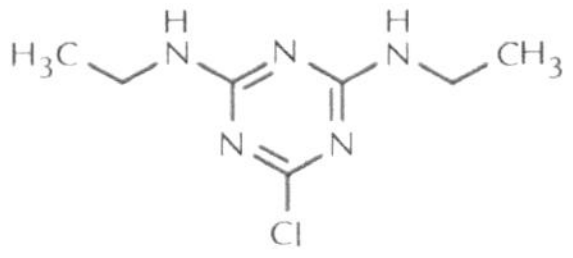

simazine

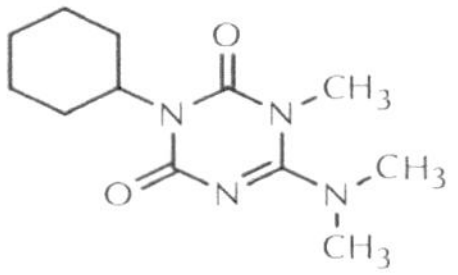

hexazinone

Regarding cultural heritage, although simazine and atrazine are both herbicides, laboratory and in situ tests applied them as algaecides. Under laboratory conditions, atrazine was the more toxic of the two agents; 0.2 mg/l completely inhibited the algal growth.[125] On the contrary, in situ results were disappointing because, even though the treated algae were inactive in photosynthesis performance, the green coloring of the stone persisted[125] (Table 4.4). Another laboratory test using triazine (trade product Irgarol 1051) at the concentration 250 µmol/l on an aero-terrestrial

TABLE 4.4 List of Herbicides Including Active Ingredients and Trade Products, Concentrations of Use, Efficacy on the Target Organisms, Indications about the Experimental Approach (In Situ or under Laboratory Conditions), Side Effects and Notes, Bibliographical References.

Commercial product/Active ingredient	Concentration	Efficacy	Organisms/Substrate	Notes/Side effects	References
Tordon 101 (a.i. picloram)	Unknown concentration. Spraying on leaves	Efficient on ruderal plants after 3 days	Plants, mosses, liverworts, ferns on sandstone		Magadán et al., 2007, 2011
Toterbane 50F (a.i. diuron)	3%, 5%, 8% in water, brushing	Effective on mosses (*Tortula muralis*, *Tortula* sp., *Bryum capillare*, *B. caespiticium*)	Mosaics, Ostia Antica, Italy	It required a very long time (4 months) to eliminate the mosses. No different effects related to the concentrations	Bartolini et al., 2006, 2007
Toterbane 50F (a.i. diuron)	3%, 5%, 8% in water, brushing	No efficacy on cyanobacteria and green algae	Mosaics, Ostia Antica, Italy		Bartolini et al., 2006, 2007
Amega 360SL (glyphosate)	3% in water, spraying on the leaves	Efficient on plants	Roman floor mosaic, mausoleum on the Appian Way, Rome (Italy)		Alberti et al., 2006
Roundup (a.i. glyphosate)	3% in water, sprayed twice with a week interval	Efficient in killing all the mosses	17 taxa of mosses growing on mosaics (Baths of Caracalla, Rome, Italy)		Altieri et al., 1996
Roundup (a.i. glyphosate)	Injection in the roots after cutting the plants	Efficient on plants			Magadán et al., 2007
Roundup (a.i. glyphosate)	Injected inside the cambium of the plants	Efficient on shrubs and small trees	Archaeological site of Eleusis (Greece)		Papafotiou et al., 2010
Glyphosate	Applied four times a year within an experimental maintenance program	Effective on weeds	Archaeological site of Pompei (Italy)		Ciarallo, 2001

TABLE 4.4 *(Continued)*

Commercial product/Active ingredient	Concentration	Efficacy	Organisms/Substrate	Notes/Side effects	References
Rodeo gold (a.i. glyphosate)	3%, 5%, 8% in water, brushing of 40 ml of the solution	After 30 days, low reduction of viability. Only after 4 months, mosses dead. The different concentrations have no different results	Archaeological site at Ostia Antica (Rome). Extensive growth of mosses on mosaics	The efficacy results on mosses are positive only after 4 months	Bartolini et al., 2006
Irgarol 1051 (a.i. 2-(*tert*-butylamino)-4-(cyclopropylamino)-6-(methylthio)-*s*-triazine)	250 μmol/l	Even though the herbicide inhibited growth and photosynthetic performance, structural properties (e.g., membrane integrity) were unaffected; hence, it did not kill the algal cells	Laboratory tests. Aero-terrestrial microalgal strain of the genus *Stichococcus*		Gladis et al., 2010
Simazine	20 mg/l, 100 mg/l	Efficient in the laboratory test. Ineffective in situ. The green coloring of the stone persisted	Laboratory and in situ tests. Biofilms		Grobbelaar, 2000
Atrazine	0.2 mg/l	Very efficient in the laboratory test. Ineffective in situ. The green coloring of the stone persisted	Laboratory and in situ tests. Biofilms		Grobbelaar, 2000

microalgal strain of the genus *Stichococcus* showed that the agent inhibited growth and photosynthetic performance but structural properties (e.g., membrane integrity) were unaffected; hence, this herbicide did not kill the algal cells[280] (Table 4.4).

Hexazinone is an a.i. of the triazine class used on many annual, biennial, and perennial weeds as well as some woody plants. For its activation, it needs rainfall or irrigation water. It has low toxicity. Regarding conservation of cultural heritage, it exhibited good efficacy on herbaceous weeds, shrubs and mosses.[255] Velpar L (a commercial product containing hexazinone) showed efficacy also on crustose lichens (*Caloplaca aurantia* and *Diploicia canescens*).[295] More applications at regular intervals were necessary.

4.3.3.14 URACIL HERBICIDES

Uracil derivatives containing a diazine ring are used as photosynthesis inhibitors herbicides, destroying weeds. A study on the removal of plant in Thai sanctuaries reported the successful use of the a.i. bromacil (5-bromo-3-sec-butyl-6-methyluracil) present in the trade product Hyvar X.[168]

Structural formulas of

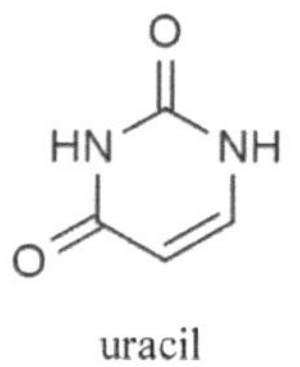

uracil

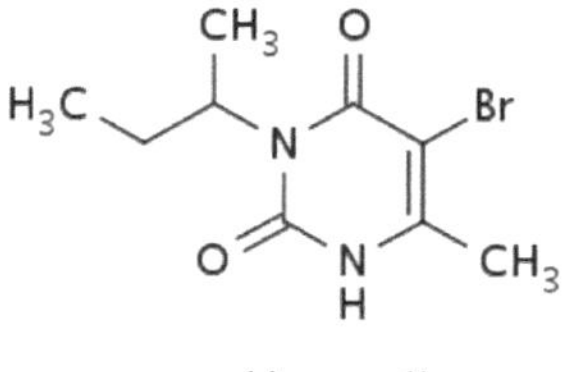

and bromacil

4.3.3.15 UREA HERBICIDES AND ALGAECIDES

Urea herbicides act by inhibiting Photosystem II, part of the photosynthesis pathway. They were introduced in 1952 and are now used as pre and postemergence herbicides for general weed control in agricultural and nonagricultural practices. Chemically, the urea herbicides contain a urea bridge substituted by triazine, benzothiazole, sulfonyl, phenyl, alkyl, or other groups.

Structural formulas of

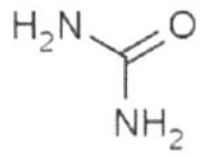

urea

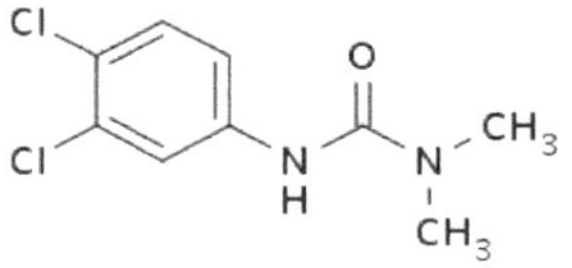

and diuron

Diuron (1,1-dimethyl, 3-(3′,4′-dichlorophenyl) urea) is a broad-spectrum residual herbicide and algaecide used in agriculture. It is also a component of marine antifouling paints to control algae and weeds. Diuron is highly toxic to aquatic organisms. Other forms of wildlife may also be harmed following exposure to high levels, but they are not so sensitive. In shallow surface waters and in the atmosphere, sunlight breaks down Diuron within a few days. However, Diuron can bind strongly to soils and sediments in water and thus it can take from months to years to break down. Many countries introduced recently restrictions of usage of diuron because of these adverse environmental concerns (toxicity at low concentrations and persistence).

Although herbicides are supposed to be efficient also on mosses, they can show unexpected performances. It was the case of Toterbane 50F (a.i. diuron) that, used to control mosses on mosaics, showed good efficacy only after 4 months[256,257] (Table 4.4). This is a drawback because the time the biocide requires to kill mosses can exceed the time available for the restoration. Moreover, no different effects occurred using three different concentrations meaning that a higher amount of a.i. did not offer a higher efficacy.

Diuron, at concentrations 0.01 and 1 mM, and a trade product containing it, Preventol A6, showed good efficacy on photoautotrophic microorganisms.[266,296]

4.3.3.16 COMMERCIAL PRODUCTS CONTAINING BIOCIDES AND WATER REPELLENTS OR CONSOLIDANTS

A few studies deal with the application of compounds and commercial products that, besides biocides, contain water repellents or consolidants. The authors used these products as biocides. An example is the product Koretrel.[260] It contains two active ingredients (alkylaminotriazine and a urea derivative) and alkyleneoxide, an epoxy resin. It is worth noting that the product contains 97.7% of epoxy resin and only 0.98% of each a.i.. It is quite hard to define it a "biocide." It is indeed a resin mixed with two biocides. The authors reported the product was very successful in killing lichens, but it considerably decreased capillary water absorption of stones and altered the color of Aurisina limestone, Muggia sandstone, and a gray granite.

Another example is that of the mixtures of Acticide CF and Acticide IOG with a water repellent.[38] The authors affirm that the mixtures were effective in reducing around 80% of the lichens that grew on a stone quarry in Spain.

In these cases, it is difficult to understand whether the effect is due to the biocides, to the resin, or to their combined action.

4.4 MECHANISMS OF ANTIMICROBIAL ACTION

This section serves as a source of information to develop an initial or general understanding of the chemistry and mode of action of a group of microbicides and herbicides. Therefore, the different compounds are arranged in groups of chemically related substances. Chemicals belonging to the same chemical class frequently have similar toxicological modes of action.

For a detailed discussion of the mechanism of action and of the characteristics of microbicides, the reader is referred to the books

- Russell, Hugo & Ayliffe's principles and practice of disinfection, preservation and sterilization, Wiley-Blackwell Pub., 2013;
- Directory of microbicides for the protection of materials: A handbook, Springer, 2008.

A brief overview is provided here. The cell offers three broad regions for biocide interaction—the cell wall, the cytoplasmic membrane, and the cytoplasm.[297] Extracellular material, cell morphology, and cellular chemical composition affect the access of biocides to these regions. Variation in cell physiology may lead to intrinsic resistance.

The mechanisms of interaction between biocides and cells depend on the chemical diversity of the biocides although the final damaging outcomes may show considerable similarity. Damage may manifest in the following ways:

- Disruption of the membrane proton motive force leading to inhibition of active transport across the membrane;
- Inhibition of respiration or catabolic/anabolic reactions;
- Disruption of replication;
- Loss of membrane integrity resulting in leakage of essential intracellular constituents such as potassium cations, inorganic phosphate, nucleotides, nucleosides, and proteins;
- Lysis of the membrane;
- Coagulation of intracellular material.

According to their mechanism of action, the microbicides are classified as membrane-active, electrophilic active, agents with oxidizing activity, and chelate formers.

The herbicides act in many ways. At the physiological level, they control plants by inhibiting photosynthesis, mimicking plant growth regulators, blocking amino acid synthesis, inhibiting elongation and division of cells, etc.[285]

4.4.1 MEMBRANE-ACTIVE MICROBICIDES

The membrane is an essential component of prokaryotic and eukaryotic cells. It has an important role in cell structure, division, and metabolism (Fig. 1.1). It is composed of phospholipids, glycolipids, sterols and proteins. The type and content of sterols differ among plant, animal, and fungal cells. Ergosterol is the predominant sterol in many fungi. Several classes of antifungal agents currently in use have exploited this difference in sterol content as the target of antifungal action.

Due to the presence of anionic groups (e.g., carboxyl and phosphate) in their membranes, most microorganisms have a negative surface charge. The membrane-active microbicides, positively charged, interact with the cell wall and/or the outer membrane, attracted by the negatively charged surface. Ionic interaction with phospholipids of the cell wall leads to their partial adsorption. This process causes loss of the permeability barrier function and modifies the membrane potential and electron transport chain. Therefore, the cell wall and the membrane lose their integrity, and the microbicides can enter the cytoplasm. At this point, they can release their lethal effects: disarrangement of the properties of the cytoplasmic membrane, inhibition of the enzymes localized on cytoplasmic membrane, leakage of cytoplasmic components, and eventually disintegration of the cell.[261] The leakage of metabolites (such as K^+ ions and inorganic phosphate), the lysis of the cell, and the disappearance of membrane enzymes are among the reported impacts of QACs on bacteria. Membrane-active microbicides such as QACs are governed in their interactions with cells by the balance between their hydrophobic and polar groups. Membrane-active microbicides include quaternary ammonium salts, alcohols, phenols, salicylanilides, carbanilides, dibenzamidines.[297]

4.4.2 ELECTROPHILIC ACTIVE COMPOUNDS

The greatest number of microbicides in use today belongs to the group of electrophilic active compounds whose molecule is characterized by the presence of a carbonyl group. The reactivity of these compounds depends on their low steric hindrance and strong electron withdrawing capability.[297] In fact, the electron deficiency at the carbonyl carbon atom enables these substances to react with nucleophilic cell components. Examples of nucleophilic entities are amino (NH_2), hydroxyl (OH), carboxyl (COOH), and thiol groups (SH), as well as the amide groups of amino acids of proteins. Thiol groups are present in proteins as part of the amino acid cysteine, where they play an important role in maintaining proteins structure and function. Many enzymes have thiol groups at the site that perform the enzyme function, and these thiol groups may participate in the enzyme reaction. If the biocide reacts with the thiol groups, the activity of the enzyme is inhibited.[279] Amide groups are also components of enzymes that are inactivated by the reaction with this kind of microbicides. Electrophilic active biocides have the advantage of being intrinsically not persistent. Therefore, they do not accumulate in the environment, characteristic that often outweighs the disadvantages imposed by their limited stability and duration of activity. Moreover, most of them have the disadvantage of a slow microbicidal effect.

This kind of microbicides includes aldehydes, compounds with activate halogen atoms (chlorine, bromine, iodine), compounds with an activated

N–S bond. When electrophilic active substances have an activated halogen atom, their efficacy increases.

4.4.3 CHELATE FORMERS

The antimicrobial activity of chelate formers is partly due to their ability to compete for the complexion of metal cations necessary for the functional cell metabolism. They include azoles and the dithiocarbamates. However, membrane activity of the compounds also plays a role.

4.4.4 UNSPECIFIC ACTIVITY (OXIDIZING AGENTS)

The oxidizing agents (listed in this book hydrogen peroxide and chlorine-containing compounds) have a microbicide effect due to their strong oxidizing power, their strong affinity to electrons that is also directed, unspecifically, toward organic matter including, of course, microorganisms. The effect is not selective and covers bacteria, algae, yeasts, fungi, but also spores. Oxidizing agents generate hydroxyl radicals that are highly reactive because they can attack membrane lipids, proteins, DNA, RNA and are thus responsible for the antimicrobial action. The pH for optimum efficacy is in the acidic range because in the alkaline media these compounds decompose too quickly. They have an intrinsic limited chemical stability. A drawback is that high reactivity and low persistence can leave some microorganisms unharmed, especially those that proliferate in biofilms.

H-O·

HO – O·

Structural formulas of a hydroxyl radical and a hydro-peroxide radical

4.4.5 PHOTOSYNTHESIS INHIBITORS

Herbicides that are photosynthesis inhibitors take electrons from the normal photosynthetic pathway resulting in production of reactive oxygen species (ROS) with reactions in excess respect to those normally tolerated by the cell, leading to plant death. Classical photosystem-II (PSII) inhibitors bind to D1 protein to prevent photosynthetic electron transfer.[285] This type of herbicides includes triazines, urea derivatives, and uracil derivatives.

4.4.6 AMINO ACID INHIBITORS

Acetohydroxyacid synthase (AHAS), also called acetolactate synthase (ALS, is an enzyme that catalyzes the first common step in the biosynthesis of amino acids valine, leucine, and isoleucine. It is the target of several classes of herbicides including imidazolinone herbicides.

Aromatic amino acid synthesis inhibitors deactivate the enzyme 5-enolpyruvylshikimate 3-phosphate (EPSP) synthase involved in the synthesis of the amino acids tyrosine, tryptophan, and phenylalanine. The herbicide glyphosate exerts this mechanism of action.

4.4.7 GROWTH REGULATORS

Compounds acting as growth regulators, also known as synthetic auxins, include many commonly used plant hormone-type herbicides.

4.5 STAINS AFTER TREATMENTS

The pigments contained in treated organisms can adhere to the substrate once released out of the cells forming green, black and/brown spots after the removal of died biomass. The phenomenon gets worst when endolithic biological growths colonize the object. The treatments can kill and remove microorganisms but are often unable to remove the pigments deposited on and adhering to the substrate. A kind of pigment, the dark brown melanins, is in fact insoluble even in strong alkaline or acidic conditions. Moreover, the effect of some biocides is cells' lysis with a consequent release of pigments in the stone. A research[298] on color changes due to biocides applied on colonized substrates showed that, among the tested biocides, quaternary ammonium salts applied on cyanobacteria caused the main discolorations.

In the case of the Jeronimos cloister in Lisbon (Portugal), the biocide application resulted in the elimination of the living organisms, but it was insufficient to eliminate the remaining dark gray spots left even after several operations of wet brushing.[299] Only the removal of a thin layer of the surface, either through abrasion or by acid attack, could take out the staining. The application of a proprietary solvent (SOLUENE 350 by Packard, a mixture of toluene, dimethyl dialkyl quaternary ammonium hydroxide and methanol) was successful in removing completely and uniformly the discoloration. The authors[299] obtained the best results applying the solvent SOLUENE 350 for 1–3 h by a poultice covered with a polyethylene sheet.

Other researches,[242] facing the same problem of organic staining on limestone substrates, solved it with the product Arte Mundit, a proprietary latex poultice composed of natural rubber and EDTA + D2 (Table 4.1). Again, EDTA showed very good performance on marble monuments of the historic cemetery of Drapano (Kefalonia Island, Greece).[240] A poultice of EDTA at 12% for 2 h removed the brown and yellow discolorations. Moreover, layers of yellow ferrous oxides were present in some areas after the cleaning. A compress with a solution of thioglycolic acid at 5%, neutralized with NH_3, for 1–2 min was enough to remove the layers.

A laboratory study tested various substances (acetone, ethanol, hydrogen peroxide, diethyl ether, mixture of ammonium carbonate +EDTA, and enzymes produced by the fungus *Aspergillus flavus*) to remove colored spots left on marble samples by fungi.[300] Results showed that microbial enzyme preparation was the most effective, followed by H_2O_2 and ammonium carbonate + EDTA solution, while the other chemicals did not have any effect on the colored spots.

4.6 BIOCIDES NO MORE ON THE MARKET, BIOCIDES THAT CHANGED THE NAME OR THE PRODUCER, VERY HAZARDOUS BIOCIDES AND STRANGE INFORMATION REPORTED IN THE LITERATURE

Some studies reported the application of commercial products that over time disappeared from the market. The reasons are diversified. Governments banned a product for environmental concerns; a manufacturer does not produce the products anymore; the product's name changes because a different company produces it but the components do not change, or vice versa, the product keeps the trade name but has new components.

Professionals involved in the field of conservation of cultural heritage experimented compounds that are not biocides affirming that they act killing biofilms and plants. Unfortunately, the authors did not report any scientific evidence proving the efficacy of these compounds. These results have been also reported hoping the future researches on treatments of biological growth on cultural heritage objects will report scientific experimented results considering what the international literature demonstrated so far.

This clause accounts for some examples and provides useful information when facing the decision of treating objects with biocides.

In terms of toxicity and long-lasting effect, the most successful microbicides in the conservation of cultural heritage have been those containing organometallic compounds (e.g., tributyltin oxide (TBTO)).[248] Organometallic compounds act producing cationic radicals that react with nucleophilic parts of the cell damaging its metabolism. An example is their reaction with SH groups causing inhibition of both endogenous and exogenous enzymes. Among organometallic compounds, TBTO was a very efficient component of antifouling coatings that prevented the growth of marine organisms on the hulls of ships. Laboratory tests showed that concentrations as low as 20 ng/l could damage nontarget aquatic organisms, particularly mollusks, negatively affecting important biological processes such as larval development or reproduction. Although highly toxic, TBTO is not persistent in the environment.[301] These concerns have led to a cessation in Europe of the manufacture of organotin compounds, which have been withdrawn from use as surface biocides. Thus, organic–tin or –mercury and other heavy metal components are no longer in use in restoration.

One of the commercial products containing TBTO was Metatin N5810/101. Many researchers reported the effectiveness of the product on several biodeteriogens such as bacterial infestations on wall paintings,[266] actinobacteria, autotrophic microorganisms, fungi,[265,292,287] lichens.[295,288] Moreover, the product did not have effects on marble[265] and on sandstone.[302] Another product containing TBT is Rocima 110. It was efficient in killing a biofilm of cyanobacteria that grew on frescoes,[94] as well as cyanobacteria and green algae on wall paintings of a crypt.[113] Its application on endolithic lichens showed acceptable results.[293] Bioestel was a mixture of a consolidant (tetraethyl orthosilicate) and two biocides, TBTO + dibutyltin dilaurate. Used to prevent biofilms' growth in the archaeological area of Fiesole (Italy), it was effective for 2 years.[29] Because of the above-mentioned limitations, the Italian supplier (CTS Company) replaced it with a product, Bioestel New, which contains the same consolidant but different biocides (terbutryn, 2-octyl-2*H*-isothiazolin-3-one, biphenyl-2-ol, 3-iodo-2propynyl butyl carbamate, dibutyltin dilaurate).

A product whose a.i. [2,3,5,6-tetrachloro-4(methylsulphonyl pyridine)] has been withdrawn from use for its environmental negative effects is Algophase. It showed a good efficacy on microorganisms but not on endolithic cyanobacteria.[288,303] Its solubility only in organic solvents gave it a "positive score" because in some cases the use of water is harmful. For the same reason, it was also used as a preventive treatment in combination with water repellents and consolidants miscible in the same solvents. The product did not have effects on marble, sandstone, and travertine[302] while it changed the color of two pigments (cobalt blue and yellow) of fresco samples.[292] The commercial product, distributed by Phase Restauro srl, is still on the market with the same name (Algophase) but, and here there is the trick, the components are very different. The new a.i. is *n*-butil-1,2 benzoisotiazolin-3-one.

European Community withdrew from usage the a.i. dichlorophen. It is a fungicide, bactericide and algaecide with a low aqueous solubility. Dichlorophen has a low mammalian toxicity but is moderately toxic to fish, aquatic invertebrates and algae. The trade product Panacide M that contains dichlorofen as a.i. showed good performance on algae, mosses, and lichens.[304]

Although pentachlorophenol and sodium pentachlorophenate (good bactericides and fungicides) have been banned in many countries for their toxicity, they are still used in some countries, such as Thailand, India, Egypt.[168,243,300] This information shows that restorers are at times not concerned about the dangerous effects of the products on humans and the environment, acting to get an "easy" result even if they accomplish it with highly toxic compounds. On the other hand, many countries are less sensitive in imposing limitations to the use of toxic compounds. In general, phenolics are now regarded as "old technology" in conservation and preservation of cultural heritage.[25]

The company Wykamol does not produce any more the product Green Murosol 20, efficient on algae, mosses, and lichens.[304] Nuodex 87 is no more on the market as well.

The product Abicide by Langlow[304] contains a different a.i.. The previous a.i. was gloquat (dodecylbenzyltrimethylammonium chloride) while the present a.i. is benzalkonium chloride.

Bayer Company no longer produces the famous Preventol R80. Lannxess now manufactures it with a slightly different name, Preventol RI80. The components (alkyl dimethyl benzyl ammonium chloride and isopropanol) did not change. A similar fate was that of another frequently used product, Hyamine 3500, which is now manufactured by Lonza Company, no longer by Rohm & Haas. It maintains 50% content of alkyl dimethyl benzyl ammonium chloride. Rhodia Company no more manufactures the product Rhodaquat RP 50 that contained 50% of benzalkonium chloride. Another example is New Des or Neo Desogen that contains 10% of benzalkonium chloride. Neo Desogen survives, but it is used in the medical field as a topical antiseptic to prevent infection. On the contrary, New Des disappeared. In Italy CTS no longer markets it. The company instead maintained the name, added a number, and sells New Des 50 that is completely different because it contains 50% of a diverse a.i., didecyl–dimethyl ammonium chloride.

Bizarre is the use of zinc silicofluoride (2–5%) as a fungicide.[243] Almost everyone involved in conservation of stone is aware that this compound is a consolidant used in the 1950s that caused serious damages to the substrates. Even though this is obviously an unbelievable misuse, the news is also reported because products with no biocide action are still applied as biocides. Hopefully, the conservation field will get rid of these empirical operations.

4.7 RESISTANCE OF ORGANISMS TO BIOCIDES

The repeated application over time of microbicides and herbicides causes selection pressure on microbes and weeds. This may eventually select for resistant individuals that will multiply and become the dominant population, resulting in biocides that are no longer effective for the control.[305,285] Resistance is the temporary or permanent ability of an organism and its progeny to remain viable and/or multiply under conditions that would destroy or inhibit other members.[306] Intrinsic resistance is a natural chromosomally controlled property or adaptation of an organism.[307] The intrinsic resistance of spores to heat, dryness, and many microbicides is due to their external envelope. Under favorable conditions, spores germinate, leaving the dormancy to acquire a metabolically active state. Despite the spores' resistance, some microbicides—mostly highly reactive substances such as aldehydes—are effective sporicidal.[308] The intrinsic resistance of Gram-negative bacteria to antimicrobial compounds is due to their outer membrane that acts as a permeability barrier able to exclude macromolecules and hydrophilic substances.[309] Testing reported that *Pseudomonas* species are especially resistant to many biocides and antimicrobial agents.

The increased tolerance of bacteria to biocides relates to phenotypic changes brought about in the surviving population.[310] These changes might involve the induction of multiple resistance operons or of other global regulatory systems that respond to subinhibitory concentrations of biocides.[282] Moreover, microorganisms embedded in the biofilm matrix are remarkably more tolerant to biocides than free-floating dispersed cells of the same bacterial strains.[94] The resistance to antimicrobials of cyanobacteria, in particular those belonging to the genus *Gloeocapsa*, can be due to their gelatinous sheath functioning as barrier to penetration of the chemicals.[113] Another factor favoring biofilms recovery is their ability in limiting nutrients and growth rate in the deeper layers of their structure following biocidal treatments.[282] For all these reasons, traditional doses of biocides can become insufficient to destroy all of the biofilm population. For example, some researchers suggest that the repeated use of subinhibitory concentrations of benzalkonium chloride induces a significant reduction of microbial sensitivity to biocides and leads extreme-tolerant fungi toward acquiring the missing virulence factors and further enhancing their stress tolerance.[311,312]

Also for weed populations, there is a general agreement that selection applied for a long enough period eventually leads to resistance. Plants have developed resistance to atrazine and more recently, to glyphosate herbicides. Simply rotating herbicide active ingredients is not enough to prevent the development of herbicide-resistant weeds. Rotating herbicides choosing different modes of action, along with other weed control methods, is necessary to prevent or delay herbicide-resistant weeds.

4.8 TOXICITY AND LEGISLATIVE REGULATIONS

Biocides are products used to control harmful organisms. They protect health, improve product performance and prevent spoilage, and are increasingly important to modern life, as consumers demand safe, long-lasting, and effective products from cosmetics to paints and from drinking water to swimming pools. For all reasons, it is important to know the properties of biocides related to their potential negative effects. Among them, a very important characteristic is toxicity that represents a hazard to the environment and human or animal health. The toxicity of a chemical often depends on its concentration, the target organism, and the matrix surrounding the cell. Use of biocides having a low potential to cause harm or applications leading to low exposures can reduce risks to human health and environment.

Common measures of toxicity are lethal dose (LD_{50}) and lethal concentration (LC_{50}) that cause death (resulting from a single or limited exposure) in 50% of the treated animals (mostly rats and rabbits). LD_{50}, orally or dermally administered, is generally expressed in milligrams (mg) of chemical per kilogram (kg) of body weight. LC_{50}, administered by inhalation, is often expressed as mg of chemical per volume [e.g., liter (l) of medium (air or water)] the organism is exposed to.[313] Chemicals are highly toxic when LD_{50} and LC_{50} are small and practically nontoxic when the value is large. However, the values of LD_{50} and LC_{50} do not reflect any effects from long-term exposure (e.g., cancer, birth defects, or reproductive toxicity) which may occur at doses below those used in short-term studies. In general, biocides can cause discomfort or ill health mainly through skin contact, eye irritation, and inhalation of fumes or mist.

The biocides are assigned to a toxicity class by means of regulatory systems created by national governments. All these systems are based on standard acute toxicity tests (several of them including chronic assays as well). European countries established a common system that classifies the biocides as very toxic (T+), toxic (T), harmful (Xn), and irritant (Xi) (see Table 4.5).

TABLE 4.5 EU Classification Criteria—Acute Toxicity

	Very toxic	Toxic	Harmful
Oral LD_{50} (mg/kg)	≤ 25	$>20 \leq 200$	$>200 \leq 2000$
Dermal LD_{50} (mg/kg)	≤ 50	$>50 \leq 400$	$>400 \leq 2000$
Inhalation LC_{50} gases, vapors (mg/l)	≤ 0.5	$>0.5 \leq 2$	$>2 \leq 20$
Inhalation LC_{50} aerosols, particulates (mg/l)	≤ 0.25	$>0.25 \leq 1$	$>1 \leq 5$
EU indication of danger (symbol)	Very toxic (T+)	Toxic (T)	Harmful (Xn)

The safety data sheet and the label of a commercial product supplied by the manufacturer (who has the legal duty to provide it) deserve close attention. The safety data sheet gives the user of a substance important information regarding its hazard, the precautionary measures to prevent risk to the environment, and the health and safety of the user. It includes the following:

- Symbols and indications of danger highlighting the most severe hazards;
- Standard risk phrases specifying hazards arising from the substance properties;
- Standard safety phrases giving advice of necessary precautions.

Different regulatory jurisdictions have different definitions for biocides and varying requirements for their registration (i.e., the process that a product must undergo to be allowed to be sold). In Europe, the recent legislation prohibits the use of highly toxic pesticides including those that are carcinogenic, mutagenic, or toxic to reproduction, those that are endocrine disrupting, and those that are persistent. Environmental issues have a big influence on the current and future use of many biocides. Major issues regarding the problems associated with pesticides in the environment are biodegradability, environmental persistence, occurrence and fate in surface waters, wastewaters and groundwater contamination, bioaccumulation and effects on biota in the aquatic environment, human health impacts, and treatments for pesticide removal.

REACH, which stands for Registration, Evaluation, Authorization and Restriction of Chemicals, is a regulation of the EU, adopted to improve the protection of human health and the environment from the risks posed by chemicals. It also promotes alternative methods for the hazard evaluation of substances to reduce the number of tests on animals. It entered into force on June 1, 2007. REACH applies to all chemical substances, not only those used in industrial processes but also in our day-to-day lives, for example in cleaning products, paints as well as in articles such as clothes, furniture, and electrical appliances. Therefore, the regulation has an impact on most companies across the Europe.

The European Biocidal Product Regulation (Regulation 528/2012) concerns the placing on the market and use of biocidal products. It replaces the Biocides Directive 98/8/EC and aims at providing a high level of protection for humans, animals and the environment. Annex I of this Regulation lists active substances identified as presenting a low risk under Regulation 1907/2006 or Directive 98/8/EC, substances identified as food additives, pheromones and other substances considered to have low toxicity, such as weak acids, alcohols and vegetable oils used in cosmetics and food. Other active substances may be added in case there is evidence that they do not give rise to concern. Since September 1, 2015, a biocide cannot be placed on the EU market if the substance supplier or product supplier is not included in the list of the BRP for the type to which the product belongs.

Among the possible effects of active ingredients on the environment, their leaching from outdoor treated materials is relevant to the field of conservation of cultural heritage.[47] The Regulation requires the evaluation of the risk assessment of the emissions following the authorization procedure. A study[314] investigated several marketed formulations of textured coatings and paints with a mixture of commonly used active ingredients (OIT, DCOIT, iodopropynyl butylcarbamate, carbendazim, isoproturon, diuron, terbutryn, and Irgarol 1051). The results showed that leach ability is due to water solubility and *n*-octanol/water partition coefficient* of the active ingredients. Leaching of biocides from façade coatings is mainly a diffusion-controlled process.

The legislation in the USA and Europe is broadly equivalent. All commercial microbicides and herbicides must be extensively tested prior to approval for sale and labeling by the Environmental Protection Agency (EPA) in the USA. The EPA has strict guidelines that also require testing of pesticides for their potential to cause cancer.

---

*K_{ow} (*n*-octanol/water partition coefficient) is defined as the ratio of a chemical's concentration in *n*-octanol divided by its concentration in water. Values of K_{ow} are usually expressed as log K_{ow}. The water–insoluble organic solvent *n*-octanol is used as a surrogate for soils and organisms to simulate the accumulation of organic molecules like pesticides into those materials. The partition coefficient is an indicator of the environmental fate of a chemical since it gives a general idea of how a chemical will be distributed in the environment. For example, chemicals with large K_{ow} values are of great concern since they can be adsorbed in soils and living organisms.

Some pesticides, considered too hazardous for sale to the public, are designated as "restricted use pesticides." Only certified applicators, who have passed an exam, may purchase or supervise the application of this kind of pesticides.

As countries regulate the use of biocides by their own national authorization procedures, it can occur that a biocide, legal in a country, results banned or restricted in use in another country. The approach to regulatory controls and safety assessment methods of biocides differs among countries depending on many factors and reasons that are beyond the context of this book. The example of DDT is emblematic to explain the situation. DDT (acronym of dichloro diphenyl trichloroethane) is an organochlorine insecticide that was first synthesized in 1874. It was very effective for insect control until the banning in 1972 by the United States Environmental Protection Agency because it can cause adverse health effects on wildlife. Moreover, it is highly persistent in the environment. Its soil half-life (i.e., the time required for half of the compound to degrade) is 2–15 years and its half-life in an aquatic environment is about 150 years. The residues in crops at levels unacceptable for export have been an important factor in recent bans in several tropical countries. While many countries no longer used it, its use continues in some parts of the world including India, China, South America, Africa, and Malaysia mainly to control malaria-carrying mosquitoes. Indoor spraying with DDT is one of the tools used to control malaria around the world. In rare cases, unfortunately, it is the most effective choice.

The consequences of different regulations among countries are also evident in the literature on biocides applied to cultural heritage. A recent study[300] suggests the use of sodium azide (NaN_3) to treat fungal biofilms on marble monuments in Cairo (Egypt) because this compound was very effective at a very low concentration (100 ppm) in a laboratory test in comparison with dichloroxylenol, thymol, penta-chlorophenol, *p*-cresol. Sodium azide is a severe poison. The toxicity of this compound is comparable to that of soluble alkali cyanides and the lethal dose for an adult human is about 0.7 g. Therefore, in most countries its use, including restoration practice, is illegal.

As already mentioned, biocides are also applied after the removal of biofilms and lichens to prevent a possible recolonization. Two studies[125,243] report the use of two surfactants, triethanol amine and lauryl sulfate, which are not used in Europe.

Scientific studies reflect the different sensitivity of countries in imposing limitations to the public use of hazard compounds. For example, pentachlorophenol (PCP) and its sodium salt, sodium pentachlorophenate, are good bactericides and fungicides but pentachlorophenol is extremely toxic to humans from acute (short-term) ingestion and inhalation exposure. Chronic (long-term) exposure to this chemical by inhalation in humans resulted in effects on the respiratory tract, blood, kidney, liver, immune system, eyes, nose, and skin. US EPA has classified pentachlorophenol as a probable human carcinogen. Since the early 1980s, the purchase and use of PCP in the USA has not been available to the public. Other countries, on the contrary, allow its use (see Section 4.6).

Nanomaterials are a new challenging type of chemicals whose application as biocides is growing. They are substances manufactured and used on a very small scale. Their structures range from approximately 1 to 100 nm in at least one dimension. Nanoparticles provide the inherent properties of the material they derive from. For example, nanoalumina maintains the properties of alumina, such as hardness and scratch resistance, but on a nanoscale. Similarly, nano silica provides hardness, nano titanium dioxide provides a high refractive index and UV stabilization, and nano zinc remains a UV light absorber. Nanomaterials though have more pronounced characteristics compared to the same material without nanoscale features. In most studies, the effectiveness of nanomaterials is much higher or longer than that of the bulk material itself.[315] This can be due to several aspects, including nanomaterial size-dependent properties, high surface-to-volume ratio of ultrafine particles, and features related to the presence of surface stabilizers. For example, capping agents are able to control the nanoparticle ionic release and, consequently, its antibiofilm properties (see a review in Longano et al., 2012[315]).

Nanotechnology is rapidly expanding. Many products containing nanomaterials are already on the European market (e.g., coatings, antibacterial clothing, cosmetics, and food products). While offering technical and commercial opportunities, the rapid increase in their use raises concerns about their potential effects on health and the environment.[316,280] Literature on this subject is still emerging.[280] Just as an example, some studies suggested that covalently linked or copolymerized nanomaterials provide a sort of confinement of nanoparticles in a dispersing and/or supporting matrix, which conversely acts an immobilizing component inhibiting nanoparticles' leaching.[315]

4.9 RECOLONIZATION AFTER TREATMENT AND MAINTENANCE PLANS

A growing number of studies considered the topic of recolonization after treatment as one of the most challenging aspects nowadays. There are two aspects connected to the topic, namely the recolonization of bare surfaces after removal of biofilms and lichens, and the recolonization of surfaces treated with consolidants and/or water repellents. Although the researches are accurate, they do not allow drawing homogeneous conclusions regarding the times of recolonization because the results are still scarce and related to different environmental conditions. However, in general, many factors influence the mode and timing of recolonization. Examples are higher plants that increase relative humidity, wind, chemical air pollution, bioreceptivity of stones, and climate[30] (Fig. 4.6).

(a)

(b)

FIGURE 4.6 Acropolis of the archaeological site of Marzabotto (Italy) (a). Recolonization by algae of an area after 1 year from the treatment (b). The type of stone (travertine) and the environmental conditions were the main factors affecting the regrowth.

This section reports the results of scientific experiences that can be of help in planning correctly both treatments and post treatments maintenance.

In humid or temperate climates, recolonization of cleaned building facades can be quite rapid, within one or 2 years.[248] The first signs of recolonization were observed 2 years after the drastic cleaning (sandblasting) of marble statues (gardens of the National Palace of Queluz, Portugal).[241] Five years later, green algae were widespread on the north- and northeast-facing areas, and lichens were already present on the south-facing areas. On the other hand, a study[245] reported that no recolonization occurred after 5 years on a 3000-year-old deer stone placed in a natural environment (Mongolia). Supposing that biological growth was completely removed in both cases, the obvious idea emerging from these results is that the microclimate more than macroclimate is responsible for the recolonization. A further cause can be the presence of remains of dead biomass as suggested in a study[38] on stones recolonized by fungi only after 16 months from the biocide treatment.

Climatic conditions and the substrate's structure are factors influencing the biological regrowth as it happened on mortars of San Roque church (Campeche, Mexico).[317] After the removal of cyanobacteria and lichens with mechanical methods, the restorers applied a yellow lime paint on mortars. After only a few months, a phototrophic-based colonization, composed of cyanobacteria and bryophytes, developed mainly beneath the restored mortars causing exfoliation of the paint layer. The high temperature and moisture annual conditions ($T > 25°C$ and RH > 70%), the pore structure that ensured a high rate of water capillary condensation and water retention, and the absence of a biocide treatment in the restoration protocol caused the very short-term recolonization of the church façade.

The biocide Parmetol DF12 was effective at inhibiting biofilms regrowth for 4 years in the extremely favorable environment of Angkor Wat monuments, Cambodia.[99] The same study reports also the application of "algal wash," a formulation containing a low concentration of a copper complex. Its long-term growth-controlling effect was effective even after 7 years. Unfortunately, the study did not specify the components of the formulation.

A study[247] reported that after 2 years of outdoor exposure, there was no regrowth of biofilms and lichens on a sarcophagus in the garden of the Aphrodisias Museum (Turkey). The restorer treated the object by brush with a biocide and wrapped it with wet cloth and plastic for 45 min.

Cleaning was undertaken using a low-pressure water jet (maximum pressure 90 bars).

The microclimatic conditions are very relevant in the case of tombs and caves.[138,318] Microbial communities of cyanobacteria and algae similar to those present before restoration, colonized only 8 months after cleaning the mural paintings of the Necropolis of Carmona (Sevilla, Spain). The same happened in a Japanese tomb, Takamatsuzuka Tumulus, where fungi recolonized the wall paintings after 2 years from the treatment with formaldehyde and alcohol.[319]

Other researches[320–322,30] studied the lichen recolonization of some limestone statues located in parks of three Venetian villas in Italy. The conservators applied consolidants and water repellents on the statues after cleaning. The authors attributed to those products the significant differences observed in recolonization times rather than to environmental conditions and to the applied biocides. A thorough rinsing in fact eliminated most residues of them. Recolonization started even after a few weeks on statues treated with a fluorinated polymer, whereas it took several years for those treated with a silane and/or polysiloxane. The phenomenon can be due to the different durability of organic products applied to stone as a study showed.[323] In fact, after exposure in a natural environment for 5 years, the products applied on a sandstone had different performances. Acrylic resins and fluoroelastomers degraded in a short time, whereas silicate-based products maintained their good performance over time.

Seven years after the cleaning and the application of a water repellent on marble statues in the gardens of the National Palace of Queluz (Portugal), a severe recolonization of the surfaces by green algae and lichens occurred.[169,241] Disfiguring dark patterns developed on the hydrophobic surfaces because the water repellent inhibited the spreading of dew or raindrops, thus favoring the formation of water stripes as the drops roll down the stone surface (Fig. 4.7). Therefore, according to the authors, the application of a water repellent after a biocidal treatment can be disadvantageous in terms of the subsequent recolonization. Furthermore, the biocide may interfere with hydrophobicity of the water repellent (see Chapter 5).

On the contrary, a study[168] reports very good results about the recolonization of stones cleaned and then treated with a water repellent: Eight years lasted the effects of a mechanical cleaning followed by the application of methyl silane on a Thai sanctuary. Algae and lichens colonized the cleaned but untreated areas in direct contact with damp soil after 5 years.

FIGURE 4.7 A marble statue in the gardens of the National Palace of Queluz, Portugal. The hydrophobicity provided by water repellents inhibited the spreading of raindrops over the surfaces. Thus, dark streaking appeared resulting from the localized accumulation of algae and cyanobacteria in the preferential runoff. Photo courtesy of Ornella Salvadori.

Over time, biological growth will tend to reappear, particularly on porous substrates, in shaded areas, and in areas that remain moist over time. Therefore, reapplication of biocides at regular intervals may be necessary when a cyclical maintenance is planned to prevent reappearance of biological growth[253] on outdoor objects.

How to face the unavoidable recolonization? An example of *good practice* and *good maintenance* is that of the preservation of marble statues in the gardens of the National Palace of Queluz, Portugal.[169] In such a context, it is opportune to avoid the application of high amounts of biocides. The authors selected two biocides: Preventol R80 (1.5–3% v/v) and zinc chloride solution (1.5–3% w/v). Preventol R80 at 1.5% was sufficient to eradicate lichens and grayish biofilms while zinc chloride solution 1.5% proved effective mainly on the biofilms. Therefore, all the statues were treated by brushing first with Preventol R80 and, 6 months later, with zinc solution. After 2 years, no recolonization occurred even in areas with damp conditions. The conservators treated again the sculptures only when visible recolonization changed from incipient to moderate intensity. After the treatment, the conservators decided to leave the statues untouched for several months. Then they brushed them lightly to remove any detaching remnants reducing the intervention to a minimum and diminishing the probability of inducing biocide-resistant species.[169] In most statues, brushing was not even necessary. This approach to maintenance well reduced the cost of periodic cleaning.[169] Moreover, the authors reported an interesting aspect about biological growth in parks and gardens. They suggested that a limited amount of biocolonization gave the garden an aged appearance in harmony with the buildings and other decorative elements of the palace. In fact, the intervention kept the statues free of the most invasive colonization while not trying to give them their original appearance. A few studies agree with this position and are a plea for greater consideration when treating exterior stone covered in lichens. A report[165] even favors minimal intervention proposing nondestructive documenting/recording of the monuments and letting the lichens contribute to the esthetics of churchyards and cemeteries.

Regarding the routinely control of spontaneous vegetation in monumental and archaeological areas, the repeated application of herbicides is not a good choice. The periodical mechanical mowing and the landscape gardening can be even more effective.[238,42] Low-impact maintenance implies the avoidance of unnecessary and destructive measures, ensuring that natural tree establishment and growth conditions could be sustained.

However, when the context needs the use of herbicides, the maintenance plan should include a selective treatment that kills just some plants and spares others. This approach entails the knowledge of the variable sensitivity of the species to different herbicides, the selectivity of the various agents, and the appropriate dosages and techniques for their application.[255]

The conservation plan of a Roman earthwork complex in the United Kingdom allowed for felling of trees and clearing of ferns because the tree coverings potentially damaged the site through root penetration.[324] After a decade, the earthworks appeared to have attained a stable state. However, continual monitoring was required because the site is potentially vulnerable due to its remoteness and the lack of staff. Improvements to the site's management included the use of sheep grazing for controlling vegetation for limited periods and the creation of a permanent trail to prevent erosion caused by visitors walking on unmarked paths.[324]

4.10 BIOCIDES' EFFECTS ON STONES

The biocides should not induce any chemical–physical variations in the substrate, nor interact with products used for cleaning, or with consolidants and protective coatings. Many resins applied as consolidants or water repellents may represent an additional nutritional source for microorganisms and therefore stimulate, under favorable environmental conditions, microbial growth. Moreover, quaternary ammonium salts may alter the performance of water repellents when applied after the biocides (*vide infra*). It is therefore opportune to remove quaternary ammonium salts from stone if the conservation process implies the following application of a water repellent.[225]

Observed negative effects of biocides on stone surfaces have stressed the need to evaluate their possible interaction on test materials prior to application in situ. The chemicals must in fact be compatible with the substrate and not cause any unwanted color changes or alter the properties of the materials. The assessment of any chemical–physical variations of the substrates induced by biocides is usually carried out in laboratory tests on noncolonized stone samples.[260] The results allow evaluating either the risks related to the application of the products on stones or the effects of a frequent use and/or over dosage.

The evaluation of the interference implies the analysis of the following parameters before and after treatment: dry weight, color, water absorption by capillarity, calcium ion concentration, and surface micromorphology observed under optical and scanning electron microscopes. The products are usually characterized measuring pH, conductivity, and ion concentration.

A possible negative effect can be due to the composition of the biocide. In fact, trade products contain compounds, called coformulants, which combine with the a.i. to make up the finished product. An example is

Rocima 103, blend of two active ingredients (2-octyl-2*H*-isothiazolin-3-one and didecyldimethylammonium chloride) in propanol and formic acid. The product Koretrel[260] is another example. It contains two active ingredients (alkylaminotriazine and a urea derivative) and alkyleneoxide, an epoxy resin. The product considerably decreased capillary water absorption and altered the surface color of Aurisina limestone, Muggia sandstone, and a gray granite.

Color and water absorption of stones are two relevant parameters that can be measured also in situ.[29] The availability of instruments for in situ measurements is very important. For example, portable spectrophotometers can easily measure color in situ (see Chapter 7). There is not a standard value but the scientific community and conservators accept ΔE^* values below 4 as a tolerable limit for the visual impact of stone surface treatments.[325]

Some types of sandstone are considerably more susceptible to the loss of minerals when treated.[85] The damage can be due to exchange mechanisms with the biocide, or to the pH of the product. Sandstone samples treated with amine or dichlorophenol-based biocides lost a high number of mineral elements to solutions in laboratory tests. Two products containing derivatives of phenol as active ingredients caused a strong color change because they yellowed (Δb^* +9.14) and darkened (ΔL^* −10.40) limestone samples.[326] As reported by the authors, the effect could be due to a chemical reaction between the phenolic compounds and the iron oxides and hydroxides contained in the limestone.

Samples of a quartz feldspathic sandstone from Tuscany (Italy) embedded for 48 h in 3% aqueous solutions of two biocides (Arsenal, Roundup) showed different patterns of alteration.[302] The application of Roundup, producing a thick net of crystals deposit on surfaces, caused a strong whitening of stone color (ΔE^* 16.12). The crystals contained calcium indicating a possible interaction of the biocide with stone minerals. Other results—a decrease in samples dry weight and an increase in capillary water absorption—confirmed the action of Roundup on sandstone. The acidic pH of Roundup solutions and the phosphonic group of the active ingredient may be the cause respectively of calcium mobilization and of calcium phosphonate formation. The application of Arsenal caused a release of calcium ions as well. Testing the same biocides on samples of Carrara marble and travertine from Tivoli (Italy),[302] Roundup damaged the marble surfaces forming many signs of corrosion and caused an increase in capillary water absorption of travertine. Arsenal produced a release of calcium ions of marble. All the biocides induced chromatic variations. On the other hand, Roundup, tested at 3% on a polychrome mosaic floor (Baths of Caracalla, Rome, Italy) did not cause any significant macroscopic alteration on the lithotypes meaning that the tests in laboratory are often far from what happens in the real cases.[271]

A study[298] tested three biocides (Preventol R80, Biotin R, Biotin T) on marble samples to check possible color changes. The authors applied the products at 2% in deionized water, brushing them twice at interval of 7 days. They did a further application embedding the samples for 24 h in a solution containing the biocides. Preventol R80 and Biotin T did not show any interaction with marble in both applications while Biotin R caused a variation of marble color (ΔE^* −3.85, Δb^* +1.25) when applied by immersion while it did not interact with marble when applied by brush. A confirmation of these results came from another study that tested Preventol R80 and Umonium 38 effects on calcarenite samples.[113] Both biocides did not modify significantly the stone color and water absorption. Laboratory testing of another quaternary ammonium-salt-based microbicide applied undiluted (D/2 Biological Solution) showed no color changes of treated dolomitic limestone compared to untreated areas.[259]

The product Preventol A6 (a.i. diuron),[266] applied on samples twice by brush with 6 days' interval, caused a slight darkening of limestone surfaces, and a less detectable change on granite surfaces.

Biocides can form chemical bonds with siliceous stones. For example, quaternary ammonium salts bind to silicates, retaining their efficacy while bound. Some studies considered it a likely factor in causing differences in the performance of quaternary ammonium salts on sandstone, limestone, and mortar test samples.[85]

Besides traditional biocides, other kinds of substances such as nanoparticles, subject of growing interest by conservation scientists, can have negative effects. Silver nanoparticles for example alter strongly the color of two types of concrete samples producing a pronounced darkening (ΔE^* 8.7 and 15.9) of the surfaces.[327]

In conclusion, there is scope for detailed further studies to elucidate the interaction between biocides and stones, what mineral components become vulnerable to biocide attack, and what long-term damages, including color changes, occur when biocides are regularly reapplied.[326]

4.11 EFFECTIVENESS OF BIOCIDES AND THE PERSISTENCE OF THEIR EFFECT

Many factors affect the performance of biocides on outdoor and indoor stone objects.[252] They vary with the composition of the biocide, the method of application, the porosity of the surface, the physical–chemical properties of the substrate and the degree of exposure of the surface to rain.[252] Other factors affecting biocides' performance are contact time, interfering substances, temperature, and concentration. Climate and microclimate affect biocides' efficacy as a testing of paints mixed with biocides and exposed to different environments in Brazil showed.[57] The researchers exposed painted concrete samples in equatorial, tropical, and temperate climatic locations for 4 years. The biocides (carbenzadim and benzisothiazolinone) were effective only in the temperate location where samples showed just very little biological discoloration. *Cladosporium* was the most frequent genus of fungi in all specimens. Samples at tropical location had the highest fungal colonization while those at equatorial place showed the greatest number of phototrophs. According to the authors, the strong colonization in the equatorial climate was due to the leaching of biocides by rainfall and to high temperatures. In fact, the equatorial location presented the highest rainfall with an annual average of 3340 mm and the highest temperature, varying between 23 and 32°C.

The presence of clays in stones may prolong the effectiveness of a biocide. On the other hand, components of stones may adsorb and inactivate the biocides. Toxic agents also undergo chemical or photodegradation. In addition, wind-blown dust, organic debris, paint pigments, and plaster layers may compromise toxic activity. The methods applied to evaluate the effectiveness of biocides should consider all these factors.

When a large amount of organic material is present, including that of the microbial colony itself, the toxicity of a biocide may decrease. A biocide can even induce the production by biofilms of extracellular polymers and soluble chemicals into the surrounding medium, conferring a protective effect to all biotic components against the action of chemicals. It is now widely accepted that microorganisms adhering to surfaces, particularly after long periods of attachment, are more resistant to disinfectants.[328] The stage of development and the presence of biofilms in the bulk of the stone strongly affected the treatment efficacy as a study on the removal of lichens and cyanobacteria present on the Segovia cathedral cloister (Spain) showed.[268] Two biocides (Biotin T soluble in water and Biotin R soluble in white spirit) were applied by brush on dolostones and granite. They were effective at killing epilithic lichens and cyanobacteria, but they had different effects on endolithic microbiota. Biotin T application resulted in a cleaner stone surface, free of epilithic lichen remains, but living fungal hyphae were detected in fissures of the stone after 2 months. Biotin R application left more superficial lichen remains but it was more efficient at eliminating mycobiont hyphae that penetrate the stone, as well as endolithic cyanobacteria.

4.12 NOVEL BIOCIDES AND ALTERNATIVE METHODS FOR THE CONTROL OF BIOLOGICAL GROWTH

As already highlighted, many are the concerns on the use of biocides in the conservation of cultural heritage—and not only in this field, because their toxicity poses risks to humans and the environment. For example, environmental alarms on indoor air quality place more demands for the development of "greener" products on paint formulations than ever before. Recently, formulators have faced restrictions on the level of volatile organic compounds that they can incorporate in their paints.

Biocides lose efficacy over time and require constant retreatment that is both costly and inconvenient, and their repeated applications have potentially harmful effects on stone heritage objects and create resistant species.[329] Moreover, their compatibility with conservation treatments is an issue. As the control of biological growth is often deemed necessary, researchers are attempting to design alternative methods and new formulations with biocidal properties for the conservation of stone objects. While in the past the studies dealing with this issue were scarce, nowadays many scholars consider attractive this new field of research. The recent findings suggest potential natural sources for the control of biological growth on stones in restoration and conservation programs.

As it is unlikely that "new" microbicidal molecules might be produced in the future, novel products might concentrate on synergistic effects between microbicides and the combination of a microbicide and permeabilizers or other non-microbicidal chemicals. The attempt is toward a more effective use of biocides and an increase in antimicrobial activity decreasing the concentration of the biocide required to get toxic effects.[330]

Recent subjects of extensive investigations in other fields regard the ways microbicides are delivered (e.g., use of polymers for the slow release of microbicidal molecules), and the application of light-activated microbicides.[330]

Suggestions that can result suitable are found in the literature related to fields different from that of cultural heritage. Ethylenediaminedisuccinate (EDDS), a biodegradable chelator defined as a green biocide enhancer, improved the efficacy of glutaraldehyde in the treatment of bacteria biofilms on carbon steel coupon surfaces.[263] EDDS reduced the glutaraldehyde dosages considerably.

In agriculture, fungicides are routinely applied in larger quantities normally as a preventive measure. When it rains, they are washed into rivers and lakes, where they can result in harmful effects on aquatic organisms. For this reason, new fungicides called "paldoxins" were developed for a sustainable agriculture.[331] The paldoxins, short for phytoalexin detoxification inhibitors, are potent inhibitors of fungal enzymes. These green biocides help to protect crops that are valuable not only for food, but also to make biofuels.

Biofouling is a major concern for underwater hull of ships and for submersed manmade structures. After the banning of TBT, widely used in coatings for its antifouling capacity, the need for bioinspired antifouling strategies has been a challenge for the scientific community.[332] Among substances recently studied, there are the extract of the red seaweed *Chondrus crispus*, and usnic acid and juglone isolated from a lichen and the black walnut, respectively.[332] All these substances proved to have good efficacy on two biofilms forming bacteria.

A laboratory study[333] on human pathogenic bacteria evaluated the activity of two selected isothiocyanates, natural substances of plant origin, on biofilms formed by *Escherichia coli*, *Pseudomonas aeruginosa*, *Staphylococcus aureus*, and *Listeria monocytogenes*. Isothiocyanates had preventive action on biofilm formation and showed a high potential to reduce the mass of biofilms at low concentrations (circa 1 mg/ml).

In the conservation field, a step toward this direction was the EU project "Inhibitors of Biofilm Damage on Mineral Materials" (BIODAM) that evaluated the combination of biocides with permeabilizers, special slime (EPS) blockers, pigment inhibitors, and photodynamic treatments.[329] The scientific approach aimed at a drastic reduction of the concentration of applied biocides. The more promising substances are the permeabilizers.

Increasing the permeability of the cell to biocides, making cells and cellular EPS envelopes penetrable by biocides, is an efficiency-increasing measure. The project studied the effect of the combination of two permeabilizers [ethylenediaminetetraacetic acid (EDTA) and polyethyleneimine (PEI)] with benzalkonium chloride on *Pseudomonas* strains, algae, cyanobacteria, and fungi isolated from stone monuments.[89,291] All substances alone had varying antimicrobial activity. *Pseudomonas* species can degrade chloride compounds, and therefore, they are not very sensitive to QACs. EDTA and PEI enhanced the activity of benzalkonium chloride toward microorganisms in the laboratory experiment. A subsequent field trial involved two sandstone objects with natural biofilms that were treated with multiple combinations of chemicals. Although treatments proved successful under laboratory conditions, field trials were inconclusive and further testing will be required to determine the most effective treatment regime.

A novel approach in the conservation of caves involved the interaction of light with photosensitizers that can exchange electrons or protons with adjacent molecules to generate ROS, highly oxidative and toxic to living cells.[214] One of these compounds is D-aminolevulinic acid (D-ALA), which is colorless. Cyanobacterial cells are abundant in phycobilisomes and chlorophyll *a*, compounds that produce ROS upon irradiation with strong red light (620–650 nm). The study[214] tested the effectiveness of D-ALA upon irradiation on cyanobacteria and biofilms collected from hypogea and catacombs in Rome (Italy). D-ALA showed ability to enhance the treatment with light because cyanobacteria transformed it into protochlorophyllide that, excited by red light, generated ROS inside the cells.

As mentioned above, the need for environmentally benign antifouling technologies has led to renewed interest in the ways that many marine organisms manage to stay free from fouling. They produce secondary metabolites that prevent the adhesion of microorganisms to their surface. The study of metabolites from these organisms may form the basis for the development of natural nontoxic antifouling compounds and their application on cultural heritage objects.[95] Among them there are zosteric acid (p-sulfoxy cinnamic acid, a natural extract from eelgrass *Zostera marina*), N-vanillylnonanamide, poly-alkyl pyridinium salts, *Ceramium botryocarpum* extract.[334,95,335] Other antifouling compounds with terrestrial origin are capsaicin, cinnamaldehyde, and carvacrol that is an essential oil of plant *Origanum vulgare*.[334–336]

Laboratory tests of some of these compounds (*Ceramium botryocarpum* extract, poly-alkyl pyridinium salts, zosteric acid, capsaicin, cinnamaldehyde) against the growth of microorganisms showed that they are more efficient on cyanobacteria than on algae. Only cinnamaldehyde exhibited a good performance on fungi.[334] Other laboratory tests used N-vanillylnonanamide, zosteric acid, and carvacrol[95] on bacteria and fungi. The results of the first compound were not completely satisfactory in inhibiting the adhesion of microbes to glass slides. Regarding the antibiofilm activity of zosteric acid, it is species specific, causing more than 90% reduction of bacteria *Escherichia coli* and *Bacillus cereus* adhesion, whereas fungi *Aspergillus niger* and *Penicillium citrinum* coverage was affected by 57%. Calvacrol was efficient[336] in inhibiting the growth of some fungi with MIC values ranging 0.1–2.0 µl/ml.

Another promising natural biocide, composed by cell filtrates of the fungus *Trichoderma harzianum*, was tested under laboratory conditions.[337] It was efficient on phototrophic microorganisms that developed on limestone samples. *Trichoderma harzianum* is a filamentous fungus (Ascomycota division) found in the soil of many climatic zones. It produces natural bioactive products such as extracellular enzymes, and many volatile (e.g., pyrones, sesquiterpenes) and nonvolatile secondary metabolites (e.g., peptaibols). The extracellular lytic enzymes (cellulase, protease, glucanase, and chitinase) play a basic role in the cell-wall degradation of a broad range of fungi and bacteria. In addition, the volatile organic compounds produced by *Trichoderma harzianum* have antibiotic activity.[337]

The results of studies[95,335] of mixtures of antifoulants with water repellents and consolidants at preventing microbiological growth are discussed in Chapter 5.

Other substances have been the subject of studies focused on "green" biocides. Metabolites (usnic acid, norstictic acid, parietin) produced by saxicolous lichens are natural competitors of rock dwelling microorganisms.[312] Their effects *in vitro* (approx. 10^{-2} mM) on microcolonial fungi, coccoid cyanobacteria, and green algae (all commonly occurring on stonework) showed high efficacy inhibiting the growth of the species. The study detected a negligible chromatic alteration ($\Delta E^* < 0.5$) caused by lichen metabolites to the white Carrara marble samples. As reported by the authors, the research is worth to be extended to evaluate the interaction of lichen secondary metabolites with stones.[312]

A promising application alternative to biocides is that of *nanoparticles*. When used as a prefix, the term "nano" implies 10^{-9}. A nanometer (nm) is one billionth of a meter or roughly the length of three atoms side by side. A DNA molecule is 2.5 nm wide, a protein approximately 50 nm, and a flu virus about 100 nm.[338] A nanoparticle is a microscopic particle with at least one dimension less than 100 nm.

Nanoparticles have photocatalytic and catalytic effects on the surfaces. Catalysis is the acceleration of a reaction by the presence of a catalyst that acts accelerating the chemical transformation of a substrate. The catalytic activity depends on the ability of the catalyst to create electron–hole pairs, which generate free radicals, primarily hydroxyl radicals, and ROS, like superoxide and hydrogen peroxide. ROS are highly reactive and, at high concentrations, are extremely harmful for all organisms as they easily oxide essential molecules such as lipids, polysaccharides and proteins. The use of nanoparticles is a new trend in the building and paint industry. Prevention of biofilm growth on surfaces is of economic importance, and hence, the industry produces photocatalytic surface coatings using their self-cleaning properties, primarily to remove unwanted organic pollutants.[280] Nanoparticles of zinc oxide, titanium dioxide, and silicon dioxide act as photocatalytic agents by absorbing radiation energy, mostly UVA radiation (315–400 nm). In contrast to water-soluble biocides, nanoparticles can be fixed more effectively in the coating matrix when added to consolidants and water repellents, and the threat of leaching out into the surrounding environment is low, thereby resulting in lower ecological risks.[280,339] Some studies have already applied photocatalysis with titanium dioxide nanoparticles to roof tiles, paints, concrete, and glass panels.[340,341] Titanium dioxide nanoparticles have deactivating effects on bacterial endospores[342] and affect bacterial and algal adhesion.[343]

The algaecide activity of nanoparticles of zinc oxide (20–60 nm) was the subject of a laboratory study. A green algal strain of the genus *Stichococcus*, isolated from a German building, was the test organism. Nanoparticles without UV radiation exposure only slightly affected the algal cells. On the contrary, nanoparticles exposed to UV radiation inhibited the algal growth. Therefore, according to the authors, photocatalysis caused this positive effect and not ZnO itself. Even if ZnO nanoparticles acted on membrane integrity and algal autofluorescence, they killed only 20–40% of the *Stichococcus* cells. Other studies confirmed these results.

TiO_2 exists in three crystalline phases, namely rutile (the most stable form), anatase, and brookite.[344] TiO_2, in the mineral form anatase, was efficient in eliminating cyanobacteria and green algae inoculated on mortar slabs exposed outdoor.[283] It was also good at removing in two weeks almost all algae and lichens that grew on buildings in Sintra (Portugal) when applied by spraying at 1% (v/v) in distilled water. TiO_2 reduced algal growth on concrete surfaces[327] while other authors[345] experimenting with ceramic materials in outdoor shaded contexts, found only a reduction in fungal growth and practically no influence on the colonization by cyanobacteria, green algae, and lichens. A recent study confirms this last result.[346] The study applied a commercial product based on photocatalytic titanium dioxide, Photocal, on samples located in a forest. After 2 years, Photocal treatment did not prevent biological colonization indicating that titanium dioxide is not efficient in all situations. The samples were shaded in summer because of the forested location, and the lack of full sun affected the photocatalytic activity. These and other results, discussed in Chapter 5, showed that UV intensity has a great impact on the photocatalytic degradation or prevention of microorganisms' development and thus TiO_2 can be efficient only in particular environments.[30,347]

The removal of microorganisms using a microorganism was the goal of a recent research[348] that proposes the fungus *Aspergillus allahabadii* as a cleaning agent of the biofilms formed on surfaces of a sandstone temple in Angkor Thom, Cambodia. Even if this method can be an alternative to biocides, as the authors affirm, the problem of the following growth of *Aspergillus allahabadii* on stone surfaces remains an unresolved issue.

Other studies[349,223] experimented hydrolase enzymes, used so far to remove biological growth on canvas, paper and polychrome artifacts. The researchers treated biofilms (alga, cyanobacteria, and heterotrophic bacteria) with glucose oxidase enzyme (GOx) homogeneously spreading it on the surfaces of travertine, peperino, and marble samples. This enzyme produces H_2O_2, the actual biocide agent, from the catalytic reaction of glucose. The samples were treated under laboratory conditions with an aqueous solution containing GOx enzyme (0.5 mg/ml) and different concentrations of glucose (ranging from 2 to 400 mM) for different contact times at room temperature. The enzymatic approach, compared with the direct application of H_2O_2, produces smaller amounts of H_2O_2, avoiding the etching of stones materials. The travertine samples showed better results in terms of cleaning efficiency and absence of etching effect on the surface. According to the authors, the proposed cleaning procedure showed good results because the enzyme controls the concentration of H_2O_2 in situ and retains it preferentially on the surface, depending on the porosity of the substrata.

The application of "algal wash," a formulation containing low concentration of a copper complex, showed high efficacy in killing lichens, algae, and fungi that colonized monuments in Angkor Wat, Cambodia.[99] The authors reported that the formulation, nicknamed Melange d'Angkor, is an eco-friendly compound, but unfortunately, they did not specify its components.

KEYWORDS

- **physical control methods**
- **mechanical and water-based control methods**
- **biocides**
- **types of microbicides and herbicides**
- **mechanisms of antimicrobial action**
- **stains after treatments**
- **biocides no more on the market**
- **resistance of organisms to biocides**
- **toxicity and legislative regulations**
- **recolonization after treatment**
- **biocides' effects on stones**
- **effectiveness of biocides and the persistence of their effect**
- **novel biocides and alternative methods for the control of biological growth**

CHAPTER 5

PREVENTION OF BIODETERIORATION

CONTENTS

ABSTRACT

The chapter focuses on the methods and on the more suitable techniques and products that are useful for the prevention of the development of micro- and macroorganisms that grow on artificial and natural stone objects, including wall paintings. It details the different strategies performed for the prevention of biological growth on indoor and outdoor stone heritage objects. The concept of planned maintenance, which is closely related to preventive conservation, has reached recently growing consideration and several techniques and diagnostic tools have been developed for the detection of possible risks, thus facilitating maintenance and management decisions.

Whether any control method has been applied or not and even if effective, it is advisable to consider a prevention approach as well, because biological growth occurs whenever the favorable conditions do not change. Aim of prevention is to slow down or inhibit biological growth by acting on the factors that help and determine its development. Preventive actions include a limitation of possible nutrients for microorganisms as well as the change of conditions (e.g., temperature, humidity, and light) favorable for microbial growth.

5.1 INDOOR ENVIRONMENTS

In indoor environments—for example museums, archives, libraries, churches—prevention implies regular dusting (without the use of water to provide stable RH conditions) and restrictions to eating and drinking for visitors and staff. Dust and dirt can contain spores and microorganisms and provide the nutrients required for microbial growth. Cleaning using vacuum cleaners equipped with high efficiency particulate arrestants (HEPA filters) can remove most spores.[144]

The stabilization of climate conditions is very important to limit any microbiological colonization of art collections.[237]

Further actions involve building maintenance (e.g., avoidance of leaks, maintenance of properly functioning heating and ventilation systems).[350] Ventilation is a simple method that prevents microbe colonization because it reduces humidity, impedes water condensation on surfaces, and diminishes the deposition of atmospheric particulate, vegetative cells, and spores.[351,98] The installation of filter systems in storage rooms avoid the ingress of fungal spores, plant pollen, and dirt particles.

Many churches have no climate control systems. In some cases, it is possible to improve indoor conditions just with simple actions. For example, a device for air circulation, a dehumidifier and an air-to-air heat pump were installed in the church of the Holy Cross (Harju Risti, Estonia). Ventilation performed well at in preventing microbial growth, and dehumidification was mostly effective in the cold period of the year. Annual energy consumption for the heating with a heat pump was the lowest compared with other systems.[352]

Only recently important European museums have started to establish microbiological monitoring programs including the measurement of fungal airborne spores by air samplers and by surface contacts examinations on objects and museum shelves.[144] Air is the most important medium for the transportation of microorganisms that reach objects' surfaces and can start developing. In museums, libraries, archives, churches and caves, there is a wide range of microorganisms that represent potential risks for the degradation of objects. Airborne spores and vegetative structures may develop on different substrates interacting with other factors, such as the microclimatic conditions, the composition and state of conservation of the objects, their location. In some environments, factors as inappropriate air conditioning systems, visitors, windows and doors in contact with the external environment, are sources of airborne particles that deposit on the objects.[353] Moreover, a restored object may be susceptible to biological growth because of the substances used in the restoration operations.[354,355] Aerobiology is the scientific study of the sources, dispersion, and effects on living systems and on the environment of airborne biological materials, such as pollen grains, bacteria, fungal spores that are passively transported by the air. Aerobiologists have traditionally been involved in the measurement and reporting of airborne pollen and fungal spores as a service to allergy sufferers. Aerobiology applied to the conservation of cultural heritage serves to evaluate the potential risks of degradation of objects located in indoor environments by airborne microorganisms and to plan interventions.[356–359,353] Indoor monitoring of airborne particles (spores and vegetative cells) allows determining their changes over time in a building or a room. This method makes it possible to know either the biological risks of objects or the hygienic safety and health protection for operators and visitors.[353] Thus, aerobiology is relevant not only for preventive conservation but also for public health. Aerobiological monitoring is periodically carried out (usually in different seasons of the year) to determine over time the air quality both quantitatively (e.g., degree of contamination) and qualitatively (e.g., presence of potentially damaging species or allergens). It often includes also measurements of aerosol particles, pollutant gases and microclimatic parameters.[98] The sampling locations should depend on the critical areas where the risk of biodeterioration is high (e.g., parts close to doors and windows, rooms with high number of visitors, etc.).

Air sampling for bacteria and fungi has intrinsic problems. The main difficulty is the variability of airborne spore concentrations that can fluctuate dramatically over relatively short distances and times.[360] This degree of variation often requires the collection and analysis of many samples to permit reliable interpretation of the results.[360]

For the analysis of the airborne viable fungal particles, two sampling systems are used, a passive and an active method. The passive sampling utilizes the impact sedimentation of particles on Petri dishes. This simple method can contemporaneously monitor different areas but permits only qualitative analyses. As an example, open Petri dishes containing a sterile cultural medium are exposed one time a week for 1 h. The active method utilizes devices like Volumetric Air Sampler that traps spores and particles on glass slides. The method allows a simple counting of particles providing a first approximation of a microbial contamination of the air. Another device used in cultural heritage field is the impact sampler multi-stage Andersen Cascade Impactor that measures concentration and particle size distribution of bacteria and fungi in ambient air. Each stage impactor contains Petri dishes with sterile agar. Multiple orifices are present on each stage. Air to be sampled enters the inlet cone and cascades through the succeeding orifices through the different stages. Viable particles can be collected on a variety of agar and incubated for counting and identification. A continuous vacuum pump provides constant air sampler flow. Active sampling can be performed in different times depending on the monitoring's aims (e.g., one time a week with 10 min' air aspiration).

A study[353] analyzed the air quality of the crypt of the Basilica of St. Peter in Perugia (Italy) sampling airborne particles weekly in the period March–July 2011. It showed wide variations in the bioaerosols with the highest values in June and July, heterogeneous spore distributions and different peak concentrations in the studied areas. Different fungal genera were present.

The study of indoor atmospheres of a museum in Antwerp (Belgium) included analysis of aerosol particles, pollutant gases, bacteria, fungi and different microclimatic parameters.[356] It proved that poorly balanced heating and air conditioning affected microclimates in the exhibition rooms and the daily flux of visitors produced rapid changes and marked thermohygrometric gradients. Microbial loads were higher in summer than in winter. However, the proportion of microorganisms capable of damaging the objects did not increase inside the museum and thus was not a risk.

Some studies examined also outdoor atmosphere because outdoor sources largely determined the composition of indoor aerosol and biological airborne particles.[359]

Diagnostic characterization and distribution of biofilms colonizing an object is a relevant step for the definition of preventive interventions. For example, the study of cyanobacteria and green algae that grew on mural paintings of the Crypt of the Original Sin (Italy) demonstrated that humidity and sunlight exposure were the main causes of the colonization.[113] Thus, the preventive measures focalized on the reduction of these two parameters. On the other hand, the critical ecological characteristics of the site required periodical examination of the state of conservation of the painted surfaces, thus every 6 months the researchers carried out the chemical and biological monitoring.

An interesting research[361] suggests a prevention strategy based on the use of a fungal detector that provides a fungal index to assess the conditions of a museum. The authors experimented the device in the storerooms of historical buildings in Higashiomi (Japan). The sealed detector contains fungal spores and nutritional sources. The spores were of aerial fungi differing in sensitivity to RH, that is, moderately xerophilic *Eurotium herbariorum*, strongly xerophilic *Aspergillus penicillioides*, and hydrophilic *Alternaria alternata*. If the microclimate of the room is favorable to fungal growth, spores will germinate. The index is the value obtained by the ratio between the growth responses of fungi (called "response units") in the detector and the exposure period. The researchers exposed the detectors for 4 weeks at survey points in the storerooms. Then, they measured the length of hyphae under a microscope and defined three levels of microclimates: level A when no germination occurred, level B when hyphae length was circa 500 μm, and level C when hyphae length was >2600 μm. In the examined storerooms the corresponding index values were <1.8, 1.8–18, and >18, respectively. Consequently, a room at level A is free of contamination; a room at level B can have fungal contamination. In a room at level C, fungal contamination is unavoidable, and countermeasures should be taken promptly.

Peculiar are the actions attempted to prevent biofilms development in caves and catacombs. Some authors proposed the improving of illumination management by minimizing lighting time, using weak-intensity lamps or light-emitting diodes (LEDs), or at least by establishing drastic illumination management when opening sites to tourists. Unfortunately, such actions are often insufficient in inhibiting algal proliferation in caves.[362] Sometimes, wrong decisions adopted for managing tourism's impact in caves resulted in fatal errors that marked the future of these fragile sites. This was the case of the worldwide famous Lascaux cave (France) that attracted a large audience since discover in 1940.[118] To meet tourists' needs, a lighting system was installed. The light caused the growth of a biofilm, called "le maladie verte," formed by green algae as already mentioned. Besides biological colonization, visitors' breath strongly affected the cave ecosystem altering the microclimate.[118] For all these reasons, a drastic decision led to its closure in 1963.

However, there are also successful results[65] at preventing biological growth in caves. The installation of lamps emitting monochromatic blue light (emission peak around 490 nm) in the Roman Catacombs of St. Callistus and Domitilla, Rome (Italy) inhibited cyanobacterial growth because cyanobacteria cannot use this spectral emission for photosynthesis and growth.

Other useful steps in decision making for preserving caves and catacombs are the following[363]:

- Improving the ventilation and air circulation to favor the decrease in moisture content in the air and walls;
- Reducing or eliminating the penetration of the natural light inside caves and catacombs;
- Periodic monitoring of environmental parameters (e.g., air temperature, relative humidity, surface temperature, air mass circulation);
- Evaluating the best treatment for the removal of biofilms considering the microclimate.

5.2 OUTDOOR ENVIRONMENTS

Although prevention measures are considerably more difficult to undertake on outdoor monuments than indoors, in some cases effective maintenance operations can be carried out. Positive results can be achieved simply through the control of surface wetness by repairing drainage, reducing the sheltering effects of closely located vegetation or other structures, removing accumulated soil and debris, controlling nutrient availability and avoiding excessive water retention.[13,104] Other, more complex, preventive interventions include the cutting of walls to insert isolating materials, and the injection of organic silicon-based compounds to provide a chemical barrier against water capillary raise.[364,365] Removal of accumulated soil and debris and prevention of excessive water retention are often adequate measures that discourage the growth of mosses.[71] Their presence is an indication of persistently damp conditions that are probably more damaging than the organism itself.

The reduction of stone water content is effective at preventing biological growth. For this purpose, temporary or permanent protective shelters are suitable tools. Mainly used in archaeological contexts, they can reduce the deterioration rate of stone artwork only if their design and construction are appropriate allowing opportune ventilation and shadowing.[71,366,104] When not properly constructed, they may even increase the deterioration rate.[30] In fact, their effectiveness depends on several factors, for example shape and material. Transparent shelters enhance biocolonization because they increase the temperature and do not halt light hitting surfaces. The well-known mosaics of the major Roman site of Villa del Casale at Piazza Armerina, Sicily (Italy) that covers an area of 2000 m^2, are a paradigmatic example. The historical shelter constructed in 1957 was a transparent structure of plastic and glass, meant to suggest the original volumes[367] (Fig. 5.1). The condition of the shelter in the late 2010s showed serious problems both in the metal structures (extensive and advanced oxidation) and in the transparent glass and plastic sheets (water infiltration, yellowing of the Perspex, etc.). The translucent roof panels created extreme fluctuations in temperature and humidity through their greenhouse-like effect. Air did not circulate well. Internal environment (peaks in temperature, high relative humidity, frequent condensation, etc.) not only hindered the appreciation of the site, but also compromised the preservation of the mosaics causing biological colonization as well.[367]

Another example is that of hoods, installed to protect sculptures from snow in wintertime, which can create a greenhouse effect. In such situations, ventilation is an important measure to prevent rapid colonization.[235]

FIGURE 5.1 Roman site of Villa del Casale, Piazza Armerina (Italy). A detail of the historical shelter constructed in 1957, a transparent structure of plastic and glass meant to suggest the original volumes.

Temporary shelters were installed to protect the excavated walls and ruins located under the ground level in the historical center of Milan (Italy). In a short time, the lack of ventilation, the water penetrating inside the site and the consequent greenhouse effect generated an extended biofilm development (Fig. 5.2). A positive example instead is that of the Mayan Hieroglyphic Stairway in the archaeological site of Copan (Honduras). In 1987, the monument was covered with a permanent tarpaulin shelter to protect the stairway from rain. In 2000, a study showed a drastic reduction of biological colonization. Lichens completely disappeared; just a small number of cyanobacteria and green algae remained.[30,71]

After the restoration of some wall paintings in Pompei (Italy), the conservation scientists who performed the diagnostic analyses suggested to protect them with polycarbonate slabs that filtered the ultraviolet and infrared rays.[368] Shape memory alloys (SMA) fixed the slabs to the walls allowing their movement when temperature changes. Thus, SMA induced a good air circulation and avoided an excessive increase intemperature on the surfaces of the wall paintings.[367] A shape-memory alloy "remembers" its original shape and, if deformed, returns to its predeformed shape when heated.

(a)

(b)

FIGURE 5.2 Temporary shelter installed to protect excavated walls and ruins located under the ground level in the historical center of Milan (Italy) (a). Lack of ventilation, water penetrating inside the site and the consequent greenhouse effect caused an extended green biofilm development (b).

External protective modern glass sheets proved effective at preventing weathering and biofilm formation on ancient stained-glass windows of three European churches, as the European project VIDRIO showed.[10] The glass sheets were installed outside, facing the external side of the original stained windows, creating a narrow chamber (interspace). The sheets had slots for the free passage of air inside the chamber. The microbial impact on the internal surface of protected windows was similar to that of the internal surface of unprotected windows thus excluding the risk of an acceleration of microbiological activity in the interspace.[10]

In some circumstances, it is possible to solve problems due to biological alteration making just few changes. It was for example the case of the Marlborough Pavilion, Kent (UK), built in the mid-1920s.[369] Open on two sides, the internal walls are painted. Wide disfiguring green patches developed on wall paintings because of the highly favorable microclimate inside the building especially during winter months. The use of a temporary modular enclosure in winter was a successful solution. With the addition of a dehumidifier, thermal buffering reduced condensation events and freeze thaw cycles, resulting in significant reduction of microbiological growth and flaking, with a relatively low energy input.[369] By using this system, the green microbiological growth did not reappear.

Another approach in architectural conservation, alternative to repeated application of biocides to inhibit biological growth, involves the installation of copper or zinc strips and wires in the length of the masonry. When rainwater washes over the metal strips, they release toxic copper or zinc ions leached by water and carried on the lower areas of stone.[370] The technique is mostly effective on new or freshly cleaned materials and on objects with a regular shape and design. These characteristics ensure even distribution of rainwater over the surfaces.[370,371] The application of the system is instead a challenge when operating on ruins at archaeological sites.[372] Metal strips appear to remain effective in controlling biological growth for years, thus eliminating the need for reapplication.[371] However, some drawbacks emerge from in situ observations. The ions released by metal strips are not in sufficient concentration nor are effective to act as biocides. This explains why it is difficult to eradicate algae, mosses, and lichens using this technique without an initial cleaning of the surfaces. For example, zinc oxides eventually kill mosses and algae but it takes a very long time (8–10 years).[371] Therefore, metal strips show good performance in inhibiting biological growth rather than killing it. Metal strips do not work on shaded surfaces.[371] Moreover, copper strips tend to produce light green staining on pale-colored calcareous stones that may be unsightly and difficult to remove. There are also some concerns about toxicity and environmental pollution. Concentrations of zinc and copper are toxic to freshwater and marine invertebrates and to freshwater fish.[371] This has led countries to establish limits for acceptable amounts of heavy metals in water bodies.[371] Industrial pollution and runoff from streets and roofs have been identified as a source of metals in water.[371] However, it is reasonable to assume that the application of metal strips on buildings would have minimal effects on the overall water quality for a given area as they would infinitesimally increase the amount of metals.[371]

Zinc strips were installed[370] on the ridge's roof of the Stanford Mausoleum, Barre (USA), over both cleaned and uncleaned areas of stone. After 12 years, no biological recolonization and no staining occurred. Only minor soiling was found (Table 5.1).

A test evaluated the performance of metal strips on an object with irregular shape, that is, a situation unfavorable to the system's efficiency.[372] On the top of a wall in San Ignacio Mini Mission (Argentina) the researchers laid out three different strips and meshes made of lead, zinc, and brass (58% copper, 40% zinc, and 2% lead). They divided the north vertical side of the wall into three sections corresponding to the ones on top of the wall with the metals. Each section was divided in areas where the abundant biological was removed with a biocide, and areas left uncleaned as a control. After 16 months, they observed just a slight growth of cyanobacteria in the cleaned areas. On the contrary, new shrubs developed in the uncleaned ones (Table 5.1).

A quite common practice aimed at decreasing the rate of a recolonization on outdoor objects is the application of biocides to inhibit the return of biofilms and lichens. This step is the last in a conservation process because it occurs after the cleaning procedures.[258,243,274,242,225,373] A study[146] assessed the performance of the product Acticide MBL5515 (active ingredients 1,2-benzisothiazolin-3-one 5%, 2-bromo-2-nitropropane-1,3-diol 15%, 2-methyl-3(2*H*)-isothiazolone 5%) when added at 0.1% to a white water-based paint (Table 5.1). Painted gypsum plasterboard panels underwent accelerated biofouling according to the standard paint test method BS3900. The researchers inoculated the painted side with spores of the most common fungi present on damp walls, that is, *Acremonium*, *Alternaria*, *Aspergillus*, *Aureobasidium*, *Cladosporium*, *Chaetomium*,

TABLE 5.1 List of Products and Methods Used for the Prevention of Biological Growth on Stone Objects. The Table Includes Concentrations of Use, Efficacy, and Indications about the Experimental Approach (In Situ or Under Laboratory Conditions), Side Effects and Notes, Bibliographical References.

Product/Method	Concentration/ Application	Efficacy in inhibiting biological growth	Substrate	Notes	Reference
Biotin R (mixture of iodopropynyl butylcarbamate, *n*-octyl-isothiazolone and 2-2′-oxydiethanol)	5% in ethanol	After 1 year, no biological growth occurred	Sandstone of the Roman Theater in Trieste (Italy)		Borgioli et al., 2006
Biotin T (alkyl-benzyl-dimethyl-ammonium chloride and isopropyl alcohol)	2% (v/v) in distilled water, brushing	Efficient after 4 months	Laboratory test, mortar samples inoculated with cyanobacteria and green algae		Fonseca et al., 2010
Anios D.D.S.H. (*n,n*-didecyl-*n*-methyl-poly(oxyethyl) ammonium propionate with alkyl-propylene-diamineguanidium acetate)	No dilution, spraying	Not efficient after 4 months	Laboratory test, mortar samples inoculated with cyanobacteria and green algae		Fonseca et al., 2010
ACTICIDE® MBL5515 (a.i. 1,2-benzisothiazolin-3-one, 2-Bromo-2-nitropropane-1,3-diol, 2-methyl-4-isothiazolin-3-one)	0.1% in water	Effective, no discoloration on surfaces	Laboratory test, approximately 100% humidity and 20 ± 2°C for 12 weeks. Painted gypsum plasterboard panels inoculated with 14 fungal species		Gaylarde et al., 2015
TiO_2 (anatase)	1% (v/v) in water, spraying	Efficient after 4 months	Laboratory test, mortar samples inoculated with cyanobacteria and green algae		Fonseca et al., 2010
TiO_2 (anatase)	1) 1% w/v 2) 0.5% w/v water	TiO_2 nanocoating was not able to prevent biofouling under weak UV exposure conditions	Laboratory test, treated clay brick samples inoculated with the green alga *Chlorella mirabilis* and the cyanobacterium *Chroococcidiopsis fissurarum*. The samples were exposed to different UV intensities		Graziani et al., 2013

TABLE 5.1 *(Continued)*

Product/Method	Concentration/ Application	Efficacy in inhibiting biological growth	Substrate	Notes	Reference
TiO_2 (anatase), commercial solution	1% w/v, water	TiO_2 nanocoating was not efficient in preventing biofouling on specimens with high porosity and roughness, while those with low porosity had minor growth	Laboratory test treated clay brick samples inoculated with green alga *Chlorella mirabilis* and the cyanobacterium *Chroococcidiopsis fissurarum*	Whitening of the samples caused by TiO_2 treatment	Graziani et al., 2016
Commercially available TiO_2-coated glasses	TiO_2 layer, thickness ca.15 nm	Not efficient	Laboratory test, commercially available photocatalytic glasses inoculated with two algal species (*Chlorella luteo-viridis* and *Coccomyxa* sp.)		Gladis & Schumann, 2011
Commercially available TiO_2-coated glasses		Considerable reduction of biofilm formation (mainly fungi) in comparison with non-TiO_2-coated glasses	Five-month exposition of samples to the atmosphere in São Paulo city (Brazil)		Shirakawa et al., 2016
Photocal®, commercial product composed of titanium dioxide	No dilution	Not efficient	Limestone samples, located outdoor in a park in Belgium for 2 years	Strong change in color for coloniza-tion and deposits	Eyssautier-Chuine et al., 2014
Silver nanoparticles in demineral-ized water	250 and 100 mg/m^2, brushing	Efficient in preventing algal fouling (*Chlorella vulgaris*)	Laboratory test, algal fouling on white architectural and autoclaved aerated concrete specimens (cementitious building materials)	They alter strongly the substrates' color producing darkening (ΔE* 8.7 and 15.9)	De Muynck et al., 2009
Alkoxy silane	40% (w/w) in isopropyl alcohol	After 12 weeks, no fouling developed on the white concrete samples. On the contrary, 65% of autoclaved aerated concrete samples were covered with algae	Laboratory test, algal fouling on white architectural and autoclaved aerated concrete specimens (cementitious building materials)		De Muynck et al., 2009

TABLE 5.1 *(Continued)*

Product/Method	Concentration/ Application	Efficacy in inhibiting biological growth	Substrate	Notes	Reference
Water repellents (silane and fluorinated compounds)		The treatment notably reduced the progression of colonization	Laboratory test (160 days). Treated mortar slabs inoculated with alga *Graesiella emersonii* through water capillary ascent and water run-off		Martinez et al., 2014
Rhodorsil® H224, alkyl polysiloxane oligomer, water repellent	10% in white spirit	Efficient	Limestone samples, located outdoor in a park in Belgium for 2 years		Eyssautier-Chuine et al., 2014
Protectosil® SC concentrate, organofunctional silane, water repellent	10% in water	Efficient	Limestone samples, located outdoor in a park in Belgium for 2 years		Eyssautier-Chuine et al., 2014
Application of the biocide 3,5,6-tetrachloro-4-methylsulfonyl-pyridine in isopropyl alcohol followed by alkyl alkoxy silane	Water repellent 40% (w/w) in isopropyl alcohol, biocide as supplied Products were applied at a 24-h interval	After 7 weeks of accelerated tests, only about 20% of the surface was covered with algae. No green streaks occurred on surfaces.	Laboratory test, algal fouling on white architectural and autoclaved aerated concrete specimens (cementitious building materials)		De Muynck et al., 2009
Mixtures of Cu nanoparticles (stabilized with tetra-octyl-ammonium) with a consolidant (Estel 1000) containing tetraethyl orthosilicate	Cu nanoparticles, concentration 0.22 mg/ml	On marble, after 3 years, fungal growth occurred. On sandstone and plaster, no biological growth	Plaster, marble, and sandstone of archaeological area of Fiesole, Firenze (Italy)		Pinna et al., 2012
Mixtures of Cu nanoparticles (stabilized with tetra-octyl-ammonium) with a water-repellent (Silo 111 containing methylethoxypolysiloxane 12% in white spirit)	Cu nanoparticles, concentration 0.22 mg/ml	On marble, after 3 years, a slight fungal growth occurred. On sandstone, no biological growth	Marble and sandstone of archaeological area of Fiesole, Firenze (Italy)	On marble, the treatment showed no changes in water repellency throughout the entire monitoring period	Pinna et al., 2012

TABLE 5.1 *(Continued)*

Product/Method	Concentration/ Application	Efficacy in inhibiting biological growth	Substrate	Notes	Reference
Mixtures of Cu nanoparticles (stabilized with tetra-octyl-ammonium) with an acrylic resin (Acrilico 30 containing 30% Paraloid™ B72 in ethyl acetate) diluted to 7% w/w in ethylacetate/ buthylacetate 50/50	Cu nanoparticles, concentration 0.22 mg/ml	After 3 years, no biological growth occurred	Plaster, archaeological area of Fiesole, Firenze (Italy)	The treatment induced an increase in plaster's brightness (ΔL^* ~7)	Pinna et al., 2012
Tetraethoxysilane (TEOS) mixed with chitosan (poly-acetyl-glucosamin, antimicrobial biopolymer)		Not efficient	Limestone samples, located outdoor in a park in Belgium for 2 years	Strong change in color for colonization and deposits	Eyssautier-Chuine et al., 2014
TEOS mixed with chitosan (poly-acetyl-glucosamin, antimicrobial biopolymer), silver nitrate, and hydrophobic silica		Efficient	Laboratory test (4 weeks). Treated limestone samples inoculated with *Chlorella vulgaris*		Eyssautier-Chuine et al., 2015
Emulsions of TiO_2 nanoparticles and silane/siloxane-based water repellents in water	TiO_2 nanoparticles <0.1 wt%, brushing	Efficient in reducing significantly biofilm surface coverage in comparison with control samples	Laboratory test (8 weeks). Treated mortar slabs inoculated with algal and cyanobacteria	Nanoparticulate treatments improved water repellence of the control emulsion	Zhang et al., 2013
Emulsions of ZnO nanoparticles and silane/siloxane–based water repellents	ZnO nanoparticles <0.1 wt%, brushing	Efficient in reducing significantly biofilm surface coverage in comparison with control samples	Laboratory test (8 weeks). Treated mortar slabs inoculated with algal and cyanobacteria	Nanoparticulate treatments improved water repellence of the control emulsion. No modifications of color	Zhang et al., 2013

TABLE 5.1 *(Continued)*

Product/Method	Concentration/ Application	Efficacy in inhibiting biological growth	Substrate	Notes	Reference
Mixtures of silver nanoparticles with polydimethylsiloxane	0.1, 0.3, 0.5 wt%	Not efficient	Laboratory test (4 weeks). Treated mortar slabs inoculated with *Chlorella vulgaris* and *Synechococcus* sp.	Considerable change of color with ΔE^* varying 5.35–8.40 respectively for the three concentrations of nanoparticles	MacMullen et al., 2014
Mixtures of TiO_2 with silicate/ acrylic-based compounds	TiO_2 2 wt%	Not efficient	Laboratory test (160 days). Treated mortar slabs inoculated with alga *Graesiella emersonii* through water capillary ascent and water run-off. The lamps used to activate photocatalysis varied in UV intensities (1.5, 3, 7.5, 12 W/m^2)		Martinez et al., 2014
Mixtures of silver nanoparticles (0.1, 0.3, 0.5 wt%) with the consolidants Primal AC33 and Wacker BS 1001	3% of polymers in water, silver nanoparticles concentration 40 mg/ml. 25 ml of the mixture coated the surfaces	Wacker BS 1001 + silver nanoparticles efficient	Laboratory test, sandstone, and limestone samples. The treated samples were inoculated with the bacterium *Streptomyces parvulus* and the fungus *Aspergillus niger*		Essa & Khallaf, 2014
Zinc strips		After 12 years, no biological recolonization and no staining occurred. Only minor soiling was found	Ridge's roof of the Stanford Mausoleum, Barre (USA), over both cleaned and uncleaned areas of stone		Wessels, 2011

TABLE 5.1 *(Continued)*

Product/Method	Concentration/ Application	Efficacy in inhibiting biological growth	Substrate	Notes	Reference
Lead, zinc, and brass strips and meshes		After 16 months, a slight growth of cyanobacteria occurred in the cleaned areas. On the contrary, new shrubs developed in the uncleaned ones	Wall in San Ignacio Mini Mission (Argentina). North vertical side divided into three sections. Each section divided in areas where the abundant biological was removed with a biocide, and areas left uncleaned as a control		Magadán et al., 2001

Penicillium, and *Ulocladium*. After 12-week long testing in a chamber at approximately 100% relative humidity and 20 ± 2°C, no discoloration was present on the surfaces.

Biotin R, a product containing a blend of carbamates and other biocides, was effective at preventing biological colonization for 1 year on sandstone of the Roman Theatre in Trieste (Italy)[267] (Table 5.1).

Prevention of algal growth on surfaces is of economic importance as it is a visual defect.[280,374,375] Hence, the manufacturers of paints and rendering systems adopt different ways to solve the problem. They develop products with an enhanced resistance against algal growth adding biocides that have also the scope of protecting products in-can.[57,376] These paint mixtures can be suitable also for cultural heritage objects, for example, mortars. Another strategy proposed by the manufacturers of paints to prevent or reduce biofouling after application is the hydrothermal approach[376] that creates an algae–hostile environment by minimizing the humidity films at the surface. The products are a new generation of energy saving paints that reflect the radiant heat of the sun, reducing the surface temperature of painted materials. Pigments used for this purpose include Complex Inorganic Colored Pigments (CICPs) or Mixed Metal Oxides (MMOs). The consequent reduction in duration of dawn condensation on the surfaces decreases the growth of microorganisms.[377] A study showed that the paints maintained this behavior over almost 2 years.[378] Unfortunately, in tropical and subtropical climates dirt and biofilms rapidly covered them changing their initial appearance and reflecting properties. After less than 2 years, cool painted metal, clay and concrete specimens exposed at various sites in California (USA) lost about 6% of the initial solar reflectance. One of the causes was the formation of biofilms made of cyanobacteria and fungi.[378]

Copper, either as a powder or as fibers, was very efficient at preventing biological growth when added to a hydraulic mortar.[379] Mortar specimens were exposed in the town of Sintra (Portugal) that has excellent climatic conditions for biological development. Over 9 years, no colonization occurred on samples containing 0.35% by weight of copper powder while control samples were colonized after 2.5 years of exposure. As a drawback, the presence of copper changed the mechanical properties of mortar. Compressive strength was reduced to 65% of the original value, flexural strength was reduced to 85%, while the elasticity modulus was the least affected being reduced only to 93%. However, as reported by the authors, these changes are not critical as the mechanical strengths are still above those usually required. Moreover, the presence of copper reduced the porosity and the capillary absorption coefficient of samples. This can be an advantageous effect because less water penetrates the mortar prepared with copper.

A study[380] reported the ability in inhibiting fungal growth of amorphous carbon films doped with copper that formed clusters sized some nanometers. The clusters were homogeneously distributed in the carbon matrix. The result is promising for the practical application of amorphous carbon films as protective media against biodeterioration.

Recent researches studied the antifouling properties of nontoxic compounds such as *N*-vanillylnonanamide and zosteric acid that affect cell-to-cell communication of microorganisms altering quorum sensing (QS) system. It has emerged as one of the most important mechanisms for controlling the development of highly structured and cooperative biofilms, on both organic and inorganic substrata.[95] The preventing action of *N*-vanillylnonanamide was not completely effective at inhibiting the adhesion of bacteria and fungi to glass slides. The activity of zosteric acid is species specific, causing more than 90% reduction of *Escherichia coli* and *Bacillus cereus* adhesion, whereas *Aspergillus niger* and *Penicillium citrinum* development was affected by 57%.[95]

TiO_2 has been the subject of growing interest and attention by scientists working in the field of cultural heritage conservation. This compound is indeed chemically stable, nontoxic, high photoreactive, cheap, and has biocide properties. The advantage of photocatalytic substances over biocides lies in their inexhaustible catalytic action, which provides long-term effectiveness.[375] Photocatalytic surfaces are environment-friendly, as they operate using only sunlight and rainwater, without any chemicals.[375] Despite all these positive features, the performance of TiO_2 is very controversial and shows that there is scope for detailed further studies to assess the efficacy of this product. Protection mechanisms of algae, bacteria and fungi should be investigated in more detail to evaluate photocatalysis as a disinfection measure.[375] Furthermore, other important topics need examination such as the interaction between nanotreatments and substrates, including the effects of the application procedures.[381] For example, a study[382] on the compatibility of TiO_2-based finishing for renders in architectural restoration showed that the presence of titanium decreased the surfaces' contact angle and remarkably increased the initial water absorption rate, which could be a drawback in façades exposed to wind driven rain. The treatment with titanium dioxide nanoparticles of clay-fired bricks caused surfaces whitening.[383] Moreover, the environmental impact and health issues related to the release of nanoparticles, as well as the cost benefits versus traditionally used products should be subjects of future studies.[381]

Some of the much relevant results on application of TiO_2 nanoparticles are here mentioned. Titanium dioxide (anatase) applied as a coating or by mixing it within mortars, demonstrated efficiency in a 4-month-long laboratory testing as it inhibited the growth of cyanobacteria and green algae.[283] Moreover, the results showed that anatase was better at preventing colonization than two conventional biocides, Biotin T and Anios D.D.S.H. (Table 5.1).

Five-month exposure to the atmosphere in São Paulo city (Brazil) of TiO_2 coated modern glasses (self-cleaning glass) resulted in a considerable reduction of biofilm formation (mainly fungi) in comparison with non-TiO_2-coated glasses.[32] The authors reported that TiO_2 inhibits the fungal spore germination.

TiO_2 nanoparticles efficaciously prevented the adhesion of an alga and a cyanobacterium (*Chlorella mirabilis* and *Chroococcidiopsis fissurarum*) on treated clay brick specimens but only under exposure to optimal UV irradiation.[384] The study clearly demonstrated the limitations of this "fashionable" compound, and showed that, under weak UV radiation, the inhibitory effect of TiO_2 nanocoatings is very low (Table 5.1). An accelerated biofouling test using treated clay brick samples inoculated with the same microorganisms showed that porosity and roughness highly affected the algal growth.[383] There was no difference between treated and untreated specimens with high porosity and roughness, while bricks with low porosity had minor growth confirming the inhibitory efficiency of TiO_2.[383]

Algae can resist the effects of TiO_2 even after 48 h of irradiation with UVA. A study[375] examined algal growth prevention using commercially available photocatalytic glasses (TiO_2-coated). Two algal species (*Chlorella luteoviridis* and *Coccomyxa* sp.) were chosen as model organisms as they grew on test specimens at 100% humidity and low UVA radiation. No effects on algal growth were detected, although the coated surfaces were photocatalytically active. The authors reported that aeroterrestrial algae survived because their cells are protected against hydroxyl radicals and cope with oxidative stress when oxygen radicals are formed during electron transport in photosystems.

Similar were the results of a study on the performance of the commercial product Photocal, composed of photocatalytic titanium dioxide, which was not efficient at preventing biofilms growth on limestone samples located outdoor in a park in Belgium.[346] After 2 years' exposure, strong change in color occurred for biocolonization and deposits (Table 5.1).

In the attempt to solve the limited application of TiO_2, a study examined TiO_2 doped with alkaline metal ions that have a larger photocatalytic spectrum.[385] The modification should improve the electron mobility of TiO_2 so that it extends its wavelength response toward the visible region. Hence, TiO_2 would exhibit high reactivity under visible light even with poor UV lighting. The study applied nondoped TiO_2 and TiO_2 doped with Ag, Fe, and Sr mixed with nano silica (dilution 1/10) on marble samples at different powder/binder ratios: 1/1, 1/10, and 1/100. All Fe-doped TiO_2 showed strong color variations while only Ag doped TiO_2 at the ratio 1/100 was suitable as it caused low color alteration. TiO_2 doped with Ag alone or in combination with Fe and Sr was effective at preventing the growth of selected Gram-positive and Gram-negative bacteria.

According to some authors[327] silver nanoparticles are not appropriate for application on stone to inhibit biological growth because they alter strongly the color of two types of concrete samples producing a pronounced darkening (ΔE^* 8.7 and 15.9) of the surfaces (Table 5.1). ΔE^* values below 4[325] are generally considered acceptable for surface treatments on stone.

A study[386] proposed the experimentation on cultural heritage objects of parvomes that are small molecules produced by living organisms ranging from bacteria to plants.[387] Many years of natural product research, coupled with recent advances and applications of genetic and genomic techniques, have revealed the presence of an enormous collection of unique small molecules that are the products of cellular metabolism. The past literature studied plant parvomes mainly for their lethal effects, disregarding concentrations and ecologically relevant functions of these molecules in the natural context.[386] Although there is still little knowledge of their functions and, in most cases, of their routes of biosynthesis, the authors proposed their application at sublethal concentrations as an antifouling strategy to affect the multicellular behavior of microorganisms.[386]

Many studies deal with the application on stones of *water repellents* and *consolidants* that are useful to prevent biofouling. These substances can inhibit microbial growth by reducing the saturation of the stone and limiting the amount of water available to organisms. A study showed that films composed by *n*-octyltriethoxysilane/tetramethyl orthosilicate (TEOS/TMOS) increased the hydrophobicity of the surface and reduced the settlement of fouling organisms.[388] The same authors developed also a coating for façades containing silica nanoparticles in an organic polymer dispersion. The coating resists "mold formation" because the hydrophilic surface allows rapid spreading of raindrops, producing a washing effect, as well as a rapid drying. The high silica content reduces adhesion of microorganisms and other particles.

Limestone samples sprayed with different consolidant solutions (acrylic emulsion, polyurethane dispersion, acrylic-epoxy polymers, alkoxysilane) were exposed outdoor in a tropical climate (Maya site at Xunantunich, Belize).[286] After 1 year, just some consolidated samples located in a sunny area had occasional black fungal deposits and white lichens. In the shade, all consolidant-treated samples showed evidence of substantial black and gray fungal growth and, in some cases, green deposits. The results indicate the importance of climatic conditions in microflora development. Similar conclusions emerged from fungal soiling of painted concrete panels exposed to various environments in Brazil for 4 years.[57] Although the paint over the panels contained the biocides carbenzadim and isothiazolinone, they could not inhibit biofilm growth in these tropical climates.

As previously mentioned, the application of methyl silane after the cleaning of a Thai Sanctuary inhibited the regrowth for more than 8 years. Instead, algae and lichens colonized the cleaned but untreated areas in direct contact with damp soil after 5 years.[168]

Two water repellents, the products Rhodorsil H224 and Protectosil SC Concentrate, efficiently prevented biological colonization on limestone samples located outdoor in a park in Belgium for 2 years, while tetraethoxysilane (TEOS) mixed with the antimicrobial agent Chitosan (*vide infra*) was not effective[346] (Table 5.1).

A 160-day-long laboratory test evaluated the ability of water repellents (silane and fluorinated compounds) at preventing colonization of alga *Graesiella emersonii* on mortar samples (Table 5.1). The treatment notably reduced the progression of colonization compared to that obtained with the control and the photocatalytic specimens (*vide infra*).[347]

Drawbacks can occur after the application of this kind of products. A study[169] observed that the hydrophobicity provided by water repellents on the surfaces of marble statues inhibited the spreading of raindrops over the surfaces. Consequently, preferential water paths formed that favored the accumulation of biomass and eventually the development of localized biofilms. The formation of these preferential water routes depends strongly on the structural characteristics of the stone surfaces. In fact, autoclaved aerated concrete samples treated with alkyl alkoxy silane showed streaking patterns of colonization when tested with an accelerated fouling experiment while white concrete specimens treated with the same water repellent did not show any streakings.[327] According to the authors, the low roughness of the latter substrate allowed an even distribution of products impeding the formation of preferential water flow paths.

Several papers suggested the application of *mixtures of hydrophobic compounds and biocides* as effective on microbial recolonization of stones.[389,390,249,391] The application can be done in a single step when the water repellent and the biocide are mixed together or in two steps when the biocide is applied before or after the water repellent.[389,391,392] While the application of a water repellent and then of a biocide on a sandstone portico prevented the regrowth of phototrophic microorganisms for 6 years, the same products did not prove effective when applied to bricks under continuous rising damp conditions.[390]

The combination of a water repellent with a biocide had a protective and preventive effect, especially on algae and bacteria, when used on mortar samples exposed to outdoor environments.[391] Some researchers reported a decrease in the performance of these treatments because of interactions between substances.[249,392] There are evidences that biocides interact with water repellents and affect the treatment's performance whether they are applied before or after them. When applied *before* water repellents, some biocides apparently hinder the polymerization and, hence, decrease the hydrophobic properties of the products.[389] However, also the application of quaternary ammonium-based biocides *after* a water repellent treatment had negative effects in terms of loss of hydrophobicity and modification of the water capillary absorption of the stone.[392] As reported by the authors, the result relates to the surfactant properties of QACs. On the contrary, the previous study[389] did not observe any effects when biocides were applied after the hydrophobic treatment. Results like this latter emerged from a study[327] focused on the performance at preventing algal fouling (*Chlorella vulgaris* var. *viridis*) on white architectural and autoclaved aerated concrete (AAC) specimens (cementitious building materials) (Table 5.1). The selected products were water repellents, biocides and their combinations—silane/siloxane mixture in water, alkyl alkoxy silane 40% (w/w) in isopropyl alcohol, 2,3,5,6-tetrachloro-4-methyl sulfonyl-pyridine in isopropyl alcohol, 3-trimethoxy silyl propyl dimethyl octadecyl ammonium chloride, and silver nanoparticles in demineralized water. The study applied a water run-off laboratory test 12-week long that produced an accelerated fouling of concretes. The run-off period started every 12 h and ran for 90 min. Porous concrete specimens treated with a silane-based water repellent followed by a chlorinated pyridine-based biocide exhibited a faster rate of drying compared to the untreated specimens. The result demonstrated that the water repellent retained its property even when combined with a biocide. Moreover, this treatment showed the best performance at preventing algal fouling on autoclaved aerated concrete samples. Only limited fouling was observed after 8 weeks of tests. For this combination, no green streaks were observed along the surfaces. Other relevant conclusions came from this study. The treatments' performance depended on bioreceptivity of the substrates. In fact, at the end of the test, no fouling developed on the white concrete samples treated with biocides and/or water repellents while algae colonized AAC specimens. The latter substrate has rough surfaces, high macroporosity, and high number of anchoring sites for microorganisms, thus algae rapidly colonized it. Moreover, many treatments on this substrate showed a loss of biocidal activity and water repellence within two to five weeks. Another negative feature observed on AAC samples was that water repellent treatments resulted in a streaking pattern of colonization as other authors[169] similarly observed on marble statues. On the other hand, no streaks occurred when the biocides were applied after the water repellents. As reported by the authors, the interesting result can be due either to improved availability of the biocides or to modification of the water repellent by the biocides, decreasing the occurrence of preferential water flow paths. Combinations of water repellents followed by biocides appeared the most effective treatment for the prevention of algal fouling on concrete with a high bioreceptivity. Differences observed in the performances of the combined treatments with a different order of application might be caused by the accessibility of the biocides. Most treatments did not cause strong color changes except for the mixture alkyl alkoxy silane + tetrachloro-4-methyl sulfonyl-pyridine on AAC samples (ΔE^* 7.9).

An innovation of this kind of mixtures was the subject of a research on water repellents chemically bound to a biocide. A study experimented trimethoxysilane coatings containing chemically bound quaternary ammonium salt (QACs) moieties.[393] Twenty different trimethoxysilane-functional QACs were synthesized and their antimicrobial activity was proved on *Escherichia coli*, *Staphylococcus aureus*, and *Candida albicans*. Compositional variables that produced heterogeneous surface morphologies provided the highest antimicrobial activity suggesting that it was primarily due to the relationship between coating chemical composition and self-assembly of QACs moieties at the coating/air interface.

Another innovation regards mixtures of polymers with natural nontoxic antifouling agents. Some derived from plants, such as capsaicin and cinnamaldehyde, and some from marine animals, such as 3-poly-alkyl pyridinium salts, *Ceramium botryocarpum* extract, and zosteric acid. They were mixed with two commercial silicone-based coatings (Silires BS OH 100, a consolidant, and Silires BS 290, a water repellent).[394,335] The researchers treated marble samples with the mixtures and tested them in the laboratory on cyanobacteria, algae, fungi, and protozoa formation. Unfortunately, the results were unsatisfactory because the mixtures did not show any preventive action on biofilm formation. Therefore, further detailed studies for defining the optimal concentration and the best condition for a constant releasing and long lasting of incorporated natural substances will be a key point for similar researches.

The treatment of limestone slabs with the product Bio Estel (a consolidant mixed with a biocide) under laboratory conditions showed very low efficacy at preventing the growth of the fungus *Ulocladium* sp. inoculated on the surfaces.[266]

A laboratory testing[395] mixed tetraethoxy silane (TEOS, frequently used by restorers as a consolidant) with chitosan (a biopolymer used for its antimicrobial potential), silver nitrate and hydrophobic silica. Treated limestone samples were inoculated with the green alga *Chlorella vulgaris*. The combination of silver nitrate and hydrophobic silica, both at high dosages, provided the best biocide effect. The addition of chitosan induced a synergistic effect, resulting in the best biocide performance at the lowest concentrations of $AgNO_3$ and hydrophobic silica.

Some studies experimented nanoparticles mixed with water repellents and consolidants, to inhibit the biological growth. Although primarily addressed to buildings, they report that the polymers functionalized with nanoparticles can be useful also for the preservation of historic monuments.

Silver nanoparticles mixed with polydimethylsiloxane were applied on mortar samples to prevent the colonization of *Chlorella vulgaris* and *Synechococcus* sp.[396] Aqueous algal and cyanobacteria cultures were pumped onto mortar through a culture-streaming test. Unlike other similar studies, this one reported the concentrations of silver nanoparticles, 0.1, 0.3, 0.5 wt%. Although the treatment performed quite well, samples exhibited some colonization after 4 weeks. However, the worst aspect was the considerable color change with ΔE^* varying 5.35–8.40 for the three concentrations (Table 5.1).

Again, silver nanoparticles (40 mg/ml) mixed with two types of consolidants were applied on sandstone and limestone samples[397] (Table 5.1). The consolidants were Primal AC33 (a mixture of methylacrylate and ethylmethacrylate) and Wacker BS 1001 (50% silane/siloxane emulsion). Both polymers were diluted to 3% final concentration with water. The researchers inoculated the samples with the bacteria *Streptomyces parvulus* and the fungus *Aspergillus niger*. Stones treated with the silicon polymer loaded with silver nanoparticles showed an elevated antimicrobial potentiality on both the microorganisms.

Another article on silver nanoparticles mixed with a silane derivative and applied on sandstone samples showed that they bind to the stone surface exhibiting a cluster disposition that is not affected by washing treatments. A surface concentration of 6.7 $\mu g/cm^2$ reduced 50–80% cell viability of *Bacillus subtilis* inoculated on the samples.[398]

Although the proven efficacy of silver nanoparticles at preventing microbial growth, they are not feasible for cultural heritage stone monuments because they caused strong color changes even at very low concentrations.

Recent approaches in the development of novel antimicrobial agents describe the application of Cu-zeolites, amorphous hydrogenated carbon films doped with copper, and Cu nanoparticles (NPs) mixed with water repellents and consolidants.[315] A study[339] reported the synthesis of CuNPs by means of a simple and reproducible electrochemical procedure called "sacrificial anode." The technique produces core–shell NPs that are morphologically stable and capable of exerting a controlled release of metal species when exposed to aqueous solutions. The nanoparticles were mixed at various concentrations (0.02%, 0.05%, 0.14%, and 0.28% w/w) with the product Estel 1100 (ethyl silicate and polysiloxane oligomers in white spirit, CTS, Italy) commonly used as water repellent/consolidant coating on stone objects. The authors applied the colloid to different calcareous stones. The 0.02% concentration did not alter the stones' color while 0.05% concentration caused a strong change (ΔE^* 10 circa). Copper atomic percentage on the surface of the nanocomposite film increased over time, reaching a plateau value after about 40 days.

Innovative mixtures of Cu nanoparticles with water repellent and consolidants (Silo 111, Acrilico 30, Estel 1000) yielded good results in terms of preventing biological colonization in the archaeological area of Fiesole (Italy) over almost 3 years[29] (Table 5.1). The mixtures were applied in situ to sandstone, marble and plaster that had been cleaned beforehand. Silo 111 and Estel 1000 apparently did not affect the substrates' color while Acrilico 30 induced an evident and persistent increase in plaster's brightness ($\Delta L \sim 7$).[29] The researchers chose copper because it has an excellent antimicrobial activity on a wide range of bacteria, algae and fungi. This ability increases when Cu is properly nanodispersed.[315] Copper affects cells producing reactive hydroxyl radicals that cause oxidation of proteins, cleavage of DNA, and RNA molecules, and membrane damage due to lipid peroxidation.[315] Likewise, in the archaeological area of Fiesole, the recolonization of stones was strongly linked to their bioreceptivity as a laboratory test (*vide infra*) showed.[327] Although the site is very favorable for biological colonization, the materials with lowest bioreceptivity were sandstone and plaster since no biological growth was detected over the 30-month period. On the contrary, marble showed a high bioreceptivity with a strong colonization by fungi in the control area. Silo 111+nanoCu showed the best performance on this substrate. The treatment was effective on biological growth for 2 years, and the areas treated with this product showed less biological growth than the others did. Moreover, marble treated with Silo 111+nanoCu showed greater water repellency throughout the entire monitoring period than the portions treated with the other products. Therefore, mixtures of Cu nanoparticles with a consolidant and a water repellent hold great promise for preventing recolonization of stone after conservation treatments.

A result confirming the limited performance of TiO_2 (*vide infra*) was that of a mixture of TiO_2 (2 wt%) with silicate/acrylic based compounds applied on mortar samples to prevent the growth of alga *Graesiella emersonii*[347] (Table 5.1). The investigation used two accelerated tests simulating different types of humidification (water capillary ascent and water run-off) under different lighting conditions. The lamps used to activate photocatalysis varied in UV intensities (1.5, 3, 7.5, 12 W/m^2). The mixtures did not reduce or slow down the growth of algae in any lighting conditions, even in the most favorable to photocatalytic reaction kinetics (7.5 and 12 W/m^2). The colonization extended overall surfaces in less than 160 days. The authors reported that the results can relate to extra polymeric substances and dead algae that create a barrier between the photocatalytic titanium dioxide and the algal cell wall. Moreover, the algal layer diminishes the amount of UV light that reaches the photocatalytic surface limiting the diffusion of the reactive oxygen species on the algae. In addition, the effect of a possible binding of TiO_2 to silicate/acrylic based compounds should be considered because TiO_2 proved to be very efficient in other studies when properly irradiated by UV light, even after a long period.

Curiously, a very similar study employing much less amount of TiO_2 came to opposite conclusions.[399] Titanium dioxide and zinc oxide nanoparticles (both at a concentration <0.1 wt%) were mixed with silane/siloxane based water repellents combined into oil-in-water (o/w) emulsions (Table 5.1). They are a new trend in façade remediation being water-based and thus reducing volatile organic compound (VOC) emissions. The researchers pumped on treated mortar slabs algae and cyanobacteria through a culture streaming test. All treated samples showed reduced surface coverage over the 8-week-long test compared to the untreated ones. Moreover, the treatment had negligible impact on substrate's color. Nanoparticle treatments performed better than the control emulsion without nanoparticles. They also improved water repellence of the control emulsions. According to the authors, this is because the metal oxides have a high concentration of polar groups that act as bonding sites for the silane and siloxane components. This allows a more structured and uniform interfacial morphology; hydrophilic domains are attracted to the interior of the system, with hydrophobic parts of the treatment governing the interfacial characteristics. Zinc oxide tended to aggregate more than the titanium dioxide, reducing distribution potential during application. The contrasting results of the two studies[347,399] likely relate to the different time length of the tests, 160 days and eight-weeks respectively. One study[347] reported that the latency period with no sign of colonization lasted 50–100 days, depending on the lighting conditions. Therefore, it is likely that the treatment was successful in the other study[399] just because it was shorter. However, small spots started appearing after eight weeks.

Although there are many studies on the antibacterial properties of zinc oxide nanoparticles that have found application as antimicrobial agents in many fields, the mechanism of their antimicrobial action needs further examinations.[400] The antibacterial effect of zinc oxide nanoparticles (ZnO–NPs) is much greater on Gram-positive bacteria than on Gram-negative ones, probably due to their different cell membrane structure.[400] A study showed that ZnO–NPs embedded in polymers do not change the consolidant/water repellent properties. Commercially available consolidants and water repellents were selected (Estel 1000 and Silo111 supplied by CTS, Italy). The researchers applied polymers and nanostructured coatings (0.5% w/w of ZnO–NPs) by brush to samples of three calcareous stones. When in contact with artificial rain, nanomaterials released zinc in the range of ppm amounts. About chromatic changes upon curing of the consolidant, ΔE^* values did not exceed 4.5. These results represent an improvement compared to copper-based coatings, since the ΔE^* values are lower than those of copper nanoparticles, despite the amount of embedded ZnO–NPs was much higher. Moreover, ZnO–NPs showed a good performance on the fungus *Aspergillus niger*.

In the attempt to combine the antifungal properties of ZnO or Zn^{2+} ions with the consolidant ability of $Ca(OH)_2$, a study[401] applied calcium zinc hydroxide dihydrate ($Ca[Zn(OH)_3]_2 \cdot 2H_2O$) on three limestones. The performance of the protective coating, which well adhered to all the substrates, was excellent as it inhibited the growth of the fungi *Penicillium oxalicum* and *Aspergillus niger* in the dark and under illumination. On the contrary, fungi grew and penetrated all the control samples that did not have the protective coating.

An *in vitro* study tested protective nanocomposite coatings based on unsaturated polyester resin and multiwall carbon nanotubes (CNT) as antifouling compounds.[402] Polymer matrix composite materials based on a thermoset resin (like polyester) and a fiber reinforcement (such as glass, carbon), or other fillers (such as silicon, calcium carbonate, etc.) offer many superior properties in comparison with conventional materials (high strength to weight ratio, superior corrosion resistance, chemical inertia, degradation resistance to UV rays, insulation properties, stability to shrinkage, etc.). The test used a bacterial suspension of *Pseudomonas stutzeri* (Gram negative) and *Micrococcus luteus* (Gram positive). The action of the coating was bactericidal (97.3%) on *M. luteus*, but not on *P. stutzeri*, with a bactericidal effect of only 3%. Moreover, the pure polyester resin without CNT showed no bactericidal effect toward both microorganisms.

Although the results on application of nanoparticles mixed with water repellents and consolidants for preventing biological growth on stones are promising, and the topic is very important having strong connections with maintenance and minimum intervention, it is not possible to draw general conclusions because most researches were performed under laboratory conditions. Therefore, controlled in situ trials of potential for biodeterioration of these new coatings are a good subject for future applied research.

The interaction between synthetic polymers and biofilms is an important aspect connected to the performance of these substances. The issue has implications both in industrial fields and in conservation of cultural heritage.[403] Outdoor stone monuments are susceptible to damage in many environments and organic as well as inorganic products are widely used for their protection and consolidation. When developing strategies for the protection of outdoor stone monuments, it is essential to consider both the effects of weathering and the biodeterioration of these products[404,350] even if the assessment of polymers biodeterioration is a difficult process as different factors can contribute.[403] Once microorganisms develop on a treated surface, they can penetrate the polymer matrix with microbial filaments causing loss of its stability. In addition, they can indirectly swell the materials through water accumulation.[405]

Many polymers used for water repellency or for consolidation or as additives for filling materials and mortars are degradable by microorganisms.[404,235] Polymers are potential substrates for heterotrophic microorganisms including bacteria and fungi that secrete enzymes such as extra cellular and intracellular depolymerases, and organic acids during their metabolic processes.[353] These substances can react with the polymers, thus increasing the rate of degradation[406] that largely depends on the chemical structure of the polymers (e.g., C–C and other types of bonds, molecular weights, structures and configuration), as well as on the participating microorganisms and environmental conditions.[406] High molecular weight polymers are less biodegradable or degrade at slower rates than those with low molecular weights[353,235] such as monomers, dimers, and oligomers of a polymer. Microbial colonization can degrade several materials including acrylates, polyurethane, polyvinyl acetates, natural/synthetic rubbers, silicon organic plastics and epoxies.[353,406] Other factors favoring microbial colonization of polymers are changes of stone physical properties and of microclimate caused by the coatings.[235]

The hydrophobic effect of water repellents has a variable duration, in many cases years, in other cases months or even weeks. The duration depends on the type of product, its concentration, and the amount applied on stone.[30] Once films of these materials lose flexibility, microfissures occur. Consequently, water permeates through the fissures and accumulates in the bulk of the stone.[235] When surfaces lose their water repellency, biological recolonization develops on areas that are directly wetted by the rain.[321]

Although many studies examined the biodeterioration resistance of water repellents and consolidants (described later), assessments of biodeterioration have rarely been quantitative because the presence of bacteria or fungi on surfaces has generally been assumed as biodeterioration. Thus, many studies have dealt with the potential for biodeterioration rather than biodeterioration itself.[353]

Most studies involved simulation tests of microbial growth on materials under laboratory condition. Although the evaluation of polymers biodeterioration through in situ trials holds much interest, researchers did not adequately address it.

Laboratory tests were conducted on the resistance to microorganisms of synthetic polymers commonly used in conservation/restoration, that is, Mowiol 4-88 and Mowiol 4-98 (polyvinyl alcohol derivate), Mowilith (vinyl acetate), Klucel (hydroxylpropylcellulose), Primal (acrylic dispersion), and Paraloid B-72.[407] Microorganisms grew on all synthetic polymers. Mowilol 4-88 and Mowiol 4-98 gave the worst results, while there was a poor colonization of Paraloid B-72. The authors thus consider the use of Mowiol products as problematic and recommend preliminary investigation about microbiological assessment before using them.

The ability of three commercial antigraffiti products, Fluorolink P56 (a polyurethane fluoropolymer), Weather Seal Blok-Guard and Graffiti Control (a solvent-based silicone elastomer), and Protectosil Antigraffiti (a water-based fluorosilane), to provide protection from graffiti attack was evaluated on sandstone and marble.[408] To investigate the products' potential to be a source for microbial growth, the researchers contaminated them with soil microbes. No growth was present on the products except for Blok-Guard that was degraded by microorganisms.

Aged polymers are more prone to biodeterioration than fresh products. Marble surfaces of Milan cathedral were treated since 1972 with acrylic resins (poly-isobutyl methacrylate).[409] UV irradiation caused molecules' modification producing oligomers. Fungi can attack them more easily than high-molecular-weight polymers. In fact, dematiaceous meristematic fungi widely developed on treated marble of the cathedral.[409] The study revealed that biodeteriorated acrylics differed chemically from the same aged but nonbiodeteriorated resins, thereby showing that resin decay processes are enhanced by biological growth.[409] Similarly, the facade of Tempio Malatestiano (Rimini, Italy) treated with acrylic resins, presented black fungal growth in cracks and fissures.[410]

Regarding the prevention of *plants* growth, a very peculiar problem worth mentioning is that of olive tree seedlings that grew on the monuments in the archaeological site of Eleusis (Greece).[275] The seed dispersal came from olive trees that are part of the landscape of the site. Spraying the trees once in a year with 400 mg/l of naphthalinacetic acid (Apponon) resulted in complete fruit abortion.[275] The authors suggest the application of this method for olive tree spreading control, and thus for prevention of olive growth, in Mediterranean archaeological sites.

Plants can play also a positive role in conservation of stone objects. It is the case of *soft capping* that consists in the use of grass, other plants and soil to cover horizontal masonry surfaces to protect the wall below.[411] A project conducted on ruins located in the United Kingdom under a wide range of environmental conditions focused on the use of soft capping as a method for conserving historic walls[411] (Fig. 5.3).

The traditional way of preservation of ruins uses hard capping, wall tops constructed from stone and mortar. For example, sealing the top of the walls of monuments in Argentina with a hydraulic mortar prevented recolonization for a long time.[372] Although it is a useful method, it has some drawbacks. For example, where the wall tops are consolidated using cements, these tend to crack open with frosts, allowing moisture ingress, which results in freeze–thaw damage. The mentioned study[411] on ruins located in the United Kingdom showed that soft capping provided a better thermal blanket than hard capping. Soft capping reduced daily temperature variations, decreasing the threat of damage from thermal expansion and contraction specially to walls that have already been consolidated using hard mortars. It also reduced the risk of frost damage because it minimized temperature drops below 0°C in substrates under it. Moreover, it appeared to reduce short-term moisture fluctuations within walls, and water running down the face. Further research is proceeding to study in more detail the

FIGURE 5.3 Soft capping on the ruins of Hailes Abbey (UK).

variability of moisture regimes under hard and soft caps. About grasses used in the soft capping, it is advisable that they are drought tolerant and can regenerate well from their own seed. Damage to the stones from soft caps is not likely as grasses and most herbaceous plants do not produce secondary thickening and therefore cannot cause such damage. Concerns that soft capping will become easily detached from the underlying walls in windy environments have proved unfounded. However, soft caps present some drawbacks. One is that the soil can be washed off if the grass dies forming bare stone areas, or can be removed by wildlife. Moreover, at the edges of soft capping, the soil dries out easily and grass grows with difficulty.

Possibly, its use is restricted to countries where rainfall is frequent in the year or at least seasonally. According to the author,[411] soft capping may be particularly beneficial where there is a risk of surface erosion from run-off and if there is the threat of frost damage or differential thermal expansion. Under other regimes of deterioration, soft capping may be less effective. Moreover, regarding the perceived value of a monument, that is how it looks, soft capping may not always be appropriate. Maintenance is required, involving the removal of woody species whose roots might penetrate the stonework. Further research is underway by the same author[411] on the use of alternative capping, such as sedum mats and seeded mats, which may prove to be more robust and easy to install under some circumstances. On the other hand, soft wall capping is relatively cheap and easy to install and maintain, in comparison with hard capping. Of course, each individual situation needs possible modifications to any general approach, depending on the local climatic conditions, the nature of the walls, and availability of materials and personnel.[411]

Conservators carry out *reburials* as a protection of complex objects such as mosaics, plasters, and fragile masonry.[364] Discussions about the appropriateness of reburial refer to issues of its length and ease of access.[247] The term "reburial" describes a preventive in situ conservation intervention, which may include specialized techniques and materials. It can refer also to a post-excavation recovering in archaeological sites.[273] The reasons underlying this conservative choice are different. Reburial can be a response to excavated mosaics left exposed and deteriorating.[412] With resources for shelter design and construction being beyond the means of many countries, reburial has become the more practical, least expensive way for national authorities to protect and preserve in situ mosaics.[412] Reburials are also useful to protect delicate objects from frost damage and plant roots penetration. They provide, even after decades of exposure, an important degree of protection, drastically reducing the mechanical and chemical stresses that the materials suffer when they are directly exposed to the environment. Therefore, a protective strategy against both biological growth and loss of material in archeological sites often include reburials.[247]

A common reburial method consists of a layer of sand or clay over a separation membrane that is in contact with the object's surfaces.[364] The membranes are mostly made of polyester fiber geotextiles (fabrics designed to survive underground), now widely available. Their flexibility in system design makes reburial, whether temporary or long term, an attractive option for conservators.[273,247] The sand layer is either left exposed or covered with a layer of plastic screening and a layer of gravel.[412] Layers of quartz sand, LECA (light expanded clay aggregate) and gravel, applied on a mosaic floor in the archaeological site of Eleusis (Greece), successfully protected the object and totally inhibited weed development.[275] A study focused on reburials of mosaics in Tunisia[412] evaluated the overall performance of the system at preventing deterioration of the mosaics. Weed root growth was the main observable alteration. In some sites, roots accumulated under the separation membranes placed in contact with the mosaic. The main reason for this colonization was the trapping of moisture underneath the membranes. Geotextiles proved to be the most effective barrier against root penetration, as they are more resistant than other materials, although certain types of plant roots were able to pass through them. The use of more significant fill depths and of a top layer of gravels were two effective ways to reduce roots' penetration.[412] In such a context, maintenance after reburial to remove vegetation also in areas close to the objects is an important action that can include seasonal use of an appropriate herbicide.[412]

The control of feral *pigeons and birds* includes measures and methods that discourage them from roosting and nesting on buildings. As these methods aimed at preventing birds' presence, they are discussed in this section. Pest control companies offer different deterrent systems, of widely varying efficacy, for proofing buildings against birds.[206] Attention should be paid in the choice of a system or a device because some of them, allowed for use in a country, would be illegal in another country.[206] There are several different sorts of physical barriers and deterrents available on the market that are durable and unobtrusive.[203] The most effective types of protection for monuments are metal wires, synthetic fiber nets, pins, and electromagnetic systems.[205,242] Nylon nettings are installed to protect an area rather than a specific surface, excluding pigeons from favored roosting and nesting zones. Nettings are also in usage to prevent pigeons' access to a building through open windows and loggias.

Metal spikes systems consist of stainless steel wires on a plastic base. There are different systems for heavy, medium or light pressure from birds and it is important to choose the right one.[203] These systems are substantially successful, but a scheduled maintenance is necessary considering pigeons' adaptive behavioral strategies to displace the spikes. Figure 5.4 shows the ability of pigeons to roost even within metal spikes.

The electric shock system produces a low voltage electrical shock when birds land on its surface. The shock does not cause actual bodily injury to birds, but "induces" them to stay away from the area. The device is designed to frighten the birds. A drawback is that the system can fail for an inadequate supply of power to the track. Debris and leaves on the track are also a major factor that results in failure of the device because the birds do not make contact with wires. Therefore, electric shock systems require regular maintenance and inspection to ensure that debris does not compromise the track. Where the system is powered by current supplied via a battery, regular inspection is required to recharge the battery.

FIGURE 5.4 Pigeons roost even within metal spikes as the presence of guano shows.

Sticky repellent gels have been used extensively to prevent birds landing on external surfaces such as windowsills or architectural features.[203] They make the target species feel uncomfortable when landing on them because their surface remains soft and unstable for the birds to walk on. Repellent gels are transparent substances designed to be effective at normal temperatures. They are less effective at temperatures below 0°C and above 40°C. They have a limited lifespan (from 2 months up to 1 year) and can leave a dark unsightly deposit on porous stones.[203]

Other methods such as ultrasound light signals, models of birds of prey, olfactory repulses, narcotics, hunting, sterilization, or expulsion are less effective.[205]

Another method known as "territorial deterrence" uses hawks or falcons to disperse pigeons. Hawks and falcons are predators of bird species and their presence is enough to discourage these birds from nesting in the area, as they no longer consider it safe. The providers of the service regularly brought the trained raptors to the site to simulate their permanent presence. This approach has been used in waste disposal sites, landfill sites, hospitals, and airports for many years as well as in historic buildings and town centers. For example, in the archeological site of Herculaneum (Italy) the method involves intensive periods of falcon flights to relocate the pigeons, followed by regular "maintenance" flights to prevent the settling of new pigeon populations.[204]

KEYWORDS

- **aerobiology**
- **illumination managements in caves**
- **filters**
- **outdoor environments**
- **protective shelters**
- **copper or zinc strips**
- **biocides**
- **nanoparticles**
- **TiO_2**
- **water repellents**
- **consolidants**
- **soft capping**
- **reburials**
- **prevention of birds' presence**

CHAPTER 6

BIOREMEDIATION

CONTENTS

ABSTRACT

The chapter addresses biological growth from its positive side, as a potential help for conservation. It includes the revision of recent studies on bioremediation applied to conservation of cultural heritage involving the use of microorganisms to remove various kinds of materials present on the surfaces.

BIOREMEDIATION

The application of bioremediation to conservation of cultural heritage involves the use of microorganisms to remove various kinds of materials present on the surfaces. In the case of stone cultural properties, the materials to remove can be sulfates, nitrates and organic substances. The method takes advantage of the ability of microorganisms to use various organic/inorganic compounds for growth. Bioremediation as an alternative cleaning technology employs mainly bacteria belonging to the genera *Desulfovibrio* and *Pseudomonas*. The sulfate reducing bacteria *Desulfovibrio desulfuricans* and *D. vulgaris* reduce sulfate to gaseous hydrogen sulfide removing black crusts.[413,80] Rapid bacterial processes correspond to a specific optimum temperature. Microorganisms' activity has also minimum and maximum limits of tolerance for temperature. Values below the former are inappropriate for cleaning while values above the latter cause microorganisms' dead. Researchers applied successfully bacterial cleaning at temperatures ranging 6–37°C.

As the method uses poultices, it is unsuitable for substrates that can deteriorate in contact with them. The following delivery materials for in situ application have been used: sepiolite, hydrobiogel-97, cotton wool, carbogel, mortar and alginate beads, agar, laponite, and arbocel. The choice depends on several factors: water retention; compatibility with microorganisms; adhesion capabilities to the object; state of conservation of the object. Moreover, the delivery system should be quickly and easily prepared, but also easily applied and removed at the end of the treatment. Carbogel showed very good performance for the high number of viable bacteria it can retain during the application.[414] It ensures an easy application of bacteria and a good contact between the cells and the surfaces to be treated. Moreover, the cells are removed easily and completely after the treatment. Carbogel entraps bacterial cells in about 10 min during gel formation, whereas in the case of inorganic matrixes the process lasts at least 2 days because bacteria must colonize the matrix by growing and adhering to the particles' surface.

Bioremediation offers several advantages over chemical and physical cleaning methods: It is very selective in removing specific compounds; it allows a complete control of the process by the operator; the costs are often lower in comparison with those of other techniques; it is an environmental-friendly method; it is relatively simple.[415]

The treatment of black crusts present on the surfaces of two sculptures made of oolitic limestone in the courtyard of the Buonconsiglio Castle (Trento, Italy) was performed applying a poultice with sulfate-reducing bacteria, which removed the chemical alteration preserving the patina present over the stone.[416] An in situ trial applied biocleaning on black crusts present on green serpentine, red marlstone and Carrara white marble of external walls of the Cathedral of Firenze (Italy).[417] The trial compared this technique with other cleaning methods, that is, a chemical method (ammonium carbonate poultice) and laser (1064 nm, Nd:YAG laser). Chemical cleaning produced a nonhomogeneous crust removal. Laser cleaning gave the best results on red stone. Biocleaning was the most controllable process and the most efficient in sulfate removal.

A study reports the bioremoval of nitrate and sulfate salts from the tuff of Matera Cathedral (Italy) using nitrate and sulfate reducing bacteria.[418] *Pseudomonas pseudoalcaligenes* and *Desulfovibrio vulgaris* cells were applied as poultices on a vertical wall. The biological procedure efficiently and homogeneously removed deposits and salt alterations.

Another successful example of biocleaning application is the restoration of the fresco *Stories of the Holy Fathers* at Camposanto Monumentale (Pisa, Italy).[419] Traces of casein and animal glue from previous restorations were present on the surfaces and on the back. Over time, the protein materials altered the painting and became resistant to commonly used solvents. The application for 2 h of viable bacterial cells of *Pseudomonas stutzeri* removed all casein and animal glue residues while did not affect painting's materials. Data on microbial monitoring showed that viable cells were not present on the fresco after the biocleaning. Moreover, no negative effects caused by bacterial metabolism were detected.

Nonetheless, the technique has some drawbacks and limitations. It is a time-consuming process, as biocleaning of thick layers needs numerous applications.[414] To solve the problem, a study applied a strain of *Desulfovibrio vulgaris* coupled with a nonionic detergent on columns and statues altered by black crusts. The synergy of the two treatments completely removed the black crusts, with few biological applications and 70% reduction of the total cleaning time.[414]

Another limitation is that the removal of new compounds (i.e., not yet experimented as substrates for the bacteria) needs research efforts for the selection of the microbial strain efficient to this purpose. Although the technique showed interesting and promising results, its advantages and limitations over other cleaning procedures have not been completely explained.[420]

Mineralization capacities of microorganisms can be used in enforcing or repairing construction materials. A range of opportunities emerged from the use of bacteria to influence material properties. Evidence of bacteria involvement in calcium carbonate precipitation, that is, biodeposition, has led to the exploration of this process in the fields of construction materials and conservation of monuments.[421] Recently, some researchers proposed bacterially induced carbonate precipitation as an environment-friendly method to protect and consolidate decayed stone monuments and objects. The method relies on the bacterially induced formation of a compatible carbonate precipitate on limestone and concrete.[421] In addition, the researchers proposed the use of microbial-induced carbonates as a binder material, that is, biocementation, for the improvement of compressive strength and durability of cementitious materials, and for remediation of cracks.

The task is too large for this book to set out a complete vision for bacterially induced carbonate precipitation. Moreover, the subject is far from the context of the book. Thus, only recent examples are discussed here to report some aspects interesting for stone conservation. For a complete description and elucidation of this issue, the review article by De Muynk et al. (2010)[421] provides an in-depth comparison of scientific approaches by different research groups.

The hydrolysis of urea by bacteria was selected as a very suitable pathway to produce carbonate ions due to the creation of an alkaline environment that favored the precipitation of calcite.[422] The hydrolysis of urea has several advantages over other carbonate generating pathways as it is easily controlled and it can produce shortly high amounts of carbonate. Microorganisms directly participated in the calcite precipitation by providing a nucleation site.[422] In fact, in the ureolytic-induced carbonate precipitation, the bacterial cell wall, with its negative charge, attracts calcium ions present in the solution. Upon addition of urea, the bacteria dissolve and release in the microenvironment inorganic carbon and ammonium. As there is a localized supersaturation of calcium ions, precipitation of calcium carbonate on the bacterial cell wall occurs. When the cell is completely encapsulated, it dies. Urea-utilizing bacteria such as *Sporosarcina pasteurii* and *Sporosarcina ureae* are commonly isolated from soil, water, etc.[422] The ureolytic sludge containing bacteria, urea, nutrient broth and an external calcium source, showed good results when applied on porous mortar samples reducing by five times the amount of water absorbed in 200 h in comparison with untreated samples.[423] XRD analyses identified two calcium carbonate polymorphs, calcite, and vaterite.

This treatment was successful also in repairing cracks of concrete samples. *Bacillus sphaericus* cells were immobilized on diatomaceous earth or inside microcapsules to protect them from the alkaline pH conditions of the concrete. This treatment enabled self-healing of cracks -about 1 mm width- in three weeks.[424] The treatment of surfaces of concrete samples produced an increased resistance toward water absorption, carbonation, chloride penetration, freezing, and thawing.[424] Porous limestone samples were consolidated at depths up to 30 mm.

Various biocalcification treatments were applied on fiber cement samples. They were immersed in a culture of living or dead cells of *Bacillus sphaericus*. Then, they were incubated in a medium containing urea and calcium chloride or in a medium containing calcium acetate, glucose, and yeast extract. A third medium was similar to the latter with the addition of urea. The treated samples were then exposed in a wooded area of São Paulo (Brazil) for 22 months to assess their susceptibility to biofilm formation.[425] Fungi and phototrophs fouled more intensely carbonated samples than noncarbonated ones. Samples treated with the third medium and live *B. sphaericus* were those best resisting to biodeterioration. This result related to low water absorption, porosity and surface hydrophilicity, which in turn were linked to the smaller size of crystals compared to that of other biocalcifying treatments.[425] There was no apparent degradation of the crystals by microbial cells.

As reported by other authors,[426,427] the application of viable bacteria and organic nutrients can be inappropriate for cultural heritage objects. Uncontrolled growth of environmental microbes, such as airborne fungi, can be promoted by the supplied nutrients producing stained patches on treated stones, while several cells of the applied strains could be still viable in the biocalcite up to 1 year.[427] Therefore, the researchers induced $CaCO_3$ mineralization by a bacteria-mediated system in absence of viable cells. *Bacillus subtilis* dead cells and its bacterial cell wall fraction (BCF) functioned as nucleation sites of calcite crystallization in solution. BCF treatment on dolomite stone samples significantly decreased the water absorption (up to 16.7%). The treatment, tested also on walls of a church located in Angera (Italy), caused little cohesion increase showing the potential of this application, even though further improvements are needed.[427]

The results of the above-mentioned studies show that the feasibility of the biodeposition treatment in situ largely depends on the time required for carbonate production. The application of calcinogenic bacteria needs long times for carbonate precipitation to occur. This means that the building material remains wet for long periods because microorganisms require a minimum amount of water to remain active. This situation causes biofilm formation and EPS production. The aspect has important economic consequences as it increases the costs of the treatment. The metabolic pathway that involved the hydrolysis of urea appears the fastest way to produce carbonate ions and hence precipitation of calcium carbonate. This is because the hydrolysis of urea is a very rapid process and depends on one enzyme only. Therefore, no additional nutrients are necessary for the long-term maintenance of the bacterial activity.

Finally, it is worth mentioning researches on alternative biological treatments for copper alloys objects.[428] Conservators working in the field of metal conservation commonly face problems connected to corrosion of iron and copper alloys. Metal artworks are mainly composed of these materials. Corrosion is an electrochemical process where the metal is oxidized by interaction with the environment, resulting in the metal return to its most stable oxidative state. Some researchers developed specific biotechnological methods aiming at modifying existing corrosion products into more stable compounds. Taking advantage of peculiar characteristics of selected fungal strains, they produced compact green patinas of copper oxalates that have a high degree of insolubility and chemical stability even in acid atmospheres (pH 3), providing bronze surfaces with good protection. The results demonstrated a different weathering behavior of the biopatina compared to standard treatments such as waxes or chemical inhibitors of corrosion.

KEYWORDS

- **bioremediation**
- **bacteria**
- **biocleaning**
- **biodeposition**
- **biocementation**
- **biopatina on metal artworks**

CHAPTER 7

SCIENTIFIC EXAMINATIONS

CONTENTS

ABSTRACT

This chapter describes and discusses the analytical techniques—both noninvasive and micro-invasive—suitable to assess the nature of biological growth, its effects on material decay and the efficacy of control methods. A detailed description of all the analytical procedures is far beyond the scope of this book. However, for a better understanding of the results' validity, this chapter includes a summary of the most useful techniques. It links long established and widely used measurement methods to innovative new technologies, explaining how they help facilitate understanding of the biological growth and its effects on stones.

In situ monitoring of the performance of methods and products for the control of biodeteriogens is discussed.

7.1 ASSESSMENT OF THE BIOLOGICAL GROWTH AND ITS EFFECT ON STONES

An appropriate assessment of biodeterioration requires a combination of microbiological investigations, analysis of macroflora, surface examination, and materials characterization. A correct description of the deterioration patterns, including biological growth, is an essential requisite when studying stone objects to understand the problems, to identify conservation needs, and to define conservation actions.[429] Important are also techniques that analyze in situ ecological parameters favorable to biological growth. Portable unilateral nuclear magnetic resonance (NMR) was very useful at mapping the moisture distribution in a deteriorated wall painting located in St. Clement Basilica, Rome (Italy).[430,431] Another technique, a multispectral imaging, provided heat and water distribution maps of an object. A study applied this technique in a catacomb (the tomb of Pancratii Roman Family, Rome, Italy) and compared the results with the analysis of the ecological factors affecting microbes' development. The combined results allowed a valuable prediction of microbial and fungal spreading on catacomb's surfaces.[129]

The study of the alterations caused by biofilms and macroflora on stones usually involves[115,432,433]:

- Examination of the biological covering and development,

- Sampling of the alteration,
- Identification of the microorganisms present in the biofilms,
- Identification of lichens, mosses, and plants,
- Microscopic observation of the interface biofilm, lichen/material,
- Elemental and mineral analysis of the damaged material,
- Final interpretation including the correlation between the morphological and metabolic properties of the identified organisms, the decay morphology, and the composition of the altered material.

The quantification of biofilms and lichens coverings is relevant to evaluate the state of conservation of buildings and objects. A study[434] applied image analysis by color-based pixel classification on the surfaces of marble, travertine and mortar stonework colonized by lichens. High-resolution images were acquired with a scanner, thus avoiding invasive surveys, and the percentage cover of lichens was subsequently measured in the laboratory using the WinCAM Pro 2007d software (Regent's Instruments). A similar method was applied to evaluate algal growth on mortar surfaces treated with water repellent and photocatalytic substances.[347] The method is of great help because statistical analysis quantifies differences between areas and thus it avoids subjective visual judgments.[47]

A range of noninvasive techniques that use different sterile tools (brushes, direct agar, loop, needle, cotton swab, adhesive tape strips, and velvet) allow the sampling of microorganisms[435,115,436] (Table 7.1). The object's condition is often an important factor in the selection of sampling methods. Moreover, the chosen technique can be highly influential in determining the quality of a sample available for cultivation. Direct agar technique transfers microorganisms directly from the surface to a nutrient source, that is, agar plates, where the level of agar is above the rim of the Petri dish.[115] Using adhesive tape strips, the operator gently applies a strip of it on the surface. The tape is then removed and placed on sterile glass microscope slides that are kept in a sterile box until arrival in the laboratory[437,438] (Fig. 7.1). The operator cuts the strips into small pieces and arranges them for microscopic and cultural examination. Light microscopy is carried out by adding a drop of sterile water or other liquid, depending on the technique applied, on the strip and placing a glass slide on the reverse of the tape to keep it flat during examination.[437]

TABLE 7.1 Sampling Methods of Microorganisms—Natural and Artificial Stones.

	When use them	Analyses
Non-invasive methods		
Cotton swab, needle	Superficial stains, microbiological growths	Microscopic observation, qualitative cultures
Agar fingerprint, velvet	Superficial stains, microbiological growths	Qualitative cultures
Lancet (for biomass only)	Thick patinas, organisms growing inside pitting, cavities, among crystals	Microscopic observation, qualitative and quantitative cultures
Adhesive tape strip	Superficial stains, microbiological growths	Microscopic observation, qualitative cultures, identification using molecular probes (FISH)
Invasive methods		
Lancet (for biomass and substrate)	To sample small amounts of biomass + substrate	Microscopic observation, qualitative and quantitative cultures, DNA, and RNA extraction for molecular examinations
Scalpel	To sample amounts of material (50 mg–1 g or more) sufficient to perform the analytical sequence	Microscopic observation, qualitative and quantitative cultures, DNA and RNA extraction for molecular examinations, thin sections, cross sections

In many cases, minimal invasive sampling methods are necessary. The objective of such sampling is to extract representative material large enough for analysis but small enough to avoid damage to the object. The sampling includes surface scraping and removal of fragments with tools that must be sterile in the case of microorganisms' colonization (Table 7.1).

The analysis of fragments can provide information on the spatial distribution of the organisms assessing those that grow over the surfaces and those dwelling deeper in the stone. The selection of sampling sites is critical and the description of environmental conditions must detail any information useful to accomplish an appropriate interpretation of the results. The information gained by careful examination of the sample can be often extrapolated to the whole object avoiding the need for further sampling.

The laboratory tests include the application of many techniques to the samples collected from the field. The literature reports the use of microscopic and chemical analysis, culture-dependent methods of identification and function, and culture-independent genetic profiling. The techniques permit, at different levels, insights into the organisms, the role they are playing, and the structure of the community.[77]

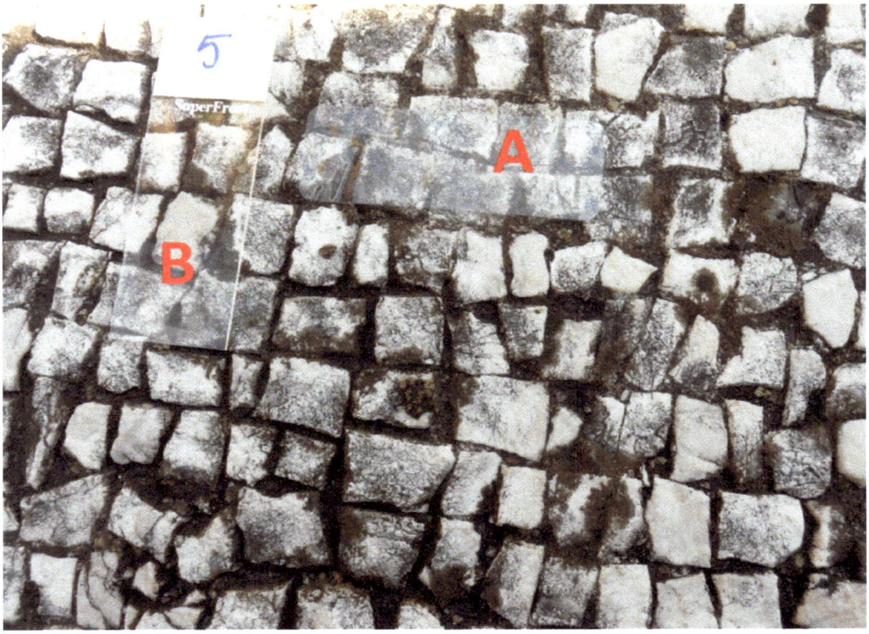

FIGURE 7.1 Sampling using adhesive tape on a mosaic (Reggio Calabria, Italy). The adhesive tape strip is applied on the mosaic surface (A). The tape is then removed and placed on the sterile glass microscope slide (B). Photo courtesy of Clara Urzí.

Microbiological studies on biodeterioration in museums or monuments are still carried out by means of classical culture-dependent tests. The methods imply the inoculation of microorganisms in appropriate cultural media (liquid solutions or solid-agar-hydrated gels containing nutrients). The growth rates in liquid media are determined by measuring the optical density with a spectrophotometer,[95] while on agar plates colonies are counted after incubation of small drops of the microbial suspension. The concentration of cells in the original sample may be calculated as colony forming units per cubic centimeter (CFU cm^{-3}).[435] When it is necessary to know the size of a microbial community, observations under optical microscope give the total number of cells (live + dead), and methods using culture-dependent tests enumerate viable cells, of course those capable of reproduction on artificial media.[435] Regarding bacteria, these traditional methods cultured and identified less than 1% of the total present in samples,[115,144,77] while they are quite effective for fungi identifying more than 70 % of them.[144] Therefore, culture-dependent approaches are still extremely useful in mycology. Analysis of specific biological molecules such as proteins, phospholipids, nucleic acids, chlorophyll, and enzymes may detect and quantify cells as well.[435]

Although only recently applied in the field on stone conservation, the advent of culture-independent techniques has been very important because they permitted dramatic increases in the number of bacteria identified from many environments and gave information about the phylogenetic composition of a microbial population without any cultivation step.[439,77,436] The molecular techniques provided also a deeper insight and understanding of fungal community structures and their consequences for the materials.

DNA-based molecular techniques analyze ribosomal DNA present in the genome of all living organisms.[440,128,435,436] Ribosomal genes contain regions where "signature sequences" have been identified for each domain of life (Archaea, Bacteria, Eucarya) as well as for microbial groups, families, genera, and species. The polymerase chain reaction (PCR) DNA amplification technique produces multiple copies of these DNA sequences by means of *in-vitro* amplification of nanograms of DNA extracted and purified from very small biofilm samples (<1 mg). Electrophoresis in denaturing gradient gels (DGGE) then separates the sequences producing characteristic band patterns (genetic fingerprints). The bands are then purified, and their sequence is compared to DNA databases to identify the microorganisms. For studies of microbial communities colonizing artworks, DGGE is the technique most often used. Another use of PCR-produced DNAs is the construction of DNA libraries that contain clones carrying ribosomal DNA fragments. After a selection of the clones, the inserted DNA is sequenced and used for identifying the microorganisms.[439] This procedure is more time-consuming but allows obtaining longer DNA sequences for a more precise microbial identification.[128]

Improvements in molecular studies showed also the advantages of RNA-based molecular analyses.[439,144] While DNA analysis shows what microorganisms are present in the sample, RNA analysis provides information on their metabolic activity, thus on their active participation in the microbial colonization. In fact, the levels of RNA in a cell are proportional to the need of that cell for synthesizing proteins required for metabolism.[439, 441,144] Methodological limitations are still a problem for a widespread application of these analyses.[439]

Adhesive tape strips serve also for the direct application of another molecular technique, fluorescence in situ hybridization (FISH). It is based on the hybridization with the complementary DNA target sequence of DNA-probes labeled with fluorochromes.[442,443] The application of FISH technique directly on microorganisms captured by adhesive tape strips added another advantage to this technique because it can visualize microbial colonization patterns and community composition "in-situ," that is directly on the tape. Since rigid fungal cell walls often impede FISH working, recently peptide nucleic acid (PNA) were applied for fluorescent in-situ detection of fungi.[144] PNA probes are synthetic DNA mimics, where a neutral polyamide backbone replaces the negatively charged DNA backbone. Therefore, PNA probes better bind to complementary targets and penetrate fungal cell walls more easily.[144] Unlike most molecular methods, FISH allows a quantification of the microbial population because fluorescent gene probes make it possible to visualize microorganisms under the microscope. For example, the technique was effective to gain insight into glass biofilm composition and density.[444]

Despites their undoubtable utility, the molecular-based methods have limitations. The DNA extraction is a critical step when applied to complex samples, and PCR-based methods inherently produce a bias in the result due to preferential amplification of some target DNA.[128] Another limitation, already mentioned, is that DNA-based methods do not give information on the organisms' physiology or viability.[128,48,439] When applied to microbes that colonize painted surfaces, the composition of the paint film, for example pigments containing heavy metals such as Pb, Hg, and Cu, may interfere with molecular techniques like DGGE altering the results of biodiversity.[47] In these cases, they should not be the only methods of detection.[47]

Worth of being mentioned are emerging scientific approaches such as proteomics and genomics that can bring new insights into the activity of microorganisms on art objects. Genomics is a discipline that applies recombinant DNA, DNA sequencing methods, and bioinformatics to sequence, assemble, and analyze the function and structure of genomes, that is, the complete set of DNA within a single cell of an organism. Proteomics studies the entire set of proteins produced or modified by an organism.

It varies with time and distinct requirements or stresses of an organism. Proteomics is more complicated than genomics because an organism's genome is quite constant, whereas the proteome differs from cell to cell and from time to time. Even studying a specific microbe, its cells may make different sets of proteins at different times, or under different conditions. Most proteins function in collaboration with other proteins, and one goal of proteomics is to identify the proteins that interact each other.[144] These methods, even though very promising, are still far from a reliable application in cultural heritage field because they are expensive, time consuming, and require highly skilled researchers.[439]

The molecular-based methods are effective for the identification of microbes but their intrinsic limitation is that they are not always available to conservation staffs. Therefore, it is worth mentioning some techniques that are reliable to assess total and viable biomass, and do not need particular skills. Among them, two techniques are remarkable for their application in situ, meaning directly on the object without needing sampling: The measure of the photosynthetic efficiency for the characterization of phototrophs (described later in this section), and the colorimetric measurements for the quantification of microbial growth.

Portable spectrophotometers easily measure color in situ. The European Standard EN15886 (2010) describes the procedure to adopt for color measurements of cultural heritage objects using the CIELAB method. The color coordinates L^*, a^*, and b^* are recorded for each selected point ($\O \sim 8$ mm). L^* values ranging from 0 to 100 represent black and white, negative and positive a^* values represent green and red, negative and positive b^* values represent blue and yellow, respectively. The total color difference ΔE^* between two measurements ($L^*_1 a^*_1 b^*_1$ and $L^*_2 a^*_2 b^*_2$) is the geometrical distance between their positions in the CIELAB color space. It is calculated using the following equation:

$$\Delta E = \sqrt{(\Delta L^{*2} + \Delta a^{*2} + \Delta b^{*2})}$$

where

$\Delta L^* = L^*_2 - L^*_1$ corresponds to the lightness difference;

$\Delta a^* = a^*_2 - a^*_1$ corresponds to the red/green difference;

$\Delta b^* = b^*_2 - b^*_1$ corresponds to the yellow/blue difference.

Colorimetric measurements, in some cases coupled with imaging analysis, have been used for the quantification of microbial growth on building materials, for the monitoring of biofilms recolonization, and for the assessment of biofilms' prevention by water repellents and consolidants.[29,346] A study[22] observed a good linear correlation between the number of cyanobacterial cells on stone and changes in color expressed as ΔE^*. Other researches[445,6] showed a direct correlation of CIELAB color parameters (L^*, a^*, and b^*) to contents of chlorophyll *a*, carotenoids, phycobiliproteins, and phycocyanin of cyanobacteria. A study[445] induced variations in color and pigments content of a cyanobacterium of the genus *Nostoc* that grew under different environmental conditions (nitrate and phosphate concentrations and light intensity). The results closely correlated the pigments content to CIELAB color coordinates enabling the formulation of predictive equations for estimating chlorophyll *a* and total carotenoid content. Monitoring biofilm development on cultural heritage monuments by means of simple nondestructive method is very important for preventive conservation. Microorganisms' assessment based on color measurements can be performed rapidly on site and is cost-effective. Future implementations should include the application of this technique to microbial communities present on stones for a nondestructive evaluation of their physiological state.

Other studies[327,446,384] proposed color measurements coupled with an image analysis method for the quantification of areas of biological coverage on surfaces. They would be a reliable tool to detect early colonization by phototrophic biofilms (greening) and to monitor their development on buildings. As reported by the authors, the shift toward lower a^* values (more green) and higher b^* values (more yellow) could be due to the main pigments responsible for the color of photosynthetic microorganisms, that is, green chlorophyll and yellow carotenoids. Parameter b^* provided the earliest indication of colonization thus being the most important in determining the total color change. The limit of perception of the greening on a granite surface was also established. Δb^* value +0.59 corresponds to 6.3 $\mu g/cm^2$ of biomass dry weight. Moreover, other researches[55,210] suggested that the values of ΔL^* and Δa^* gave useful information to determine the colonization intensity, indicating biofilm thickness.

There is another simple, rapid and highly sensitive technique useful for the determination of microbial activity, that is, the measurement of adenosine triphosphate (ATP). It is helpful for long-term monitoring of changes in microbial activity as well. Unlike the measure of the photosynthetic efficiency for the characterization of phototrophs (*vide infra*) and the colorimetric measurements, it must be carried out under laboratory conditions. Its use does not require specialized skills. All living organisms use adenosine triphosphate (ATP) molecule as an energy store.

Therefore, the presence and concentration of ATP can indicate the amount of active microbiological contamination of a sample. The analysis must be performed rapidly after sampling as ATP degrades quickly. A bioluminescence assay with firefly luciferin/luciferase measures ATP. Firefly uses the energy-rich phosphate bond of ATP to produce light with its enzyme system luciferin/luciferase. In the presence of ATP, luciferase catalyzes the oxidation of luciferin producing visible light. The amount of ATP is directly proportional to the amount of light emitted. On the market, there are devices such as the HY-LiTE Luminometer with Jet A1 Pens that is a test system designed to measure the level of microbiological contamination in fuel and water. Microorganisms contamination measurement using ATP bioluminescence is often reported in the field of food hygiene for sanitary control.[447] It takes only a few minutes to check for biocontamination, while the traditional agar incubation method takes several days. The simplicity and rapidity of the test makes it useful to estimate the distribution of biocontamination in museums.[447]

The analytical methods used in the field of cultural properties are identical to those used in modern materials science. A technique routinely performed during the initial examination is the observation of cross-sections of samples by conventional incident light microscopy. It allows studying the structure of materials, identifying their components, and assessing the interaction between biofilms/lichens and the substrates. Staining methods are very useful to detect details of this interaction. Periodic Acid-Schiff (PAS) for example visualizes hyphae because it stains in red/fuchsia polysaccharides such as glycogen, and mucous substances such as glycoproteins, glycolipids, and mucins. The application of an imaging method to stained cross-sections with the WinCAM Pro 2007d software was useful to quantify the hyphae spread in the bulk of marble, travertine and mortar.[434] The differences in hue, saturation, and intensity of the lichen thalli and PAS stained hyphae in comparison to the substrata allowed the discrimination of lichens and hyphae.

Varieties of dyes stain microbial cells making them visible under the microscope, and allowing a detailed observation of their shape. Fluorescent substances such as acridine orange and DAPI (4′,6-diamidino-2-phenylindole) permit also the visualization directly on samples using epi-illumination, without the need of biofilms removal from the substrate.[435] Some dyes allow distinguishing viable, or active, from nonviable, or inactive, cells. For example, the commercial product BacLight™ Bacterial Viability Kit, made of a mixture of propidium iodide and fluorescein dyes, discriminates between dead and viable cells because microorganisms with intact cell membranes fluoresce green while microorganisms with damaged cell membranes fluoresce red.[435,48,95] Another product useful to distinguish between dead and viable cells is SYTOX 1 Green that penetrates only the permeable membranes of dead cells.[280,220]

Photosynthetic organisms such as algae and cyanobacteria are autofluorescent. Therefore, they can be visualized without staining. When observed under optical microscope with UV illumination, the photosynthetic pigments of autotrophic organisms emit fluorescence. Live autotrophic organisms appear red whereas the dead cells appear white.[228] As the technique indicates the vitality of autotrophic organisms, it is widely used by biologists involved in cultural heritage conservation to check the effectiveness of treatments for the elimination of these organisms (Fig. 7.2).

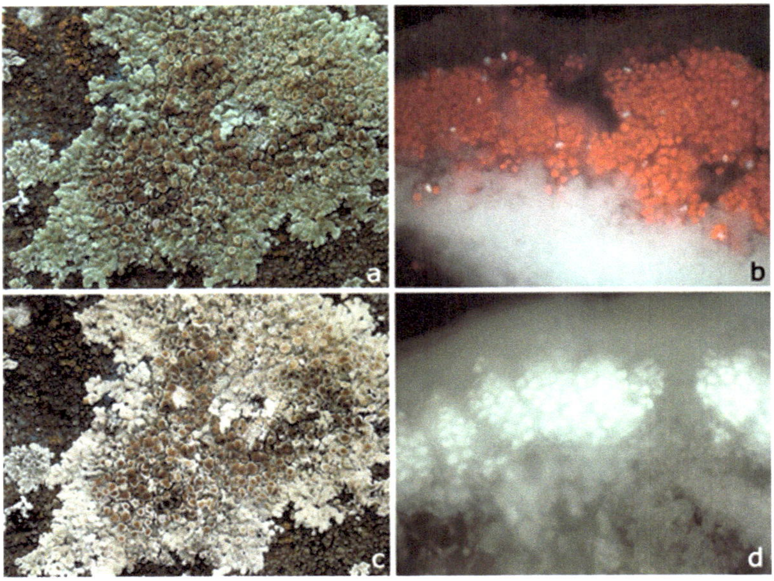

FIGURE 7.2 Crustose lichen *Protoparmeliopsis muralis* (a). Polished cross-section of the lichen showing the red fluorescence of photobiont cells, indication of cell viability and integrity of the photosynthetic apparatus. UV light imaging (b). Morphological modifications of the lichen thallus when treated with biocides (c). Polished cross-section of the treated lichen showing the white fluorescence of photobiont cells, indication of devitalization (d). (From Caneva, G.; Nugari, M. P.; Salvadori, O. *Plant Biology for Cultural Heritage: Biodeterioration and Conservation*. Getty Conservation Institute, Los Angeles, 2008. Used with permission from Nardini Editore.)

Scanning electron microscope and related methods (SEM-EDX energy dispersive X-ray analysis, ESEM environmental SEM or low-vacuum SEM) are suitable to investigate material composition and mineral structure of stone samples and patinas.[448] SEM-EDX allows the individual detection, measurement and elemental analysis of features such as particles and patinas.[449] It performs a qualitative and semiquantitative microanalysis by providing element spatial distribution maps. When carried out using backscattered electrons (BSE), the color of the maps indicates the relative concentrations of elements. Dark represents the absence of the investigated element and white indicate the maximum concentration of the element. Scanning electron microscopy with backscattered electrons (SEM-BSE) can be useful to check the vitality of biofilms and lichens that grow on and in the bulk of substrates.[38,268] The procedure implies the fixation of samples in glutaraldehyde and then in osmium tetroxide. Researchers dehydrate the samples in a series of ethanol solutions, and embed them in resin. Then, they finely polish the blocks of resin-embedded rock samples, coat them with carbon and observe them under SEM. OsO_4 staining of cytoplasmic lipids makes fungal hyphae appear white, thus well visible. This procedure also serves to detect endolithic microhabitats. When treated, the cells' collapse is evident.

Contrary to SEM, ESEM operates at low vacuum and provides fast, accurate images of biofilms and lichens, showing their spatial relationship without needing extensive manipulation of the samples. This instrument has a secondary electron detector capable of forming high-resolution images at pressures in the range of 0.1–20 torr. At these relatively high pressures, the instrument permits the observation of nonconductive biological samples without the need of metallic coating or any special preparations as in scanning electron microscope. Therefore, it provides better images of the extracellular polymeric substances (EPS) of biofilms and microbial cells.

A powerful technique to analyze a microbial community directly on the sample is confocal laser scanning microscopy (CLSM).[444,436] CLSM detects fluorescent light, but it differs from conventional epifluorescence microscopy by acquiring the fluorescent signals exclusively from the focal plane. Consecutive optical slices along the *Z*-axis of an image series ("confocal stack") can be prepared for three-dimensional reconstructions.[443] Therefore, CLSM combines the advantages of digital fluorescence microscopy with the detection of optical sections of samples by enhanced vertical and axial resolution and quality. It is possible to apply to the recorded digital images a wide range of processing and analysis routines to obtain quantitative information or three-dimensional reconstructions of the scanned samples. As a direct method to study microorganisms, CLSM avoids the PCR biases typical of molecular methods, thus it can quantify accurately microbes when suitable methods for image analysis are applied. CLSM allows the detection of three kinds of objects: (1) molecules, cells, and tissues stained with one or more fluorochromes; (2) organisms that express fluorescent proteins; (3) autofluorescent cells, tissues, and substrates.[443] Examples of fluorochromes are PicoGreen and SYTO 17 that stain nucleic acids, and Congo red that detects (1→4)-β-D-glucan, a typical component of fungal and bacterial cell wall.[444] The application of this technique is very suitable for the study of endolithic microorganisms because it allows the quantification of the volume of substratum occupied by endolithic growth, thus measuring its impact on stones.[141] Unlike other techniques, it provides information about the real spatial arrangement of the colonization in the bulk of the stone. The study[141] of marble samples using CLSM with double staining (propidium iodide for nucleic acids and Concavalin-A conjugate with the fluorophore Alexa Fluor 488 for the extra-polymeric substances) showed clearly the three-dimensional structure of endolithic microorganisms (cyanobacteria and fungi) along with the matrix of extracellular polymeric substances.

Other methods suitable to study the interaction between microflora and stones and to assess its possible protection to substrates are:

- Measurements of the capillary water uptake (discussed in the next clause) and of the surface hardness of colonized and noncolonized materials;
- Drill resistance measurements;
- Measurements of the surface porosity of a colonized stone compared with that of the rock core.

7.2 ASSESSMENT OF THE EFFICACY OF CONTROL METHODS

A very important aspect to evaluate the success of treatments concerns the effects considered relevant to assess cells' death. The literature often reports the complete inhibition of microbes' growth and the lack of physiological (e.g., photosynthetic) performance as adequate characteristics to assess biocides' efficacy. Unfortunately, these features do not automatically imply cell death because recovery is possible. Algae, for example, can become dormant, resistant, or can repair damages. Moreover, they are also able to use alternative substrates and energy sources. The loose of membrane integrity is instead equivalent to mortality.[280] This differentiation is very important when performing treatments to control biofilm growth on stones. The application of fluorescent dyes, for example, allows the discrimination between dead and viable cells because they penetrate only the permeable membranes of dead cells (*vide infra*).

The experimental approach reported by the literature to assess the efficacy of biocides includes three types of controlled application[450]:

1) Testing the biocide's efficacy *in vitro* on agar or liquid cultures of the target microorganisms isolated from the test sites;
2) Inoculation and cultivation of a microbial community on artificial and natural stone samples simulating real conditions, with a successive or simultaneous treatment to study the interaction among chemicals, microorganisms, and stones;
3) Application of biocides in situ on the object or on samples made of the same substrate and kept outdoors for long time. Often the researchers use samples prepared using different methods making it difficult to compare results across studies.[448] They also stress the samples to make them similar to the stones of the cultural heritage objects.

The first experimental approach is appropriate mainly for the microbicides. A simple laboratory method is the determination of minimum inhibitory concentration (MIC) of a biocide measuring the inhibition halo of microorganisms' growth in solid agar. MIC is the lowest concentration of a substance that inhibits the growth of a microorganism. This semi-quantitative method is based on the diffusion of the biocide through the solid medium, the incubation time and the specific growth rate of the test organism.[266,251] Based on the same principle, the method uses liquid solutions where the microorganisms are inoculated (Fig. 7.3).

Some laboratory studies perform accelerated fouling tests under optimal conditions for cells' development to assess the performance of biocides applied on stone samples. They are based on a culture-streaming test with water run-off on samples' surfaces.[327,384,396,347] These tests produce an accelerated fouling of samples. The acceleration factor of the set-up is approximately in the range 13–20 respect to specimens exposed to outdoor conditions (e.g., specimens with a biofilm coverage between 25 and 50% after four weeks of accelerated tests require about 1 year of outdoor exposure for the occurrence of the first visible discoloration).[327]

FIGURE 7.3 Test for the evaluation of biocides efficiency. A known number of microorganisms grows in a liquid nutrient medium. Different concentrations of a biocide are added to the medium. The efficiency of the various concentrations is measured. (From Caneva, G.; Nugari, M. P.; Salvadori, O. *Plant Biology for Cultural Heritage: Biodeterioration and Conservation*. Getty Conservation Institute, Los Angeles, 2008. Used with permission from Nardini Editore.)

Unfortunately, microbicides often perform differently when tested under laboratory conditions and in situ. In some cases, they are effective in the laboratory experiments while showing low or no efficiency in situ on the same microorganisms.[264] One reason is that the sensitivity of species in the lab is usually much higher than that in situ.[271,277] A second reason is that various biocides are less effective on sessile microorganisms, that is, biofilms, than on free-floating dispersed cells.[282] With their complex architecture and dynamic nature, biofilms provide a physical barrier that protects microorganisms from detrimental substances, such as biocides. Therefore, the dose needed for a microbial community on outdoor stone monuments might be much higher than the one needed to kill successfully isolated organisms growing on agar plates.[48] For these reasons researchers recognized the results obtained with the second experimental approach mentioned above as being more relevant because it simulates in situ conditions.

Conservators often apply control methods in situ testing them on a small scale, preferentially on the affected monument itself, to determine their effectiveness on the biological growth colonizing different microhabitats. It is possible to evaluate their efficiency in situ through various techniques that range from the simplest ones to others able to provide precise measures related to the killing of cells. Close empirical observation, by surveying the necrotic status of biofilms, lichens, moss cushions, leaves, and the comparison of "before and after" morphologies and reduction of biological cover, is important but not definitive and rigorous evidences of the performance and of changes. A study[259] suggested the application of a new technique that avoids subjective visual judgments. It compares areas using digital photography with a special set up. Data are analyzed with a commercially available software. Statistical analysis quantifies changes on the stone surfaces related to different control methods.

There are several instruments that allow to determine the efficacy of mechanical, physical and chemical methods under both laboratory conditions and in situ trials. Almost all the previously described techniques for the assessment of biological activity (ATP measurement, staining, autofluorescence of photosynthetic organisms, SEM-BSE, etc.) are in usage also for this scope.

The technique that measures ATP (*vide infra*) has some drawbacks when applied to check the performance of treatments for the removal of biofilms. A study showed that some biocides may interfere with luminescence measurement when the ATP concentration is low.[447] When the treatment was carried out with agents that kill cells attacking the cell wall, ATP luminescence and the survived spores were not correlated because of the elution of intracellular ATP.[447] On the contrary, in the case of treatments with ultraviolet rays or agents that interact with cellular enzymes, ATP luminescence was correlated with survived spores.[447] According to the authors, when using ATP measurements to check the efficiency of control methods, it is necessary to consider these limitations.

A laboratory technique applied to assess lichen vitality for monitoring the effects of different air pollutants overall foliose and fruticose lichens is electrical conductivity (EC).[451,452] This simple technique evaluates the integrity of lichen cell membranes measuring the amount of K^+ ions and other electrolytes leached by the organism. When damaged, lichens release ions. Thus, EC values increase considerably in comparison to healthy lichens and are proportional to the degree of damage. Lichens from polluted sites present an approximately double value in comparison to lichens from rural sites.[452] The procedure, not described in detail, implies that foliose and fruticose lichens are cleaned and soaked in demineralized water; then water conductivity is determined. A study[222] applied the technique to evaluate treatments of crustose lichens that colonized monuments. Results allowed distinguishing between reversible and irreversible damages. Lichens suffered reversible damages when EC values doubled those of controls, while the damages were irreversible if these values increased threefold or more. A limitation of the technique, when applied to crustose lichens with very thin thalli, is that the contribution of mineral substrate in EC values is high. The study showed that EC method is a valid technique to evaluate the effects of control methods on some kinds of lichens. Its use is extremely simple in testing physical control methods, while in the case of chemical treatments, it is necessary to calculate and subtract their ionic contribution.

The application of the hyperspectral imaging technique showed promising results.[231] It measures the reflectivity variations of the cleaned surfaces in a noninvasive way. The technique was applied using a system that consisted of the integration of an imaging spectrograph with a monochrome matrix array sensor.[231] It provided results similar to those obtained with traditional techniques (optical microscopy, petrographic microscopy, SEM).

To date a variety of portable instruments exists to assess in situ the efficacy of methods and chemicals. One of them is fluorescence lidar (light detection and ranging), a noninvasive, remote sensing technique that measures laser-induced fluorescence in outdoor environments.[453–457] The method can assess and characterize photoautotrophic microorganisms present on stone surfaces because it detects chlorophyll *a* fluorescence and fluorescent accessory pigments as well. Spectra obtained with this method identify green algae and cyanobacteria thanks to the typical fluorescence of phycocyanin at 660 nm and chlorophyll *a* at 690 nm.[458] The equipment has imaging capabilities providing the acquisition of hyperspectral fluorescence maps over extended areas of monuments from distances as great as 80 m and with a typical acquisition speed of about

15 min/m^2. A further improvement of the hyperspectral fluorescence lidar imaging technique was achieved with the introduction of a target pointing system for localizing the acquired images. This makes possible an extensive monitoring with suitable spatial resolution and the creation of false-color thematic maps to outline the distribution of photoautotrophic biodeteriogens.

The use of portable fluorometers, such as pulse amplitude modulation (PAM) fluorometer, that measure the photosynthetic efficiency of photosystem II (PS II) is of great help to assess the reaction and vitality of phototrophic organisms after a treatment.[260] The technique gives reliable and detailed analytical information without needing sampling of the organisms.[296,459,169,260,376,395,233] For example, PAM fluorometry quantified the resistance of a water saturated paint against algal growth with sufficient accuracy.[376] Along with lidar, it is a powerful tool in restoration treatments for rapid and noninvasive in situ control of treatments' effectiveness and for the monitoring of covering of phototrophic microorganisms. The technique measures chlorophyll *a* fluorescence parameters that can be used as a marker of the photochemical pathways of utilization of absorbed light energy. Among these parameters, QY_{max} (PSII maximum quantum yield) is expressed by

$$QY_{max} = F_V/F_{max}$$

$$F_V = F_{max} - F_0$$

where F_0 is the minimum fluorescence and F_{max} is the maximum fluorescence obtained when exposing the sample to a pulse of very intense light after dark adaptation. The optimal value of QY_{max} for most plants is around 0.83, for bryophytes and lichens is circa 0.70–0.75[293,220,232] while for microalgae the values range between 0.4 and 0.7.[460] There are also imaging systems that provide, with the help of the false color code, images of the photosynthetic activity and its spatial–temporal variations (Figs. 7.4 and 7.5). The fluorescence yield is displayed in the image window using a false color code bar, with the colors encoding for numerical values between 0, corresponding to black and 1, corresponding to purple. The imaging PAM in combination with a scanning device allowed the detailed observation of an algal colonization, even when it was below the human visual threshold.[376]

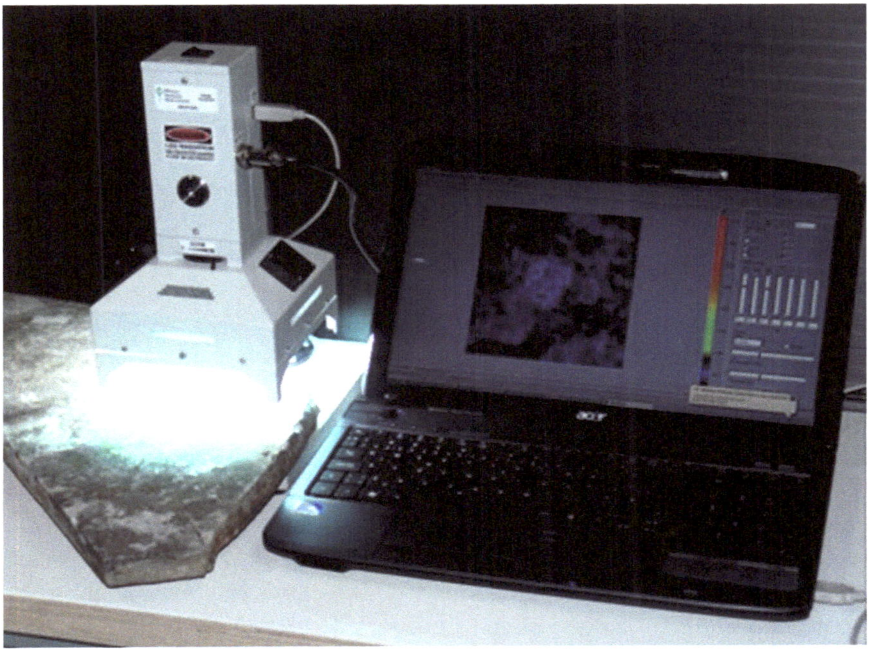

FIGURE 7.4 Pulse amplitude modulation (PAM) imaging fluorometer.

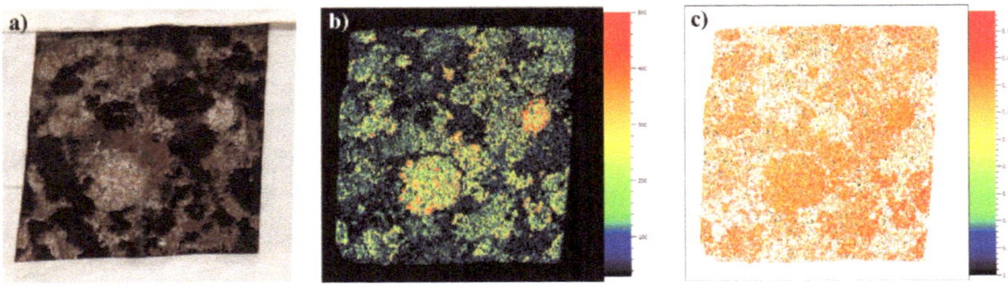

FIGURE 7.5 Images obtained applying PAM imaging fluorometer on a marble slab colonized by lichens and biofilms. The selected area—visible light imaging (a), "false color" map of the maximum fluorescence yield (F_{max}) of lichens and biofilms (b), distribution of F_V/F_m values (c). PAM measurements show the integrity of the photosynthetic apparatus.

The color of the surface after a treatment is an important parameter for the product's choice. The color coordinates L^*, a^*, and b^* are recorded before and after treatments. ΔE^* values allow an evaluation of differences in the chromatic aspect between untreated and treated areas.

Water absorption is another relevant parameter that can be measured in situ to check any changes in the water absorption capacity of stones caused by biocides or by other treatments.[29] Water absorption is the process whereby a fluid is drawn into a porous unsaturated material under the action of capillary forces. The capillary suction depends on the pore volume and geometry, and the saturation level of the stone. Water inside cavities of porous stone materials is one of the main causes of accelerated deterioration. Therefore, water absorption of porous stone materials is an indication of the degree of deterioration and its sensitivity to future deterioration. Another application of water absorption is the monitoring of stone's water repellence as a function of time. Portable devices can measure the water absorption in situ. A measuring technique is the so-called Karsten tube test that uses a glass tube filled with water and bonded to the test material (Fig. 7.6). The water exerts pressure on the surface and a graduated scale indicates, over time, the amount of water absorbed. The test can be executed in laboratory as well as in situ, but it is not always reliable for in situ analyses.[461]

Another measuring technique exists, the contact sponge method.[462] It consists of a 1034 Rodac plate (5.6 cm in diameter) containing a natural fiber Calypso sponge by Spontex that is imbibed with distilled water. This makes the sponge thicker than the rim of the plate. The plate is pressed on the stone surface for 1 min using a dynamometric spring (Fig. 7.7). Water absorption is determined by calculating the difference, in mg/cm^2, between the plate weights measured before and after contact with the surfaces.

The topic of a common scientific way to assess the application of analytical methodologies for the study of stone alterations and the effectiveness of intervention treatments acquired a growing interest by the scientific community dealing with conservation of cultural heritage. Italy has been one of the first countries to sense the necessity of *standardizing common analytical methodologies* by establishing the UNI-NORMAL (Normativa Manufatti Lapidei) Commission. Within this Commission, the Working Group Biology focused, among others, on issues related to the methods of control of biodeterioration.

Recently many European countries decided to create a commission within CEN (the European Committee for Standardization) dedicated to conservation of cultural heritage. CEN is a legal association that comprises National Standards Bodies (e.g., AFNOR—Association Française de Normalization, BSI—British Standards Institution, UNI—Ente

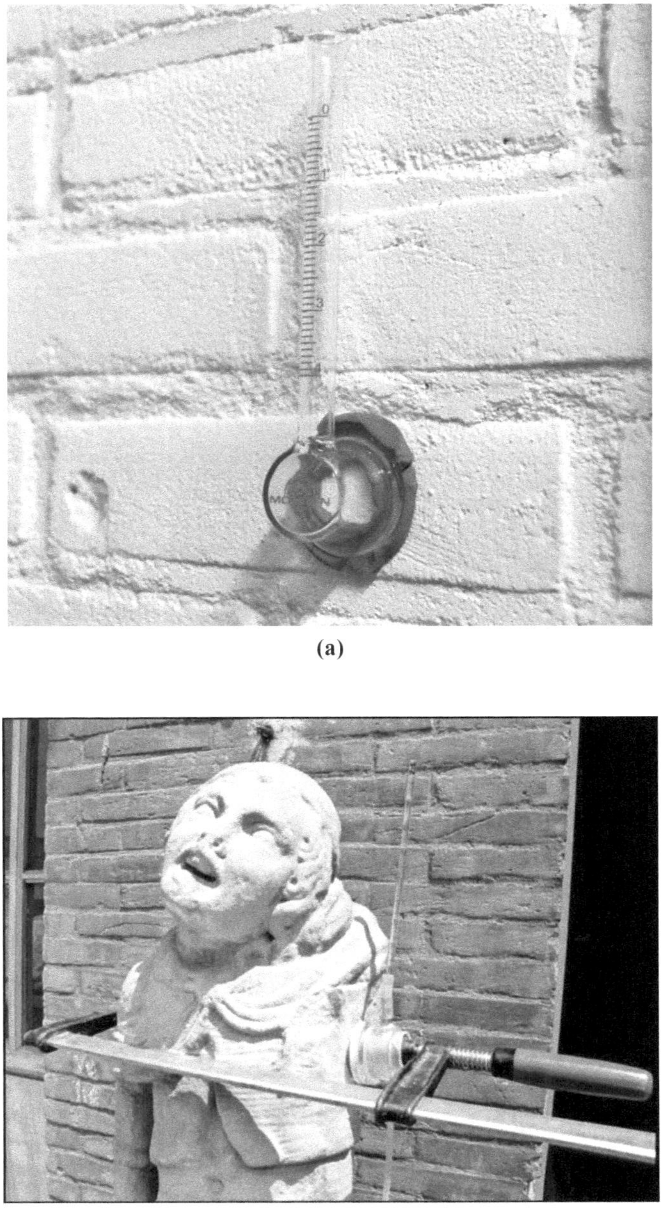

(a)

(b)

FIGURE 7.6 Application in situ of Karsten tube to measure water absorption of a wall (a) and of a statue (b). Photo courtesy of Roberto Fusco, TQC ITALIA s.r.l.

(a)

(b)

FIGURE 7.7 Contact sponge method measures in situ water absorption of a stone. It consists of a 1034 Rodac plate containing a natural fiber Calypso sponge by Spontex that is imbibed with distilled water. This makes the sponge thicker than the rim of the plate (a). The plate is pressed on the stone surface using a dynamometric spring (b).

di Unificazione Italiano, DIN—Deutsches Institut fur Normung).[328] It is responsible for the production of European Standards (ENs) and other technical documents in relation to various kinds of products, materials, services and processes (www.cen.eu). The Technical Committee CEN/TC 346 has the scope of standardizing terminology, requirements, test methods for sampling, measurements of colors in the field of cultural heritage properties. It also deals with the characterization of materials, the processes, practice, methodologies, and documentation of conservation of tangible cultural heritage to support its preservation, protection and maintenance, and to enhance its significance. The European standardization activity in the field of conservation of cultural heritage is essential to acquire a common unified scientific approach. CEN/TC 346 will hopefully focus in the future on common relevant themes such as how to use biocides (contact time/concentration/temperature), how to assess that they are working in practice, how to avoid any undesirable side effect, including selection of resistant microbes.[463]

Worth mentioning is also the *Illustrated glossary on stone deterioration patterns* by ICOMOS-ISCS (2010) that describes the weathering processes of stone surfaces (including those caused by biological growth) with the aid of a specific terminology.

KEYWORDS

- **noninvasive and micro-invasive techniques**
- **assessment of the weathering effects of biological growth**
- **assessment of efficacy of control methods**
- **image analysis**
- **sampling**
- **culture-dependent methods**
- **molecular techniques**
- **colorimetric measurements**
- **staining methods**
- **SEM**
- **ESEM**

- **confocal laser scanning microscopy**
- **accelerated fouling tests**
- **in situ trials**
- **ATP measurements**
- **electrical conductivity**
- **lidar**
- **portable fluorometers**
- **water absorption**
- **standardization**

REFERENCES

1. Martines, G. Marmo e restauro dei monumenti antichi: estetica delle rovine, degrado delle strutture all'aperto, una ipotesi di lavoro. In *Marmo e Restauro. Situazioni e Prospettive*, Museo del Marmo, Carrara, Italy, 1983; pp 83–92.
2. Ashbee, J. Ivy and the Presentation of Ancient Monuments and Building. In *Ivy on Walls*, English Heritage: 2010; pp 9–15.
3. Rossi Manaresi, R.; Grillini, G. C.; Pinna, D.; Tucci, A. La formazione di ossalati di calcio su superfici monumentali: genesi biologica o da trattamenti? *Proc. of the Conference "Le pellicole ad ossalato: origine e significato nella conservazione delle opere d'arte"*, Milano, Italy, 1989; pp 113–125.
4. Diao, M.; Taran, E.; Mahler, S.; Nguyen, A. V. A Concise Review of Nanoscopic Aspects of Bioleaching Bacteria–Mineral Interactions. *Adv. Coll. Interface Sci.* **2014,** *212*, 45–63.
5. Albertano, P.; Urzì, C. Structural Interactions Among Epilithic Cyanobacteria and Heterotrophic Microorganisms in Roman Hypogea. *Microbial. Ecol.* **1999,** *38*(3), 244–252.
6. Vázquez-Nion, D.; Sanmartín, P.; Silva, B.; Prieto, B. Reliability of Colour Measurements for Monitoring Pigment Content in a Biofilm-Forming Cyanobacterium. *Int. Biodeteriorat. Biodegradat.* **2013,** *84*, 220–226.
7. Maggi, O.; Persiani, A. M.; De Leo, F.; Urzì, C. Fungi. In *Plant Biology for Cultural Heritage: Biodeterioration and Conservation*; Caneva, G., Nugari, M. P., Salvadori, O., Eds.; Getty Conservation Institute: Los Angeles, 2008; pp 65–70.
8. Gadd, G. M. Geomycology: Biogeochemical Transformations of Rocks, Minerals, Metals and Radionuclides by Fungi, Bioweathering and Bioremediation. *Mycol. Res.* **2007,** *111*, 3–49.
9. Nimis, P. L.; Martellos, S. ITALIC–The Information System on Italian Lichens. Version 4.0. 2008. University of Trieste, Dept. of Biology, IN4.0/1 (http://dbiodbs.univ.trieste.it/).
10. Bernardi, A.; Becherini, F.; Verità, M.; Ausset, P.; Bellio, M.; Brinkmann, U.; Cachier, H.; Chabac, A.; Deutsch, F.; Etcheverry, M. P.; Geotti Bianchini, F.; Godoi, R. H. M.; Kontozova-Deutsch, V.; Lefèvre, R.; Lombardo, T.; Mottner, P.; Nicola, C.; Pallot-Frossard, I.; Rölleke, S.; Römich, H.; Sommariva, G.; Vallotto, M.; Van Grieken, R. Conservation of Stained glass Windows with Protective Glazing: Main Results from the European VIDRIO Research Programme. *J. Cult. Heritage* **2013,** *14*, 527–536.
11. Guillitte, O. Bioreceptivity: A New Concept for Building Ecological Studies. *Sci. Total Environ.* **1985,** *167*, 215–220.
12. Caneva, G.; Di Stefano, D.; Giampaolo, C.; Ricci, S. Stone Cavity and Porosity as a Limiting Factor for Biological Colonization: The Travertine of Lungotevere (Rome). In *10th International Congress on Deterioration and Conservation of Stone*; Kwiatkoski, D., Löfvendahl, R., Eds.; ICOMOS: Stockholm, 2004; Vol. 1, pp. 227–232.

13. de los Rios, A.; Cámara, B.; García del Cura, M. A.; Rico, V. J.; Galván, V.; Ascaso, C. Deteriorating Effects of Lichen and Microbial Colonization of Carbonate Building Rocks in the Romanesque Churches of Segovia (Spain). *Sci. Total Environ.* **2009,** *407*(3), 1123–1134.
14. Coutinho, M. L.; Miller, A. Z.; Macedo, M. F. Biological Colonization and Biodeterioration of Architectural Ceramic Materials: An Overview. *J. Cult. Heritage* **2015,** *16*, 759–777.
15. Cámara, B.; Álvarez de Buergo, M.; Fort, R.; Ascaso, C.; de los Ríos, A.; Gomez-Heras, M. Another Source of Soluble Salts in Urban Environments Due to Recent Social Behaviour Pattern in Historical Centres. In *Science, Technology and Cultural Heritage*; Rogerio-Candelera, Ed.; Taylor & Francis Group: London, 2014; pp 1–6.
16. Miller, A. Z.; Laiz, L.; Dionísio, A.; Macedo, M. F. Primary Bioreceptivity: A Comparative Study of Different Portuguese Lithotypes. *Int. Biodeteriorat. Biodegradat.* **2006,** *57*, 136–142.
17. Olsson-Francis, K.; Simpson, A. E.; Wolff-Boenisch, D. The Effect of Rock Composition on Cyanobacterial Weathering of Crystalline Basalt and Rhyolite. *Geobiology* **2012,** *10*, 434–444.
18. Favero-Longo, S. E.; Gazzano, C.; Girlanda, M.; Castelli, D.; Tretiach, M.; Baiocchi, C.; Piervittori, R. Physical and Chemical Deterioration of Silicate and Carbonate Rocks by Meristematic Microcolonial Fungi and Endolithic Lichens (Chaetothyriomycetidae). *Geomicrobiol. J.* **2011,** *28*, 732–744.
19. Miller, A. Z.; Sanmartín, P.; Pereira-Pardo, L.; Dionísio, A.; Saiz-Jimenez, C.; Macedo, M. F.; Prieto, B. Bioreceptivity of Building Stones: A Review. *Sci. Total Environ.* **2012,** *426*, 1–12.
20. Prieto, B.; Silva, B. Estimation of the Potential Bioreceptivity of Granitic Rocks from Their Intrinsic Properties. *Int. Biodeteriorat. Biodegradat.* **2005,** *56* (4), 206–215.
21. Wiktor, V.; De Leo, F.; Urzì, C.; Guyonnet, R.; Grosseau, P.; Garcia-Diaz, E. Accelerated Laboratory Test to Study Fungal Biodeterioration of Cementitious Matrix. *Int. Biodeteriorat. Biodegradat.* **2009,** *63*, 1061–1065.
22. Prieto, B.; Silva, B.; Aira, N.; Alvarez, L. Toward a Definition of a Bioreceptivity Index for Granitic Rocks: Perception of the Change in Appearance of the Rock. *Int. Biodeteriorat. Biodegradat.* **2006,** *58*, 150–154.
23. Eggins, H. O. W.; Oxley, T. A. Biodeterioration and Biodegradation. *Int. Biodeteriorat. Biodegradat.* **2001,** *48*(1–4), 12–15.
24. Gillatt, J. Dry-Film Biocides: The Next Generation. *Double Liaison* **2003** (535), 45–57.
25. Allsopp, D.; Seal, K.; Gaylarde, C. *Introduction to Biodeterioration.* Cambridge University Press: Cambridge, UK, 2004.
26. Caneva, G.; Nugari, M. P.; Salvadori, O. *Plant Biology for Cultural Heritage: Biodeterioration and Conservation.* Getty Conservation Institute: Los Angeles, 2008.
27. Caneva, G.; Ceschin, S. Ecology of Biodeterioration. In *Plant Biology for Cultural Heritage: Biodeterioration and Conservation*; Caneva, G., Nugari, M. P., Salvadori, O., Eds.; Getty Conservation Institute: Los Angeles, 2008; pp 35–58.
28. Shirakawa, M. A.; Beech, I. B.; Tapper, R.; Cincotto, M. A.; Gambale, W. The Development of a Method to Evaluate Bioreceptivity of Indoor Mortar Plastering to Fungal Growth. *Int. Biodeteriorat. Biodegradat.* **2003,** *51*(2), 83–92.

29. Pinna, D.; Salvadori, B.; Galeotti, M. Monitoring the Performance of Innovative and Traditional Biocides Mixed with Consolidants and Water-Repellents for the Prevention of Biological Growth on Stone. *Sci. Total Environ.* **2012,** *423*, 132–141.
30. Salvadori, O.; Charola, A. E., Methods to Prevent Biocolonization and Recolonization: An Overview of Current Research for Architectural and Archaeological Heritage. In *Biocolonization of Stone: Control and Preventive Methods*; Charola, A. E., McNamara, C., Koestler, R. J., Eds.; Smithsonian Institute Scholarly Press: Washington, 2011; pp 37–50.
31. Schabereiter-Gurtner, C.; Piñar, G.; Lubitz, W.; Rölleke, S. Analysis of Fungal Communities on Historical Church Window Glass by Denaturing Gradient Gel Electrophoresis and Phylogenetic 18S rDNA Sequence Analysis. *J. Microbiol. Methods* **2001,** *47*, 345–354.
32. Shirakawa, M. A.; Vanderley, M. J.; Mocelin, A.; Zilles, R.; Toma, S. H.; Araki, K.; Toma, H. E.; Thomaz, A. C.; Gaylarde, C. C. Effect of Silver Nanoparticle and TiO_2 Coatings on Biofilm Formation on Four Types of Modern Glass. *Int. Biodeteriorat. Biodegradat.* **2016,** *108*, 175–180.
33. Carmona, N.; Laiz, L.; Gonzalez, J. M.; Garcia-Herasa, M.; Villegasa, M. A.; Saiz-Jimenez, C. Biodeterioration of Historic Stained Glasses from The Cartuja de Miraflores (Spain). *Int. Biodeteriorat. Biodegradat.* **2006,** *58*, 155–161.
34. Marvasi, M.; Vedovato, E.; Balsamo, C.; Macherelli, A.; Dei, L.; Mastromei, G.; Perito, B. Bacterial Community Analysis on the Mediaeval Stained Glass Window "Nativita" in the Florence Cathedral. *J. Cult. Heritage* **2009,** *10*, 124–133.
35. Piñar, G.; Garcia-Valles, M.; Gimeno-Torrente, D.; Fernandez-Turiel, J. L.; Ettenauer, J.; Sterflinger, K. Microscopic, Chemical, and Molecular-Biological Investigation of the Decayed Medieval Stained Window Glasses of Two Catalonian Churches. *Int. Biodeteriorat. Biodegradat.* **2013,** *84*, 388–400.
36. Rodrigues, A.; Gutierrez-Patricio, S.; Miller, A. Z.; Saiz-Jimenez, C.; Wiley, R.; Nunes, N.; Vilarigues, M.; Macedo, M. F. Fungal Biodeterioration of Stained-Glass Windows. *Int. Biodeteriorat. Biodegradat.* **2014,** *90*, 152–160.
37. Hoppert, M.; Flies, C.; Pohl, W.; Gunzl, B.; Schneider, J. Colonization Strategies of Lithobiontic Microorganisms on Carbonate Rocks. *Environ. Geol.* **2004,** *46*, 421–428.
38. Cámara, B.; de los Ríos, A.; Urizal, M.; de Buergo, M. A.; Varas, M. J.; Fort, R.; Ascaso, C. Characterizing the Microbial Colonization of a Dolostone Quarry: Implications for Stone Biodeterioration and Response to Biocide Treatments. *Microbial Ecol.* **2011,** *62* (2), 299–313.
39. Caneva, G.; Ceschin, S.; Salvadori, O.; Kashiwadani, H.; Moon, K. H.; Futagami, Y. Biodeterioration of Stone in Relation to Microclimate in the Ta Nei Temple – Angkor (Cambodia). In *12th International Congress on the Deterioration and Conservation of Stone*, Columbia University: New York, US, 2012; pp 1–12.
40. Caneva, G.; Bartoli, F.; Ceschin, S.; Salvadori, O.; Futagami, Y.; Salvati, L. Exploring Ecological Relationships in the Biodeterioration Patterns of Angkor Temples (Cambodia) Along a Forest Canopy Gradient *J. Cult. Heritage* **2015,** *16*(5), 728–735.
41. Jim, C. Y.; Chen, W. Y. Bioreceptivity of Buildings for Spontaneous Arboreal Flora in Compact City Environment. *Urban Forest. Urban Green.* **2011,** *10*, 19–28.

42. Jim, C. Y. Drivers for Colonization and Sustainable Management of Tree-Dominated Stonewall Ecosystems. *Ecol. Eng.* **2013,** *57*, 324–335.
43. Kumbaric, A.; Ceschin, S.; Zuccarello, V.; Caneva, G. Main Ecological Parameters Affecting the Colonization of Higher Plants in the Biodeterioration of Stone Embankments of Lungotevere (Rome). *Int. Biodeteriorat. Biodegradat.* **2012,** *72*, 31–41.
44. Bellinzoni, A. M.; Caneva, G.; Ricci, S. Ecological Trends in Travertine Colonisation by Pioneer Algae and Plant Communities. *Int. Biodeteriorat. Biodegradat.* **2003,** *51*, 203–210.
45. Miller, A. Z.; Laiz, L.; Gonzalez, J. M.; Dionísio, A.; Macedo, M. F.; Saiz-Jimenez, C. Reproducing Stone Monument Photosynthetic-Based Colonization Under Laboratory Conditions. *Sci. Total Environ.* **2008,** *405*, 278–285.
46. May, E.; Zamarreño, D.; Hotchkiss, S.; Mitchell, J.; Inkpen, R. Bioremediation of Algal Contamination on Stone. In *Biocolonization of Stone: Control and Preventive Methods*; Charola, A. E., McNamara, C., Koestler, R. J., Eds.; Smithsonian Institute Scholarly Press: Washington, 2009; pp 59–70.
47. Gaylarde, C. C.; Morton, L. H. G.; Loh, K.; Shirakawa, M. A. Biodeterioration of External Architectural Paint Films – A Review. *Int. Biodeteriorat. Biodegradat.* **2011,** *65*(8), 1189–1198.
48. Scheerer, S. *Microbial Biodeterioration of Outdoor Stone Monuments. Assessment Methods and Control Strategies*. Cardiff University, Cardiff, UK, 2008.
49. Ramírez, M.; Hernández-Mariné, M.; Novelo, E.; Roldán, M. Cyanobacteria-Containing Biofilms from a Mayan Monument in Palenque, Mexico. *Biofouling* **2010,** *26*(4), 399–409.
50. Lange, O. L., Photosynthetic Productivity of the Epilithic Lichen *Lecanora muralis*: Long-Term Field Monitoring of CO_2 Exchange and Its Physiological Interpretation I. Dependence of Photosynthesis on Water Content, Light, Temperature, and CO_2 Concentration from Laboratory Measurements. *Flora* **2002,** *197*, 233–249.
51. Purvis, O. W.; Chimonides, J.; Din, V.; Erotokritou, L.; Jeffries, I.; Jones, G. C.; Louwhoff, S.; Read, H.; Spiro, B. Which Factors are Responsible for the Changing Lichen Floras of London? *Sci. Total Environ.* **2003,** *310,* 179–189.
52. Jain, A.; Bhadauria, S.; Kumar, V.; Chauhan, R. S. Biodeterioration of Sandstone Under the Influence of Different Humidity Levels in Laboratory Conditions. *Build. Environ.* **2009,** *44*(6), 1276–1284.
53. Abuku, M.; Janssen, H.; Roels, S. Impact of Wind-Driven Rain on Historic Brick Wall Buildings in a Moderately Cold and Humid Climate: Numerical Analyses of Mould Growth Risk, Indoor Climate and Energy Consumption. *Energy Build.* **2009,** *41*(1), 101–110.
54. McIlroy de la Rosa, J. P.; Casares Porcel, M.; Warke, P. A. Mapping Stone Surface Temperature Fluctuations: Implications for Lichen Distribution and Biomodification on Historic Stone Surfaces. *J. Cult. Heritage* **2013,** *14*, 346–353.
55. Cutler, N. A.; Viles, H. A.; Ahmad, S.; McCabe, S.; Smith, B. J., Algal 'greening' and the Conservation of Stone Heritage Structures. *Sci. Total Environ.* **2013,** *442*, 152–164.
56. Adamson, C.; McCabe, S.; Warke, P. A.; McAllister, D.; Snith, B. J. The Influence of Aspect on the Biological Colonization of Stone in Northern Ireland. *Int. Biodeteriorat. Biodegradat.* **2013,** *84*, 357–366.

57. Shirakawa, M. A.; Tavares, R. G.; Gaylarde, C. C.; Taqueda, M. E. S.; Loh, K.; John, V. M. Climate as the Most Important Factor Determining Anti-Fungal Biocide Performance in Paint Films. *Sci. Total Environ.* **2010,** *408*, 5878–5886.
58. Gómez-Bolea, A.; Llop, E.; Ariño, X.; Saiz-Jimenez, C.; Bonazza, A.; Messina, P.; Sabbioni, C. Mapping the Impact of Climate Change on Biomass Accumulation on Stone. *J. Cult. Heritage* **2012,** *13*, 254–258.
59. Leissner, J.; Kilian, R.; Kotova, L.; Jacob, D.; Mikolajewicz, U.; Broström, T.; Ashley Smith, J.; Schellen, H. L.; Martens, M.; van Schijndel, J.; Antretter, F.; Winkler, M.; Bertolin, C.; Camuffo, D.; Simeunovic, G.; Vyhlídal, T. Climate for Culture: Assessing the Impact of Climate Change on the Future Indoor Climate in Historic Buildings using Simulations. *Heritage Sci.* **2015,** *3*, 38.
60. Smith, B. J.; McCabe, S.; McAllister, D.; Adamson, C.; Viles, H. A.; Curran, J. M. A Commentary on Climate Change, Stone Decay Dynamics and the 'Greening' of Natural Stone Buildings: New Perspectives on 'Deep Wetting'. *Environ. Earth Sci.* **2011,** *63*, 1691–1700.
61. Albertano, P.; Bruno, L. The Importance of Light in the Conservation of Hypogean Monuments. In *Molecular Biology and Cultural Heritage: Proceedings of the International Congress on Molecular Biology and Cultural Heritage*; Saiz-Jimenez, C., Ed.; Swets & Zeitlinger, Lisse: Sevilla, Spain, 2003; pp 171–177.
62. Albertano, P.; Bruno, L.; Bellezza, S. New Strategies for the Monitoring and Control of Cyanobacterial Films on Valuable Lithic Faces. *Plant Biosyst.* **2005,** *139*(3), 311–322.
63. Zammit, G.; De Leo, F.; Albertano, P.; Urzì, C. A Preliminary Study of Microbial Communities Colonizing Ochre-Decorated Chambers at the Hal Saflieni Hypogeum at Paola, Malta. In *11th International Congress on Deterioration and Conservation of Stone*; Lukaszewicz, J. W., Niemcewicz, P., Eds.; Wydawnictwo Naukowe Uniwersytetu Mikolaja Kopernika: Torun, Poland, 2008; pp 555–562.
64. Llop, E.; Alvaro, I.; Hernández-Mariné, M.; Sammut, S.; Gómez-Bolea, A. Colonization of Maltese Catacombs by Phototrophic Biofilms. How Much Does Light Matter? In *Progress in Cultural Heritage Preservation – EUROMED*, 2012; pp 289–293.
65. Bruno, L.; Bellezza, S.; Urzì, C.; De Leo, F. A Study for Monitoring and Conservation in the Roman Catacombs of St. Callistus and Domitilla, Rome (Italy). In *The Conservation of Subterranean Cultural Heritage*, CRC Press/Balkema: Leiden, The Netherlands, 2014; pp 37–44.
66. Borderie, F.; Alaoui-Sehmer, L.; Bousta, F.; Alaoui-Sossé, B.; Aleya, L. Cellular and Molecular Damage Caused by High UV-C Irradiation of the Cave-Harvested Green Alga *Chlorella minutissima*: Implications for Cave Management. *Int. Biodeteriorat. Biodegradat.* **2014,** *93*, 118–130.
67. Portillo, M. C.; Gonzalez, J. M. Sulfate-Reducing Bacteria are Common Members of Bacterial Communities in Altamira Cave (Spain). *Sci. Total Environ.* **2009,** *407*, 1114–1122.
68. Sarró, M. I.; García, A. M.; Rivalta, V. M.; Moreno, D. A.; Arroyo, I. Biodeterioration of the Lions Fountain at the Alhambra Palace, Granada (Spain). *Build. Environ.* **2006,** *41*(12), 1811–1820.
69. Davis, K. J.; Luttge, A. Quantifying the Relationship between Microbial Attachment and Mineral Surface Dynamics Using Vertical Scanning Interferometry (VSI). *Am. J. Sci.* **2005,** *305*(6–8), 727–751.

70. (a) Hueck, H. J. The Biodeterioration of Materials – An Appraisal. In *International biodeterioration symposium*; Walters, A. H., Elphick, J. J., Eds.; Elsevier: Southampton, 1968; pp 6–12; (b) Hueck, H. J. The Biodeterioration of Materials – An Appraisal. *Int. Biodeteriorat. Biodegradat.* **2001,** *48*(1–4), 5–11.
71. Caneva, G.; Salvadori, O.; Ricci, S.; Ceschin, S. Ecological Analysis and Biodeterioration Processes Over Time at the Hieroglyphic Stairway in the Copan (Honduras) Archaeological Site. *Plant Biosyst.* **2005,** *139*(3), 295–310.
72. Warscheid, T. The Evaluation of Biodeterioration Processes on Cultural Objects and Approaches for Their Effective Control. In *Art, Biology, and Conservation: Biodeterioration of Works of Art*; Koestler, R. J., Koestler, V. H., Charola, A. E., Nieto-Fernandez, F. E., Eds.; Metropolitan Museum of Art: New York, 2003; pp 14–27.
73. Siegesmund, S.; Snethlage, R. E. *Stone in Architecture. Properties, Durability*. Springer-Verlag: Berlin Heidelberg, 2014; p 550.
74. Skinner, H. C. W.; Jahren, A. H. Biomineralization. In *Treatise on Geochemistry*; Schlesinger, W. H., Ed.; Elsevier: Amsterdam 2003; Vol. 8, pp 117–184.
75. Gadd, G. M.; Rhee, Y. J.; Stephenson, K.; Wei, Z. Geomycology: Metals, Actinides and Biominerals (Short Survey). *Environ. Microbiol. Rep.* **2012,** *4*(3), 270–296.
76. Favero-Longo, S. E.; Castelli, D.; Salvadori, O.; Belluso, E.; Piervittori, R. Pedogenetic Action of the Lichens *Lecidea atrobrunnea*, *Rhizocarpon geographicum* gr. and *Sporastatia testudinea* on Serpentinized Ultramafic Rocks in an Alpine Environment. *Int. Biodeteriorat. Biodegradat.* **2005,** *56*, 17–27.
77. Viles, H. A. Microbial Geomorphology: A Neglected Link Between Life and Landscape. *Geomorphology* **2012,** *157–158*, 6–16.
78. Konhauser, K. *Introduction to Geomicrobiology*. Blackwell, Oxford: 2007.
79. Dupraz, C.; Reid, R. P.; Braissant, O.; Decho, A. W.; Norman, R. S.; Visscher, P. T. Processes of Carbonate Precipitation in Modern Microbial Mats. *Earth-Sci. Rev.* **2009,** *96*(3), 141–162.
80. Mapelli, F.; Marasco, R.; Balloi, A.; Rolli, E.; Cappitelli, C.; Daffonchio, D.; Borin, S. Mineral–Microbe Interactions: Biotechnological Potential of Bioweathering. *J. Biotechnol.* **2012,** *157*, 473–481.
81. Miller, A. Z.; Dionísio, A.; Sequeira Braga, M. A.; Hernández-Mariné, M.; Afonso, M. J.; Muralha, V. S. F.; Herrera, L. K.; Raabe, J.; Fernandez-Cortes, A.; Cuezva, S.; Hermosin, B.; Sanchez-Moral, S.; Chaminé, H.; Saiz-Jimenez, C. Biogenic Mn Oxide Minerals Coating in a Subsurface Granite Environment. *Chem. Geol.* **2012,** *322–323*, 181–191.
82. Urzì, C.; De Leo, F.; Bruno, L.; Pangallo, D. New Species Description, Biomineralization Processes and Biocleaning Applications of Roman Catacombs-Living Bacteria. In *The Conservation of Subterranean Cultural Heritage*; Jimenez, C. S., Ed.; CRC Press/Balkema: Leiden, The Netherlands, 2014; pp 65–72.
83. Warscheid, T.; Braams, J. Biodeterioration of Stone: A Review. *Int. Biodeteriorat. Biodegradat.* **2000,** *46*(4), 343–368.
84. Gorbushina, A. A.; Krumbein, W. E.; Volkmann, M. Rock Surfaces as Life Indicators: New Ways to Demonstrate Life and Traces of Former Life. *Astrobiology* **2002,** *2*, 203–213.
85. Cameron, S.; Urquhart, D. C. M.; Young, M. E. *Biological Growths on Sandstone Buildings: Control and Treatment*. Edinburgh: Historic Scotland, 1997.

86. Martinez, M.; Martinez, P. C.; Laverde, P.; Gaylarde, C. C. Microbiological Studies of Biofilm Present on Stones from the National Museum Building, Bogotà, Colombia. In *Molecular Biology and Cultural heritage*; Saiz-Jimenez, C., Ed.; A.A. Balkema: Sevilla, Spain, 2003; pp 259–262.
87. Crispim, C. A.; Gaylarde, C. C. Cyanobacteria and Biodeterioration of Cultural Heritage: A Review. *Microbial Ecol.* **2005,** *49*(1), 1–9.
88. Perry, T. D. I.; McNamara, C. J.; Mitchell, R. Biodeterioration of Stone. In *Scientific Examination of Art: Modern Techniques in Conservation and Analysis*; National Academies Press: Washington, DC, 2005; pp 72–84.
89. Alakomi, H. L.; Saarela, M.; Gorbushina, A. A.; Krumbein, W. E.; McCullagh, C.; Robertson, P.; Rodenacker, K. Control of Biofilm Growth through Photodynamic Treatments Combined with Chemical Inhibitors: In Vitro Evaluation Methods. In *International Conference on Heritage, Weathering and Conservation*, 2006; pp 713–717.
90. Gorbushina, A. A. Life on the Rocks. *Environ. Microbiol.* **2007,** *9*(7), 1613–1631.
91. Scheerer, S.; Ortega-Morales, O.; Gaylarde, C. Chapter 5. Microbial Deterioration of Stone Monuments – An Updated Overview. In *Advances in Applied Microbiology*, Elsevier Inc.: Amsterdam, 2009; Vol. 66, pp 97–139.
92. Katharios-Lanwermeyer, S.; Xi, C.; Jakubovics, N. S.; Rickard, A. H. Mini-Review: Microbial Coaggregation: Ubiquity and Implications for Biofilm Development. *Biofouling: J. Bioadhesion Biofilm Res.* **2014,** *30*(10), 1235–1251.
93. Vega, L. M.; Mathieu, J.; Yang, Y.; Pyle, B. H.; McLean, R. J. C.; Alvarez P. J. J. Nickel and Cadmium Ions Inhibit Quorum Sensing and Biofilm Formation without Affecting Viability in *Burkholderia multivorans. Int. Biodeteriorat. Biodegradat.* **2014,** *91*, 82–87.
94. Cappitelli, F.; Abbruscato, P.; Foladori, P.; Zanardini, E.; Ranalli, G.; Principi, P.; Villa, F.; Polo, A.; Sorlini, C. Detection and Elimination of Cyanobacteria From Frescoes: The Case of the St. Brizio Chapel (Orvieto Cathedral, Italy). *Microbial Ecol.* **2009,** *57*, 633–639.
95. Cappitelli, F.; Villa, F.; Sorlini, C., New Environmentally Friendly Approaches Against Biodeterioration of Outdoor Cultural Heritage. In *Biocolonization of Stone: Control and Preventive Methods*; Charola, A. E., McNamara, C., Koestler, R. J., Eds.; Smithsonian Institute Scholarly Press: Washington, 2011; pp 51–58.
96. Villa, F.; Pitts, B.; Lauchnor, E.; Cappitelli, F.; Stewart, P. S. Development of a Laboratory Model of a Phototroph-Heterotroph Mixed-Species Biofilm at the Stone/Air Interface. *Front. Microbiol.* **2015,** *6, 1251.*
97. McNamara, C. J.; Mitchell, R. Microbial Deterioration of Historic Stone. *Front. Ecol. Environ.* **2005,** *3*(8), 445–451.
98. Ranalli, G.; Zanardini, E.; Sorlini, C. Biodeterioration – Including Cultural Heritage. In *Encyclopedia of Microbiology*, Elsevier Academic Press: Amsterdam, 2009; pp 191–205.
99. Warscheid, T.; Leisen, H. Microbiological Studies on Stone Deterioration and Development of Conservation Measures at Angkor Wat. In *Biocolonization of Stone: Control and Preventive Methods*; Charola, A. E., McNamara, C., Koestler, R. J., Eds.; Smithsonian Institute Scholarly Press: Washington, 2011; pp 1–18.
100. De Belie, N. Microorganisms Versus Stony Materials: A Love–Hate Relationship. *Mater. Struct.* **2010,** *43*, 1191–1202.

101. Polo, A.; Gulotta, D.; Santo, N.; Di Benedetto, C.; Fascio, U.; Toniolo, L.; Villa, F.; Cappitelli, F. Importance of Subaerial Biofilms and Airborne Microflora in the Deterioration of Stonework: A Molecular Study. *Biofouling* **2012,** *28*(10), 1093–1106.
102. Ortega-Morales, B. O.; Nakamura, S.; Montejano-Zurita, G.; Camacho-Chab, J. C.; Quintana, P.; del Carmen De la Rosa-García, S. Implications of Colonizing Biofilms and Microclimate on West Stucco Masks at North Acropolis, Tikal, Guatemala. *Heritage Sci.* **2013,** 1–32.
103. McNamara, K.; Konkol, N. R.; Ross, B. P.; Mitchell, R. Characterization of Bacterial Colonization of Stone at Global and Local Scales. In *Biocolonization of Stone: Control and Preventive Methods*; Charola, A. E., McNamara, C., Koestler, R. J., Eds.; Smithsonian Institute Scholarly Press: Washington, 2011; pp 29–36.
104. Doehne, E.; Price, C. A. *Stone Conservation: An Overview of Current Research.* 2nd ed.; The Getty Conservation Institute: Los Angeles, 2010; p 158.
105. Jacobson, A. D.; Wu, L. Microbial Dissolution of Calcite at T = 28 °C and Ambient pCO_2. *Geochimica et Cosmochimica Acta* **2009,** *73,* 2314–2331.
106. McNamara, C. J.; Perry IV, T. D.; Bearce, K.; Hernandez-Duque, G.; Mitchell, R. Measurement of Limestone Biodeterioration using the Ca^{2+} Binding Fluorochrome Rhod-5N. *J. Microbiol. Methods* **2005,** *61*(2), 245–250.
107. Perry, T. D., IV; Duckworth, O. W.; Mcnamara, C. J.; Martin, S. T.; Mitchell, R. Effects of the Biologically Produced Polymer Alginic Acid On Macroscopic Calcite Dissolution Rates. *Environ. Sci. Technol.* **2004,** *38*(11), 3040–3046.
108. Cappitelli, F.; Salvadori, O.; Albanese, D.; Villa, F.; Sorlini, C., Cyanobacteria Cause Black Staining of the National Museum of the American Indian Building, Washington, DC, USA. *Biofouling* **2012,** *28*(3), 257–266.
109. Keshari, N.; Adhikary, S. P., Diversity of Cyanobacteria on Stone Monuments and Building Facades of India and Their Phylogenetic Analysis. *Int. Biodeteriorat. Biodegradat.* **2014,** *90*, 45–51.
110. Wierzchos, J.; Davila, A. F.; Artieda, O.; Cámara-Gallego, B.; De los Rios, A.; Nealson, K. H.; Valea, S.; García-González, M. T.; Ascaso, C. Ignimbrite as a Substrate for Endolithic Life in the Hyper-Arid Atacama Desert: Implications for the Search for Life on Mars. *Icarus* **2013,** *224*(2), 334–346.
111. Cámara, B.; Suzuki, S.; Nealson, K. H.; Wierzchos, J.; Ascaso, C.; Artieda, O.; de los Ríos, A. Ignimbrite Textural Properties as Determinants of Endolithic Colonization Patterns from Hyper-Arid Atacama Desert. *Int. Microbiol.* **2014,** *17*, 235–247.
112. Caneva, G.; Lombardozzi, V.; Ceschin, S.; Casanova Municchia, A.; Salvadori, O. Unusual Differential Erosion Related to the Presence of Endolithic Microorganisms (Martvili, Georgia). *J. Cult. Heritage* **2014,** *15*, 538–545.
113. Nugari, M. P.; Pietrini, A. M.; Caneva, G.; Imperi, F.; Visca, P., Biodeterioration of Mural Paintings in a Rocky Habitat: The Crypt of the Original Sin (Matera, italy). *Int. Biodeteriorat. Biodegradat.* **2009,** *63*(6), 705–711.
114. Imperi, F.; Caneva, G.; Cancellieri, L.; Ricci, M. A.; Sodo, A.; Visca, P., The Bacterial Aetiology of Rosy Discoloration of Ancient Wall Paintings. *Environ. Microbiol.* **2007,** *9*(11), 2894–2902.
115. Kyi, C. The Significance of Appropriate Sampling and Cultivation in the Effective Assessment of Biodeterioration. *Zeitschrift für Kunsttechnologie und Konservierung* **2006,** *20*(2), 344–351.

116. Gittins, M.; Vedovello, S.; Dvalishvili, M.; Kuprashvili, N. Determination of the Treatment and Restoration Needs of Medieval Frescos in Georgia. In *ICOM Committee for Conservation, ICOM-CC: 13th Triennial Meeting*, Rio de Janeiro, 2002; Vol. 2, pp 560–564.
117. Zucconi, L.; Gagliardi, M.; Isola, D.; Onofri, S.; Andaloro, M. C.; Pelosi, C.; Pogliani, P.; Selbmann, L. Biodeterioration Agents Dwelling in or on the Wall Paintings of the Holy Saviour's Cave (Vallerano, Italy). *Int. Biodeteriorat. Biodegradat.* **2012,** *70*, 40–46.
118. Bastian, F.; Jurado, V.; Nováková, A.; Alabouvette, C.; Saiz-Jimenez, C., The Microbiology of Lascaux Cave. *Microbiology* **2010,** *156*, 644–652.
119. Pohl, W.; Scheider, J. Impact of Endolithic Biofilms on Carbonate Rock Surfaces. *Nat. Stone Weathering Phenomena, Conserv. Strategies Case Stud.* **2002,** *205*, 177–194.
120. Pinna, D.; Salvadori, O. Biodeterioration Processes in Relation to Cultural Heritage Materials. Stone and Related Materials. In *Plant Biology for Cultural Heritage*; Caneva, G., Nugari, M. P., Salvadori, O., Eds.; The Getty Conservation Institute: Los Angeles, 2008; Vol. 1, pp 128–143.
121. Horath, T.; Bachofen, R. Molecular Characterization of an Endolithic Microbial Community in Dolomite Rock in the Central Alps (Switzerland). *Microbial Ecol.* **2009,** *58*(2), 290–306.
122. Ariño, X.; Saiz-Jimenez, C.; Hernandez-Marine, M. Colonization of Cryptoendolithic Niches in Mortars by Phototrophic Microrganisms. In *Protection and Conservation of the Cultural Heritage of the Mediterranean Cities*; Galan, E., Zezza, F., Eds.; Swets & Zeitlinger: Lisse, 2002; pp 127–131.
123. Golubic, S.; Pietrini, A. M.; Ricci, S. Euendolithic Activity of the Cyanobacterium *Chroococcus lithophilus* Erc. in Biodeterioration of the Pyramid of Caius Cestius, Rome, Italy. *Int. Biodeteriorat. Biodegradat.* **2015,** *100*, 7–16.
124. Lombardozzi, V.; Castrignanò, T.; D 'Antonio, M.; Casanova Municchia, A.; Caneva, G. An Interactive Database for an Ecological Analysis of Stone Biopitting. *Int. Biodeteriorat. Biodegradat.* **2012,** *73*, 8–15.
125. Grobbelaar, J. U. Lithophytic Algae: A Major Threat to the Karst Formation of Show Caves. *J. Appl. Phycol.* **2000,** *12*(3–5), 309–315.
126. Albertano, P. Methodological Approaches to the Study of Stone Alteration Caused by Cyanobacterial Biofilms in Hypogean Environments. In *Art, Biology, and Conservation: Biodeterioration of Works of Art*; Koestler, R. J., Koestler, V. H., Charola, A. E., Nieto-Fernandez, F. E., Eds.; Metropolitan Museum of Art: New York, 2003; pp 302–315.
127. Cuezva, S.; Sanchez-Moral, S.; Saiz-Jimenez, C.; Cañaveras, J. C. Microbial Communities and Associated Mineral Fabrics in Altamira Cave, Spain. *Int. J. Speleol.* **2009,** *38*(1), 83–92.
128. Saarela, M.; Alakomi, H. L.; Suihko, M. L.; Maunuksela, L.; Raaska, L.; Mattila-Sandholm, T. Heterotrophic Microorganisms in Air and Biofilm Samples from Roman Catacombs, with Special Emphasis on Actinobacteria and Fungi. *Int. Biodeteriorat. Biodegradat.* **2004,** *54*(1), 27–37.
129. Fabretti, G.; Bartolini, M.; De Cicco, M. A.; Bertinetti, M.; Montella, F.; Sclocchi, M. C.; Colaizzi, P.; Pinzari, F. Multispectral Imaging Applied to the Delimitation of

the Ecological Niche of Deteriorating Microorganisms in the Tomb of the Pancratii Roman Family. In *Art '14 Non-Destructive Investigations and Microanalysis for the Diagnostics and Onservation of Cultural and Environmental Heritage*, Madrid, 2014; p IND 39.

130. Abdel-Haliem, M. E. F.; Sakr, A. A.; Ali, M. F.; Ghaly, M. F.; Sohlenkamp, C. Characterization of Streptomyces Isolates Causing Colour Changes of Mural Paintings in Ancient Egyptian Tombs. *Microbiol. Res.* **2013,** *168*, 428–437.
131. Zanardini, E.; Abbruscato, P.; Scaramelli, L.; Onelli, E.; Realini, M.; Patrignani, G. Red Stains on Carrara Marble: A Case Study of the Certosa of Pavia, Italy. In *Art, Biology, and Conservation: Biodeterioration of Works of Art*; Koestler, R. J., V. H. K., Charola, A. E., Nieto-Fernandez, F. E., Ed.; Metropolitan Museum of Art, New York, 2003.
132. Giamello, M.; Pinna, D.; Porcinai, S.; Sabatini, G.; Siano, S. Multidisciplinary Study and Laser Cleaning Tests of Marble Surfaces of Porta Della Mandorla, Florence. In *10th Inter. Congress on Deterioration and Conservation of Stone*, Stockholm, ICOMOS Sweden, 2004; pp 841–848.
133. Grissom, C. A.; Gervais, C.; Little, N. C.; Bieniosek, G.; Speakman, R. J. Red "Staining" on Marble: Biological or Inorganic Origin? *APT Bulletin* **2010,** *41*(2/3), 11–20.
134. Gervais, C.; Grissom, C.; McNamara, C.; Konkol, N. R.; Mitchell, R. Case Study: Red Staining on Marble. In *Biocolonization of Stone: Control and Preventive Methods: Proceedings from the MCI Workshop Series*; Charola, A. E., McNamara, C. J., Koestler, R. J., Eds.; Smithsonian Institution Scholarly Press: Washington, D.C., 2011; pp 97–100.
135. Konkol, N.; McNamara, C.; Sembrat, J.; Rabinowitz, M.; Mitchell, R. Enzymatic Decolorization of Bacterial Pigments from Culturally Significant Marble *J. Cult. Heritage* **2009,** *10*(3), 362–366.
136. Anonimous, In *Concise Encyclopedia of Bioscience*, McGraw-Hill Professional: New York, 2004.
137. Milanesi, C.; Baldi, F.; Vignani, R.; Ciampolini, F.; Faleri, C.; Cresti, M. Fungal Deterioration of Medieval Wall Fresco Determined by Analysing Small Fragments Containing Copper. *Int. Biodeteriorat. Biodegradat.* **2006,** *57*, 7–13.
138. Miura, S., Conservation of Mural Paintings of the Takamatsuzuka Tumulus and Its Current Situation. In *Mural Paintings of the Silk Road: Cultural Exchanges between East and West*; Yamauchi, K., Taniguchi, Y., Uno, T., Eds.; Archetype Publications Ltd.: Tokio, 2006; pp 127–130.
139. Rosado, T.; Gil, M.; Mirão, J.; Candeias, A.; Caldeira, A. T. Oxalate Biofilm Formation in Mural Paintings Due To Microorganisms – A Comprehensive Study. *Int. Biodeteriorat. Biodegradat.* **2013,** *85*, 1–7.
140. Giustetto, R.; Gonella, D.; Bianciotto, V.; Lumini, E.; Voyron, S.; Costa, E.; Diana, E. Transfiguring Biodegradation of Frescoes in the Beata Vergine del Pilone Sanctuary (Italy): Microbial Analysis and Minero-Chemical Aspects. *Int. Biodeteriorat. Biodegradat.* **2015,** *98*, 6–18.
141. Casanova Municchia, A.; Percario, Z.; Caneva, G. Detection of Endolithic Spatial Distribution in Marble Stone. *J. Microscopy* **2014,** *256*(1):37–45.
142. Hoffland, E.; Kuyper, T. W.; Wallander, H.; Plassard, C.; Gorbushina, A. A.; Haselwandter, H.; Holmstrom, S.; Landeweert, R.; Lundstrom, U. S.; Rosling, A.; Sen, R.;

Smits, M. M.; van Hees, P. A. W.; van Breemen, N. The Role of Fungi in Weathering. *Front. Ecol. Environ.* **2004,** *2*(5), 258–264.
143. Sterflinger, K., Fungi as Geologic Agents. *Geomicrobiol. J.* **2000,** *17*(2), 97–124.
144. Sterflinger, K., Fungi: Their Role in Deterioration of Cultural Heritage. *Fungal Biol. Rev.* **2010,** *24*, 47–55.
145. Gadd, G. M.; Bahri-Esfahani, J.; Li, Q.; Rhee, Y. J.; Wei, Z.; Fomina, M.; Liang, X. Oxalate Production by Fungi: Significance in Geomycology, Biodeterioration and Bioremediation. *Fungal Biol. Rev.* **2014,** *28*, 36–55.
146. Gaylarde, C.; Otlewska, A.; Celikkol-Aydin, S.; Skóora, J.; Sulyok, M.; Pielech-Przybylska, K.; Gillatt, J.; Beech, I.; Gutarowska, B. Interactions between Fungi of Standard Paint Test Method BS3900. *Int. Biodeteriorat. Biodegradat.* **2015,** *104*, 411–418.
147. Verdier, T.; Coutand, M.; Bertron, A.; Roques, C. A Review of Indoor Microbial Growth Across Building Materials and Sampling and Analysis Methods. *Build. Environ.* **2014,** *80*, 136–149.
148. Gutarowska, B.; Zakowska, Z. Elaboration and Application of Mathematical Model for Estimation of Mould Contamination of Some Building Materials Based on Ergosterol Content Determination. *Int. Biodeteriorat. Biodegradat.* **2002,** *49*(4), 299–305.
149. Görs, S.; Schumann, R.; Häubner, N.; Karsten, U. Fungal and Algal Biomass in Biofilms on Artificial Surfaces Quantified by Ergosterol and Chlorophyll *a* as Biomarkers. *Int. Biodeteriorat. Biodegradat.* **2007,** *60*(1), 50–59.
150. De Leo, F.; Urzi, C.; de Hoog, G. S. A New Meristematic Fungus, *Pseudotaeniolina globosa*. *Antonie van Leeuwenhoek* **2003,** *83,* 351–360.
151. Tesei, D.; Marzban, G.; Zakharova, C.; Isola, D.; Selbmann, L.; Sterflinger, K. Alteration of Protein Patterns in Black Rock Inhabiting Fungi as a Response to Different Temperatures. *Fungal Biol.* **2012,** *116*, 932–940.
152. Gaylarde, P. M.; Gaylarde, C. C.; Guiamet, P. S.; Gomez De Saravia, S. G.; Videla, H. A. Biodeterioration of Mayan Buildings at Uxmal and Tulum, Mexico. *Biofouling* **2001,** *17*(1), 41–45.
153. Marvasi, M.; Donnarumma, F.; Frandi, A.; Mastromei, G.; Sterflinger, K.; Tiano, P.; Perito, B. Black Microcolonial Fungi as Deteriogens of Two Famous Marble Statues in Florence, Italy. *Int. Biodeteriorat. Biodegradat.* **2012,** *68*, 36–44.
154. Onofri, S.; Zucconi, L.; Isola, D.; Selbmann, L. Rock-Inhabiting Fungi and Their Role in Deterioration of Stone Monuments in the Mediterranean Area. *Plant Biosyst.* **2014,** *148*(2), 384–391.
155. Szczepanowska, H. M.; Cavaliere, A. R., Tutankhamun Tomb, A Closer Look at Biodeterioration: Preliminary Report. In *Schimmel: Gefahr für Mensch und Kulturgut durch Mikroorganismen*; Rauch, A., Miklin-Kniefacz, S., Harmssen, A., Eds.; Konrad Theiss Verlag GmbH & Co.: Stuttgart, 2004; pp 42–47.
156. Jurado, V.; Sanchez-Moral, S.; Saiz-Jimenez, C. Entomogenous Fungi and the Conservation of the Cultural Heritage: A Review. *Int. Biodeteriorat. Biodegradat.* **2008,** *62*, 325–330.
157. Popovic-Zivancevic, M. Microbiological Contamination of Movable Cultural Property: Yugoslavian Examples. In *Schimmel: Gefahr für Mensch und Kulturgut durch Mikroorganismen. Fungi: a Threat for People and Cultural Heritage through Microorganisms*; Rauch, A., Miklin-Kniefacz, S., Harmssen, A., Eds.; Konrad Theiss Verlag GmbH & Co.: Stuttgart, Germany, 2004; pp 158–167.

158. Chen, J.; Blume, H. P.; Beyer, L. Weathering of Rocks Induced by Lichen Colonization – A Review. *Catena* **2000,** *39,* 121–146.
159. Bjelland, T. H.; Thorseth, I. H. Comparative Studies of the Lichen–Rock Interface of Four Lichens in Vingen, Western Norway. *Chem. Geol.* **2002,** *192*, 81–98.
160. de Los Rios, A.; Wierzchos, J.; Ascaso, C. Microhabitats and Chemical Microenvironments under Saxicolous Lichens Growing on Granite. *Microbial Ecol.* **2002,** *43*(1), 181–188.
161. Adamo, P.; Violante, P. Weathering of Rocks and Neogenesis of Minerals Associated with Lichen Activity. *Appl. Clay Sci.* **2000,** *16*(5–6), 229–256.
162. Edwards, H. G. M.; Seaward, M. R. D.; Attwood, S. J.; Little, S. J.; De Oliveira, L. F. C.; Tretiach, M. FT-Raman Spectroscopy of Lichens on Dolomitic Rocks: An Assessment of Metal Oxalate Formation. *Analyst* **2003,** *128*(10), 1218–1221.
163. St. Clair, L. L.; Seaward, M. R. D. *Biodeterioration of Stone Surfaces: Lichens and Biofilms as Weathering Agents of Rocks and Cultural Heritage*. Kluver Academic Publishers: Dordrecht, The Netherlands, 2004.
164. Kiurski, J. S.; Ranogajec, J. G.; Ulhelji, A. L.; Radeka, M. M.; Bokorov, M. T. Evaluation of the Effect of Lichens on Ceramic Roofing Tiles by Scanning Electron Microscopy and Energy-Dispersive Spectroscopy Analyses. *Scanning* **2005,** *27*, 113–119.
165. Sheppard, M. A Liking for Lichen. *ICON news: the magazine of the Institute of Conservation* **2007,** *13*, 22–26.
166. Bordignon, F.; Postorino, P.; Dore, P.; Laurenzi Tabasso, M. The Formation of Metal Oxalates in the Painted Layers of a Medieval Polychrome on Stone, As Revealed by Micro-Raman Spectroscopy. *Stud. Conserv.* **2008,** *53*(3), 158–169.
167. Pinna, D.; Salvadori, O. Processes of Biodeterioration. General Mechanisms. In *Plant Biology for Cultural Heritage*; Caneva, G., Nugari, M. P., Salvadori, O., Eds.; The Getty Conservation Institute: Los Angeles, 2008; pp 15–34.
168. Aranyanark, C. Biological agents in the Weathering of Sandstone Sanctuaries in Thailand. *AICCM Bull.* **2003,** *28*, 11–15.
169. Charola, A. E.; Vale Anjos, M.; Delgado Rodrigues, J.; Barreiro, A. Developing a Maintenance Plan for the Stone Sculptures and Decorative Elements in the Gardens of the National Palace of Queluz, Portugal. *Restorat. Build. Monuments* **2007,** *13*(6), 377–388.
170. Stretch, R.; Viles, H. A. Lichen Weathering on Lanzarote Lava Flows. *Geomorphology* **2002,** *47*, 87–94.
171. Scarciglia, F.; Saporito, N.; La Russa, M. F.; Le Pera, E.; Macchione, M.; Puntillo, D.; Crisci, G. M.; Pezzino, A. Role of Lichens in Weathering of Granodiorite in the Sila Uplands (Calabria, Southern Italy). *Sedimentary Geol.* **2012,** *280*, 119–134.
172. Aghamiri, R.; Schwartzman, D. W. Weathering Rates of Bedrock by Lichens: A Mini Watershed Study. *Chem. Geol.* **2002,** *188,* 249–259.
173. Prieto, B.; Edwards, H. G. M.; Seaward, M. R. D. A Fourier Transform-Raman Spectroscopic Study of Lichen Strategies on Granite Monuments. *Geomicrobiol. J.* **2000,** *17*(1), 55–60.
174. Schiavon, N., Biodeterioration of Calcareous and Granitic Building Stones in Urban Environments. *Geol. Soc. Special Publ.* **2002,** *205*, 195–205.

175. Favero-Longo, S. E.; Borghi, A.; Tretiach, M.; Piervittori, R. In Vitro Receptivity of Carbonate Rocks to Endolithic Lichen-Forming Aposymbionts. *Mycol. Res.* **2009,** *113* (10), 1216–1227.
176. Tretiach, M.; Favero-Longo, S. E.; Crisafulli, P.; Gazzano, C.; Carbone, F.; Baiocchi, C.; Giovine, M.; Modenesi, P.; Rinino, S.; Chiapello, M.; Salvadori, O.; Piervittori, R. How do Endolithic Lichens Dissolve Carbonates? In *Biology of Lichens and Bryophytes: Lichenological Abstracts*; Nash III, T., Seaward, M., Eds.; American Bryological and Lichenological Society and the International Association for Lichenology: Tempe, AZ., 2008; p 72.
177. Matthews, J. A.; Owen, G. Endolithic Lichens, Rapid Biological Weathering and Schmidt Hammer R-values On Recently eExposed Rock Surfaces: Storbreen Glacier Foreland, Jotunheimen, Norway. *Geografiska Annaler. Series A, Phys. Geogr.* **2008,** *90*(4), 287–297.
178. Chiari, G.; Cossio, R. Ethyl Silicate Treatment's Control by Image Treatment Procedure. In *I silicati nella conservazione: indagini, esperienze e valutazioni per il consolidamento dei manufatti storici*, Associazione Villa dell'arte Torino, 2002; pp 147–156.
179. Chiari, G.; Cossio, R. Lichens on a Sandstone: Do They Cause Damage? In *Botany 2001 Plants and People*, Botanical Society of America, Albuquerque, New Mexico, 2001.
180. Gazzano, C.; Favero-Longo, S. E.; Matteucci, E.; Roccardi, A.; Piervittori, R. Index of Lichen Potential Biodeteriogenic Activity (LPBA): A Tentative Tool to Evaluate the Lichen Impact on Stonework. *Int. Biodeteriorat. Biodegradat.* **2009,** *63*(7), 836–843.
181. Hoppert, M.; Konig, S. The Succession of Biofilms on Building Stone and Its Possible Impact on Biogenic Weathering. In *Heritage, Weathering and Conservation*; Fort, R., Alvarez de Buergo, M., Gomez-Heras, M., Vazquez-Calvo, C., Eds.; Taylor & Francis Group: London, 2006; pp 311–315.
182. Concha-Lozano, N.; Gaudon, P.; Pages, J.; de Billerbeck, G.; Lafon, D.; Eterradossi, O. Protective Effect of Endolithic Fungal Hyphae on Oolitic Limestone Buildings. *J. Cult. Heritage* **2012,** *13*(2), 120–127.
183. Carballal, R.; Paz-Bermúdez, G.; Sánchez-Biezma, M. J.; Prieto, B. Lichen Colonization of Coastal Churches in Galicia: Biodeterioration Implications. *Int. Biodeteriorat. Biodegradat.* **2001,** *47*, 157–163.
184. Bungartz, F.; Garvie, L. A. J.; Nash III, T. H. Anatomy of the Endolithic Sonoran Desert Lichen *Verrucaria rubrocincta* Breuss: Implications for Biodeterioration and Biomineralization. *Lichenol.* **2004,** *36*(1), 55–73.
185. Garcia-Vallès, M.; Topal, T.; Vendrell-Saz, M. Lichen Growth as a Factor in the Physical Deterioration or Protection of Cappadocian Monuments. *Environ. Geol.* **2003,** *43*, 776–781.
186. Carter, N. E. A.; Viles, H. A. Bioprotection Explored: The Story of a Little Known Earth Surface Process. *Geomorphology* **2005,** *67*, 273–281.
187. Carter, N. E. A.; Viles, H. A. Experimental Investigations into the Interactions between Moisture, Rock Surface Temperatures and An Epilithic Lichen Cover in the Bioprotection of Limestone. *Build. Environ.* **2003,** *38*, 1225–1234.
188. Ariño, X.; Ortega-Calvo, J. J.; Gomez-Bolea, A.; Saiz-Jimenez, C. Lichen Colonization of the Roman Pavement at Baelo Claudia (Cadiz, Spain): Biodeterioration vs. Bioprotection. *Sci. Total Environ.* **1995,** *67*, 353–363.

189. Wendler, E.; Prasartet, C. Lichen Growth on Old Khmer-style Sandstone Monuments in Thailand: Damage Factor of Shelter? In *12th Triennial Meeting of the ICOM Committee for Conservation*, Lyon, 1999; Vol. 2, pp 750–754.
190. Fiol, L.; Fornós, J. J.; Ginés, A. Effects of Biokarstic Processes on the Development of Solutional Rillenkarren in Limestone Rocks. *Earth Surf. Processes Landforms* **1996,** *21*, 447–452.
191. Piervittori, R.; Salvadori, O.; Isocrono, D. Literature on Lichens and Biodeterioration of Stonework. *The Lichenologist* **2004,** *36*(2), 145–157.
192. Bartoli, F.; Casanova Municchia, A.; Futagami, Y.; Kashiwadani, H.; Moon, K. H. Biological Colonization Patterns on the Ruins of Angkor Temples (Cambodia) in the Biodeterioration vs Bioprotection Debate. *Int. Biodeteriorat. Biodegradat.* **2014,** *96*, 157–165.
193. Watt, D. Managing Biological Growth on Buildings. *Historic Churches. The Building Conservation Directory: Special Report Magazine* **2006,** *13*(November), 36–38.
194. Lisci, M.; Monte, M.; Pacini, E. Lichens and Higher Plants on Stone: A Review. *Int. Biodeteriorat. Biodegradat.* **2003,** *51*, 1–17.
195. Caneva, G.; Galotta, G.; Cancellieri, L.; Savo, V. Tree Roots and Damages in the Jewish Catacombs of Villa Torlonia (Roma). *J. Cult. Heritage* **2009,** *10*, 53–62.
196. Celesti-Grapow, L. Carlo Blasi, The Role of Alien and Native Weeds in the Deterioration of Archaeological Remains in Italy. *Weed Technol. (Invasive Weed Symposium)* **2004,** *18*, 1508–1513.
197. Signorini, M. A. The IP (Impact Index): A Contribution of Botanists to Vegetation Control in Monumental Sites. *Inf. Bot. Ital.* **1996,** *28*, 7–14.
198. Dehnen-Schmutz, K. Alien Species Reflecting History: Medieval Castles in Germany. *Diversity Distribut.* **2004,** *10*(2), 147–151.
199. Caneva, G.; Pacini, A.; Celesti Grapow, L.; Ceschin, S. The Colosseum's Use and State of Abandonment as Analysed Through its Flora. *Int. Biodeteriorat. Biodegradat.* **2003,** *51*, 211–219.
200. Viles, H. Research Methods and Sites In *Ivy on Walls*, English Heritage, 2010; pp 16–21.
201. Sternberg, T. Field and Laboratory Results In *Ivy on Walls*, English Heritage, 2010; pp 22–27.
202. Sternberg, T.; Viles, H.; Cathersides, A.; Edwards, M. Dust Particulate Absorption by Ivy (*Hedera helix* L) On Historic Walls in Urban Environments. *Sci. Total Environ.* **2010,** *409*, 162–168.
203. Rees, S., Feral Pigeons: A Forgotten Pest? In *Integrated Pest Management for Collections*; Kingsley, H., Pinniger, D., Xavier-Rowe, A., Winsor, P., Eds.; James & James: London, 2001; pp 106–113.
204. Martelli Castaldi, M.; Court, S. I falchi di Ercolano. *Forma urbis: itinerari nascosti di Roma antica* **2005,** *10*(3), 44–45.
205. Hermans, T. Overlast Door Duiven. Nuisance Pigeons. *RDMZ Info* **2005,** *9*, 1–8.
206. Haag-Wackernagel, D.; Geigenfeind, I. Protecting Buildings Against Feral Pigeons. *Eur. J. Wildl. Res.* **2008,** *54*(4), 715–721.
207. Allsopp, C.; Allsopp, D. An Updated Survey of Commercial Products Used to Protect Materials Against Biodeterioration. *Int. Biodeteriorat. Biodegradat.* **2001,** *48*(1–4), 243.

208. Kigawa, R.; Sano, C.; Kiyuna, T.; Tazato, N.; Sugiyama, J.; Takatori, K.; Kumeda, Y.; Morii, M.; Hayakawa, N.; Kawanobe, W. New Measure to Control Microorganisms in Kitora Tumulus: Effects of Intermittent UV Irradiation. *Sci. Conserv.* **2010,** *49*, 253–264.
209. Berti, S.; Pinzari, F.; Tiano, P. Control of Biodeterioration. Physical Methods. In *Plant Biology for Cultural Heritage*; Caneva, G., Nugari, M. P., Salvadori, O., Eds.; The Getty Conservation Institute, Los Angeles, 2008; pp 313–318.
210. Borderie, F.; Tête, N.; Cailhol, D.; Alaoui-Sehmer, L.; Bousta, F.; Rieffel, D.; Aleya, L.; Alaoui-Sossé, B. Factors Driving Epilithic Algal Colonization in Show Caves and New Insights into Combating Biofilm Development with UV-C Treatments. *Sci. Total Environ.* **2014,** *484*, 43–52.
211. Bartolini, M.; Pietrini, A. M.; Ricci, S. Use of UV-C Irradiation on Artistic Stonework for Control of Algae and Cyanobacteria In *Of Microbes and Art. The role of Microbial Communities in the Degradation and Protection of Cultural Heritage.*, Plenum Pub Corp: Florence, Italy, 1999; pp 221–227.
212. De Lucca, A. J.; Carter-Wientjes, C.; Williams, K. A.; Bhatnagar, D. Blue Light (470 nm) Effectively Inhibits Bacterial and Fungal Growth. *Lett. Appl. Microbiol.* **2012,** *55*(6), 460–466.
213. Hsieh, P.; Pedersen, J. Z.; Bruno, L. Photoinhibition of Cyanobacteria and Its Application in Cultural Heritage Conservation. *Photochem. Photobio.* **2014,** *90*(3), 533–543.
214. Hsieh, P.; Pedersen, J. Z.; Albertano, P. Generation of Reactive Oxygen Species Upon Red Light Exposure of Cyanobacteria from Roman Hypogea. *Int. Biodeteriorat. Biodegradat.* **2013,** *84*, 258–265.
215. Albertano, P.; Pacchiani, D.; Capucci, E. The Public Response to Innovative Strategies for the Control of Biodeterioration in Archaeological Hypogea. *J. Cult. Heritage* **2004,** *5*(4), 399–407.
216. Kawanobe, W.; Kuchitsu, N.; Hayakawa, N. Controlling Vegetation Growth on the Usuki-Magaibutsu (cliff sculpture). *Hozon kagaku* **2001,** *40*, 64–68.
217. Leavengood, P.; Twilley, J.; Asmus, J. F., Lichen Removal from Chinese Spirit Path Figures of Marble. *J. Cult. Heritage* **2000,** *1*, S71–S74.
218. Abdel-Haliem, M. E. F.; Ali, M. F.; Ghaly, M. F.; Sakr, A. A. Efficiency of Antibiotics and Gamma Irradiation in Eliminating Streptomyces Strains Isolated from Paintings of Ancient Egyptian Tombs. *J. Cult. Heritage* **2013,** *14*, 45–50.
219. Tretiach, M.; Bertuzzi, S.; Candotto Carniel, F. Heat Shock Treatments: A New Safe Approach Against Lichen Growth on Outdoor Stone Surfaces. *Environ. Sci. Technol.* **2012,** *46*(12), 6851–6859.
220. Bertuzzi, S.; Candotto Carniel, F.; Pipan, G.; Tretiach, M. Devitalization of Poikilohydric Lithobionts of Open-Air Monuments by Heat Shock Treatments: A New Case Study Centred on Bryophytes. *Int. Biodeteriorat. Biodegradat.* **2013,** *84*, 44–53.
221. Olmi, R.; Bini, M.; Cuzman, O.; Ignesti, A.; Frediani, P.; Priori, S.; Riminesi, R.; Tiano, P. Investigation of the Microwave Heating Method for the Control of Biodeteriogens on Cultural Heritage Assets. In *Regional Project TDT-BioArt*, Istituto per la Conservazione e Valorizzazione dei Beni Culturali: Firenze, Italy, 2011.
222. Cuzman, O. A.; Faraloni, C.; Pinna, D.; Riminesi, C.; Sacchi, B.; Tiano, P.; Torzillo, G. Evaluation of Treatments Efficiency Against Lichens Growing on Monumental Stones by Electrical Conductivity. *Int. Biodeteriorat. Biodegradat.* **2013,** *84*, 314–321.

223. Valentini, F.; Diamanti, A.; Carbone, M.; Bauer, E. M.; Palleschi, G. New Cleaning Strategies Based On Carbon Nanomaterials Applied to the Deteriorated Marble Surfaces: A Comparative Study with enzyme based treatments. *Appl. Surf. Sci.* **2012,** *258*, 5965–5980.
224. Odgers, D. Steam Cleaning. *The Building Conservation Directory,* Cathedral Communications Limited, Tisbury, UK, 2013.
225. Bouichou, M.; Marie-Victoire, E., *Le nettoyage des betons anciens*. Circle des Partenaires du Patrimoine: Champs sur Marne, France, 2009.
226. Pantazidou, A.; Theoulakis, P. Cyanophytes and Associated Flora at the Neoclassical Palace of Sts. George and Michael in Corfù (Greece): Aspects of Cleaning Precedures. In *4th International Symposium on the Conservation of Monuments in the Mediterranean*, A. Moropoulou, F. Z., E. Kollias, I. Papachristodoulou, Ed. Rhodes, 1997 pp 355–363.
227. Speranza, M.; Sanz, M.; Oujja, M.; de los Rios, A.; Wierzchos, J.; Pérez-Ortega, S.; Castillejo, M.; Ascaso, C. Nd-YAG Laser Irradiation Damages to *Verrucaria nigrescens*. *Int. Biodeteriorat. Biodegradat.* **2013,** *84*, 281–290.
228. DeCruz, A.; Wolbarsht, M. L.; Andreotti, A.; Colombini, M. P.; Pinna, D.; Culberson, C. F. Investigation of the Er:YAG Laser at 2.94 μm to Remove Lichens Growing on Stone. *Stud. Conserv.* **2009,** *54*(4), 268–277.
229. López, A. J.; Rivas, T.; Lamas, J.; Ramil, A.; Yáñez, A. Optimisation of Laser Removal of Biological Crusts in Granites. *Appl. Phys. A: Mater. Sci. Process.* **2010,** *100*(3), 733–739.
230. Pozo, S.; Montojo, C.; Rivas, T.; López-Díaz, A. J.; Fiorucci, M. P.; López De Silanes, M. E. Comparison between Methods of Biological Crust Removal on Granite. *Key Eng. Mater.* **2013,** *548*, 317–325.
231. Pozo-Antonio, J. S.; Fiorucci, M. P.; Rivas, T.; López, A. J.; Ramil, A.; D., D. B. Suitability of Hyperspectral Imaging Technique to Evaluate the Effectiveness of the Cleaning of a Crustose Lichen Developed on Granite. *Appl. Phys. A* **2016,** *122:100* DOI 10.1007/s00339-016-9634-5.
232. Osticioli, I.; Mascalchi, M.; Pinna, D.; Siano, S. Removal of *Verrucaria nigrescens* from Carrara Marble Artefacts using Nd:YAG Lasers: Comparison Among Different Pulse Durations and Wavelengths. *Appl. Phys. A* **2015,** *118*(4), 1517–1526.
233. Mascalchi, M.; Osticioli, I.; Riminesi, C.; Cuzman, O. A.; Salvadori, B.; Siano, S. Preliminary Investigation of Combined Laser and Microwave Treatment for Stone Biodeterioration. *Stud. Conserv.* **2015,** *60*(Supplement 1), 19–26.
234. Sanz, M.; Oujja, M.; Ascaso, C.; de los Ríos, A.; Pérez-Ortega, S.; Souza-Egipsy, V.; Wierzchos, J.; Speranza, M.; Vega Cañamares, M.; Castillejo, M. Infrared and Ultraviolet Laser Removal of Crustose Lichens on Dolomite Heritage Stone. *Appl. Surf. Sci.* **2015,** *346*, 248–255.
235. Sterflinger, K.; Sert, H., Biodeterioration of Buildings and Works of Art – Practical Implications on Restoration Practice. In *Heritage, Weathering and Conservation*; Fort, R., Alvarez de Buergo, M., Gomez-Heras, M., Vazquez-Calvo, C., Eds.; Taylor & Francis Group: London: 2006; Vol. 1, pp 299–304.
236. Lee, C. H.; Lee, M. S.; Kim, Y. T.; Kim, J. Deterioration Assessment and Conservation of a Heavily Degraded Korean Stone Buddha from the Ninth Century. *Stud. Conserv.* **2006,** *51*(4), 305–316.

237. Drewello, U.; Röllig, A.; Heck, S.; Guggenmoos, B.; Schürer, R.; von Ulmann, A.; Drewello, R. Mikroorganismen im Museum: Erfassung und Gegenmassnahmen. In *Schimmel: Gefahr für Mensch und Kulturgut durch Mikroorganismen*, Rauch, A., Miklin-Kniefacz, S., Harmssen, A., Eds.; Konrad Theiss Verlag GmbH & Co.: Stuttgart, 2004; pp 184–193.
238. Caneva, G.; Nugari, M. P.; Salvadori, O. Control of Biodeterioration and Bioremediation Techniques. Mechanical Methods. In *Plant Biology for Cultural Heritage*, The Getty Conservation Institute: Los Angeles, 2008; p 312.
239. Twilley, J.; Leavengood, D. Scientific Investigation and Large Scale Sandstone Treatments: The Washington Legislative Building. In *9th International Congress on Deterioration and Conservation of Stone*, Venice, 2000; pp 513–522.
240. Spathis, P.; Pantazidou, A.; Mavromati, M.; Papastergiadis, E. Influence of Environmental Conditions and Application of Cleaning Methods Against Biodeterioration of Marble Monuments. In *Progress in Cultural Heritage Preservation – EUROMED 2012*, Cyprus, 2012; pp 271–276.
241. Delgado Rodrigues, J.; Vale Anjos, M.; Charola, A. E. Recolonization of Marble Sculptures in a Garden Environment. In *Biocolonization of Stone: Control and Preventive Methods: Proceedings from the MCI Workshop Series*; Charola, A. E., McNamara, C. J., Koestler, R. J., Eds.; Smithsonian Institution Scholarly Press: Washington, D.C., 2011; pp 71–85.
242. Allanbrook, T.; Normandin, K. C. The Restoration of the Fifth Avenue Facades of the Metropolitan Museum of Art. *APT Bull.* **2007,** *38*(4), 45–53.
243. Asthana, K. K.; Lakhani, R. Strategies for the Restoration of Heritage Building with Innovative Active Compatible Materials. In *Studies in art and archaeological conservation: Dr. B.B. Lal Commemoration Volume*, Bisht, A. S., Singh, S. P., Eds.; Agam Kala Prakashan: Delhi, 2004; pp 9–26.
244. Giovagnoli, A.; Nugari, M. P.; Pietrini, A. M. The Ice Clean System for Removing Biological Patina: The Case of Piramide of Caio Cestio in Rome. In *10th International Conference on Non-Destructive Investigations and Microanalysis for the Diagnostics and Conservation of Cultural Heritage*, Firenze, 2011; Vol. E8, pp 1–7.
245. DePriest, P. T.; Beaubien, H. F. Case study: Deer Stones of Mongolia after Three Millennia. In *Biocolonization of Stone: Control and Preventive Methods*; Charola, A. E., McNamara, C., Koestler, R. J., Eds.; Smithsonian Institute Scholarly Press: Washington, 2011; pp 103–108.
246. Schulte, S.; Wingender, J.; Flemming, H.C., Efficacy of Biocides Against Biofilms. In *Directory of Microbicides for the Protection of Materials*; Paulus, W., Ed.; Springer: Berlin, 2008; pp 93–115.
247. Severson, K. Formulating Programs for Long-Term Care of Excavated Marble: Removing and Suppressing Biological Growth. *Stud. Conserv.* **2010,** *55*(Supplement 2), 172–177.
248. Wakefield, R. Masonry Biocides – Assessments of Efficacy and Effects on Stone. *SSCR J.: Quart. News Mag. Scottish Soc. Conserv. Restorat.* **1997,** *8*(2), 5–11.
249. Nugari, M. P.; Salvadori, O. Biocides and Treatment of Stone: Limitations and Future Prospects. In *Proceedings of Art, Biology and Conservation: Biodeterioration of Works of Art*, New-York, USA, 2003; pp 519–535.

250. Knight, D. J., Cooke, M., Eds.; *The Biocides Business: Regulation, Safety and Applications*. WILEY-VCH Verlag GmbH: Weinheim, Germany, 2002.
251. Rodin, V. B.; Zhigletsova, S. K.; Kobelev, V. S.; Akimova, N. A.; Kholodenko, V. P. Efficacy of Individual Biocides and Synergistic Combinations. *Int. Biodeteriorat. Biodegradat.* **2005,** *55*, 253–259.
252. Koestler, R. J.; Salvadori, O. Methods of Evaluating Biocides for the Conservation of Porous Building Materials. *Sci. Technol. Cult. Heritage* **1996,** *5*(1), 63–68.
253. Slaton, D.; Normandin, K. C. Masonry Cleaning Technologies. *J. Architect. Conserv.* **2005,** *11*(3), 7–31.
254. Fraise, A. P., Maillard, J. Y., Sattar, S. A., Eds.; *Russell, Hugo & Ayliffe's: Principles and Practice of Disinfection, Preservation and Sterilization.* 5th ed.; Wiley-Blackwell: Malden, Mass., 2013; 678 p.
255. Caneva, G.; Nugari, M. P.; Salvadori, O. Control of Biodeterioration and Remediation Techniques. Herbicides Treatments. In *Plant Biology for Cultural Heritage;* Caneva, G., Nugari, M. P., Salvadori, O., Eds.; The Getty Conservation Institute: Los Angeles, 2008; pp 335–340.
256. Bartolini, M.; Ricci, S.; Fazio, F. Valutazione sperimentale di erbicidi per il trattamento di colonizzazioni di muschi su pavimentazioni musive. In *Pavimentazioni storiche: uso e conservazione*; Biscontin, G., Driussi, G., Eds.; Arcadia Ricerche: Bressanone, 2006; pp 685–691.
257. Bartolini, M.; Pietrini, A. M.; Ricci, S. Valutazione dell'efficacia di alcuni nuovi biocidi per il trattamento di microflora fotosintetica e di briofite su materiali lapidei. *Bollettino ICR* **2007,** *14*, 101–111.
258. Nugari, M. P.; Salvadori, O. Biodeterioration Control of Cultural Heritage: Methods and Products. In *Molecular Biology and Cultural Heritage*; Saiz-Jimenez, C., Ed.; A.A. Balkema: Sevilla, Spain, 2003; pp 233–242.
259. Charola, A. E.; Wachowiak, M.; Keats Webb, E.; Grissom, C. A.; Vicenzi, E. P.; Chong, W.; Szczepanowska, H.; DePriest, P. Developing a Methodology to Evaluate the Effectiveness of a Biocide. In *12th International Congress on the Deterioration and Conservation of Stone*, Columbia University: New York, 2012.
260. Tretiach, M.; Crisafulli, P.; Imai, N.; Kashiwadani, H.; Hee Moon, K.; Wada, H.; Salvadori, O. Efficacy of a Biocide Tested on Selected Lichens and Its Effects on Their Substrata. *Int. Biodeteriorat. Biodegradat.* **2007,** *59*(1), 44–54.
261. Rutala, W. A.; Weber, D. J. Registration of Disinfectants Based on Relative Microbicidal Activity. *Infect. Control Hosp. Epidemiol.* **2004,** *25*(4), 333–341.
262. Videla, H. A. Prevention and Control of Biocorrosion. *Int. Biodeteriorat. Biodegradat.* **2002,** *49*(4), 259–270.
263. Wen, J.; Zhao, K.; Gu, T.; Raad, I. I. A Green Biocide Enhancer for the Treatment of Sulfate-Reducing Bacteria (SRB) Biofilms on Carbon Steel Surfaces using Glutaraldehyde. *Int. Biodeteriorat. Biodegradat.* **2009,** *63*, 1102–1106.
264. Orial, G.; Bousta, F. Les altérations biologiques et les biens patrimoniaux: Chapitre IV. Les traitements: Définitions, sélection des produits et mise en oeuvre. *Monumental: Revue Scientifique et Technique des Monuments Historiques* **2005,** *1*, 107–112.
265. Diakumaku, E.; Gorbushina, A. A.; Krumbein, W. E. Treatments of Biodeteriorated Monuments with Biocides: Acropolis, Athens, and Other sites in the Mediterranean

Basin. In *4th International Symposium on the Conservation of Monuments in the Mediterranean: New Concepts, Technologies and Materials for the Conservation of Historic Cities, Sites and Complexes*; Moropoulou, A. I., Zezza, F., Kollias, E., Papachristodoulou, I., Eds.; Technical Chamber of Greece: Rhodes, 1997; Vol. 3, pp 99–108.

266. Blazquez, A. B.; Lorenzo, J.; Flores, M.; Gomez-Alarcon, G. Evaluation of the Effect of Some Biocides Against Organisms Isolated from Historic Monuments. *Aerobiologia* **2000,** *16*(3–4), 423–428.
267. Borgioli, L.; De Comelli, A.; Pressi, G. Indagini microbiologiche per la verifica dell'efficacia di alcuni biocidi esenti da metalli pesanti. *Progetto restauro* **2006,** *38*, 24–29.
268. de los Ríos, A.; Pérez-Ortega, S.; Wierzchos, J.; Ascaso, C. Differential Effects of Biocide Treatments on Saxicolous Communities: Case Study of the Segovia Cathedral Cloister (Spain). *Int. Biodeteriorat. Biodegradat.* **2012,** *67*, 64–72.
269. Borgmann-Strahsen, R. Comparative Assessment of Different Biocides in Swimming Pool Water. *Int. Biodeteriorat. Biodegradat.* **2003,** *51* 291–297.
270. Meyer, B., Approaches to Prevention, Removal and Killing of Biofilms. *Int. Biodeteriorat. Biodegradat.* **2003,** *51*, 249–253.
271. Altieri, A.; Poggi, D.; Ricci, S. Mosaic Pavements from the Thermae of Caracalla (Rome): Biodeterioration and Methods of Control. In *VIth conference of the International Committee for the Conservation of Mosaics*; Michaelides, D., Ed.; International Committee for the Conservation of Mosaics: Nicosia, Cyprus, 1996; pp 249–260.
272. Ciarallo, A. M., Controllo integrato della vegetazione nell'area archeologica di Pompei. In *Il governo dei giardini e dei parchi storici: restauro, manutenzione, gestione*; Canestrini, F., Furia, F., Iacono, M. R., Eds.; Edizioni scientifiche italiane: Naples, 2001; pp 280–287.
273. Alberti, B.; Giammichele, L.; Morganti, L.; Tancioni, G. Il mosaico sommitale del Mausoleo Rotondo sulla Via Appia Antica: diffusione di una tipologia e intervento conservativo. In *Pavimentazioni storiche. Uso e conservazione*; Biscontin, G., Driussi, G., Eds.; Arcadia Ricerche: Bressanone, 2006; pp 157–166.
274. Magadán, M. L.; Korth, G. M. A.; Cedrola, M. L.; Charola, A. E.; Pozzobon, J. L. Addressing biocolonization in the conservation project of the Portal of the Church at San Ignacio Mini', Misiones, Argentina. *Restoration of buildings and monuments - Bauinstandsetzen und Baudenkmalpflege* **2007,** *13*(6), 401–412.
275. Papafotiou, M.; Kanellou, E.; Economou, G. Alternative Practices for Vegetation Management in Archaeological Sites – The Case of Eleusis. *Acta Horticulturae* **2010,** *881*, 879–883.
276. Schmid, T.; Panne, U.; Adams, J.; Niessner, R. Investigation of Biocide Efficacy by Photoacoustic Biofilm Monitoring. *Water Res.* **2004,** *38,* 1189–1196.
277. Nugari, M. P.; Salvadori, O. Control of Biodeterioration and Bioremediation Techniques. Stone Materials. In *Plant Biology for Cultural Heritage*; Caneva, G., Nugari, M. P., Salvadori, O., Eds.; The Getty Conservation Institute: Los Angeles, 2008; pp 326–335.
278. Mulec, J. Human Impact on Underground Cultural and Natural Heritage Sites, Biological Parameters of Monitoring and Remediation Actions for Insensitive Surfaces: Case of Slovenian Show Caves. *J. Nat. Conserv.* **2014,** *22*, 132–141.

279. Chapman, J. S.; Diehl, M. A.; Fearnside, K. B.; Leightley, L. E. *Antifungal mechanism of dichloro-N-octylisothiazolone*; International Research Group on Wood Protection: Stockholm, Sweden, 1998; p 13.
280. Gladis, F.; Eggert, A.; Karsten, U.; Schumann, R. Prevention of Biofilm Growth on Man-Made Surfaces: Evaluation of Antialgal Activity of Two Biocides and Photocatalytic Nanoparticles. *Biofouling* **2010,** *26*(1), 89–101.
281. Shirakawa, M. A.; Selmo, S. M.; Cincotto, M. A.; Gaylarde, C.; Brazolin, S.; Gambale, W. Susceptibility of Phosphogypsum to Fungal Growth and the Effect of Various Biocides. *Int. Biodeteriorat. Biodegradat.* **2002,** *49*(4), 293–298.
282. Ludensky, M. Control and Monitoring of Biofilms in Industrial Applications. *Int. Biodeteriorat. Biodegradat.* **2003,** *51*, 255–263.
283. Fonseca, A. J.; Pina, F.; Macedo, M. F.; Leal, N.; Romanowska-Deskins, A.; Laiz, L.; Gómez-Bolea, A.; Saiz-Jimenez, C. Anatase as an Alternative Application for Preventing Biodeterioration of Mortars: Evaluation and Comparison with Other Biocides. *Int. Biodeteriorat. Biodegradat.* **2010,** *64*, 388–396.
284. Savvides, A. I.; Nikolakopoulou, T. I.; Kyratsous, N.; Katsifas, E. A.; Kanini, G.; Karagouni, A. D. Bacterial Deterioration of Marble Monuments: A Case Study of The Conservation Project of Acropolis Monuments. *Geomicrobiol. J.* **2014,** *31*(8), 726–736.
285. Jablonkai, I. Molecular Mechanism of Action of Herbicides. In *Herbicides – Mechanisms and Mode of Action*; Hasaneen, M. N., Ed.; InTech: Rijeka, Croatia, 2011; pp 1–22.
286. Ginell, W. S.; Kumar, R.; Doehne, E. Conservation Studies on Limestone from the Maya Site at Xunantunich, Belize. In *Materials Research Society Symposium*, 1995; Vol. 352, pp 813–821.
287. Tomaselli, L.; Lamenti, G.; Tiano, P. Chlorophyll Fluorescence for Evaluating Biocide Treatments Against Phototrophic Biodeteriogens. *Ann. Microbiol.* **2002,** *52*(3), 197–206.
288. Ascaso, C.; Wierzchos, J.; Souza-Egipsy, V.; De Los Rios, A.; Rodrigues, J. D. In situ Evaluation of the Biodeteriorating Action of Microorganisms and the Effects of Biocides on Carbonate Rock of the Jeronimos Monastery (Lisbon). *Int. Biodeteriorat. Biodegradat.* **2002,** *49*(1), 1–12.
289. Borgioli, L.; Pressi, G.; Secondin, S. Valutazione dell'efficacia di prodotti biocidi attraverso test microbiologici di laboratorio e saggi applicativi in cantiere. Evaluation of the Effectiveness of Biocide Products in Microbiological Laboratory Tests and Sample Applications at the Work Site. *Progetto Restauro* **2003,** *26*, 39–46.
290. Guiamet, P. S.; Rosato, V.; Gómez de Saraviae, S.; García, A. M.; Moreno, D. A. Biofouling of Crypts of Historical and Architectural interest at La Plata Cemetery (Argentina). *J. Cult. Heritage* **2012,** *13*, 339–344.
291. Young, M. E.; Alakomi, H. L.; Fortune, I.; Gorbushina, A. A.; Krumbein, W. E.; Maxwell, I.; McCullagh, C.; Robertson, P.; Saarela, M.; Valero, J.; Vendrell, M. Development of a Biocidal Treatment Regime to Inhibit Biological Growths on Cultural Heritage: BIODAM. *Environ. Geol.* **2008,** *56*(3–4), 631–641.
292. Monte, M.; Ferrari, R.; Lonati, G.; Malagodi, M. Biocidal Activity on Microbic Bio-Deteriogens and on Frescoes. In *Science and Technology for the Safeguard of Cultural Heritage in the Mediterranean Basin*, Elsevier: Paris, 1999; Vol. 1, pp 633–639.

293. Tretiach, M.; Bertuzzi, S.; Salvadori, O. Chlorophyll *a* Fluorescence as a Practical Tool for Checking the Effects of Biocide Treatments on Endolithic Lichens. *Int. Biodeteriorat. Biodegradat.* **2010,** *64*, 452–460.
294. Bastian, F.; Alabouvette, C.; Jurado, V.; Saiz-Jimenez, C. Impact of Biocide Treatments on the Bacterial Communities of the Lascaux Cave. *Naturwissenschaften* **2009,** *96*(7), 863–868.
295. Monte, M.; Nichi, D. Effects of Two Biocides in the Elimination of Lichens from Stone Monuments: Preliminary Findings. *Sci. Technol. Cult. Heritage* **1997,** *6*(2), 209–216.
296. Tomaselli, L.; Lamenti, G.; Tiano, P. Diagnostic Tools for Monitoring Phototrophic Biodeteriogens. In *Molecular Biology and Cultural Heritage: Proceedings of the International Congress on Molecular Biology and Cultural Heritage, 4–7 March 2003, Sevilla, Spain*; Saiz-Jimenez, C., Ed.,; A.A. Balkema: Rotterdam, 2003; pp 247–251.
297. Denyer, S. P.; Stewart, G. S. A. B. Mechanisms of Action of Disinfectants. *Int. Biodeteriorat. Biodegradat.* **1998,** *41*(3–4), 261–268.
298. Bartolini, M.; Ricci, S. Alterazioni cromatiche della pietra indotte dall'applicazione di biocidi su patine biologiche. *Bollettino ICR* **2009,** *18–19*, 10–22.
299. Delgado Rodrigues, J.; Valero, J. A Brief Note on the Elimination of Dark Stains of Biological Origin. *Stud. Conserv.* **2003,** *48*(1), 17–22.
300. Abdelhafez, A. A. M.; El-Wekeel, F. M.; Ramadan, E. M.; Abed-Allah, A. A. Microbial Deterioration of Archaeological Marble: Identification and Treatment. *Ann. Agric. Sci.* **2012,** *57*(2), 137–144.
301. Laughlin, R. B. J.; Lindén, O. Tributyltin: Contemporary Environmental Issues. *Ambio* **1987,** *16*(5), 252–256.
302. Altieri, A.; Coladonato, M.; Lonati, G.; Malagodi, M.; Nugari, M. P.; Salvadori, O. Effects of Biocidal Treatments on Some Italian Lithotypes Samples. In *4th International Symposium on the Conservation of Monuments in the Mediterranean: New Concepts, Technologies and Materials for the Conservation of Historic Cities, Sites and Complexes*, Techical Chamber of Greece, Rhodes, 1997; Vol. 3, pp 31–40.
303. Pinck, C.; Balzarotti-Kämmlein, R.; Mansch, R. Biocidal Efficacy of Algophase Against Nitrifying Bacteria. In *Protection and Conservation of the Cultural Heritage of the Mediterranean Cities: Proceedings of the 5th International Symposium on the Conservation of Monuments in the Mediterranean Basin*; Galán, E., Zezza, F., Eds.; A.A. Balkema: Sevilla, Spain, 2002; pp 449–453.
304. Quayle, N. J. T. Report on Investigation into Right Biocides. *Conservation News (United Kingdom Institute for Conservation of Historic and Artistic Works)* **1995,** *57*, 25–28.
305. Green, F. I.; Clausen, C. A. Copper Tolerance of Brown-Rot Fungi: Time Course of Oxalic Acid Production. *Int. Biodeteriorat. Biodegradat.* **2003,** *51*, 145–149.
306. Cloete, T. E. Resistance Mechanisms of Bacteria to Antimicrobial Compounds. *Int. Biodeteriorat. Biodegradat.* **2003,** *51*, 277–282.
307. Morton, L. H. G.; Greenway, D. L. A.; Gaylarde, C. C.; Surman, S. B. Consideration of Some Implications of the Resistance of Biofilms to Biocides. *Int. Biodeteriorat. Biodegradat.* **1998,** *41*(3–4), 247–259.
308. Paulus, W. *Directory of Microbicides for the Protection of Materials: A Handbook*. Springer: Dordrecht, 2008; p 787.

309. Alakomi, H. L.; Paananen, A.; Suihko, M. L.; Helander, I. M.; Saarela, M. Weakening Effect of Cell Permeabilizers on Gram-Negative Bacteria Causing Biodeterioration. *Appl. Environ. Microbiol.* **2006,** *72*(7), 4695–4703.
310. Heinzel, M. Phenomena of Biocide Resistance in Microorganisms. *Int. Biodeteriorat. Biodegradat.* **1998,** *41*, 225–234.
311. Gostincar, C.; Grube, M.; Gunde-Cimerman, N. Evolution of Fungal Pathogens in Domestic Environments. *Fungal Biol.* **2011,** *115*, 1008–1018.
312. Gazzano, C.; Favero-Longo, S. E.; Iacomussi, P.; Piervittori, R. Biocidal Effect of Lichen Secondary Metabolites Against Rock-Dwelling Microcolonial Fungi, Cyanobacteria and Green Algae. *Int. Biodeteriorat. Biodegradat.* **2013**, *84*, 300–306.
313. Caneva, G.; Nugari, M. P.; Salvadori, O. Control of Biodeterioration and Bioremediation Techniques. Chemical Methods. In *Plant Biology for Cultural Heritage*. The Getty Conservation Institute: Los Angeles, 2008; pp 318–321.
314. Schoknecht, U.; Gruycheva, J.; Mathies, H.; Bergmann, H.; Burkhardt, M. Leaching of Biocides Used in Façade Coatings Under Laboratory Test Conditions. *Environ. Sci. Technol.* **2009,** *43*(24), 9321–9328.
315. Longano, D.; Ditaranto, N.; Sabbatini, L.; Torsi, L.; Cioffi, N. Synthesis and Antimicrobial Activity of Copper Nanomaterials. In *Nano-Antimicrobials. Progress and Prospects*; Nicola Cioffi, N., Rai, M., Eds.; Springer: 2012; pp 119–150.
316. Ju-Nam, Y.; Lead, J. R. Manufactured Nanoparticles: An Overview of Their Chemistry, Interactions and Potential Environmental Implications. *Sci. Total Environ.* **2008,** *400*, 396–414.
317. Jurado, V.; Miller, A. Z.; Cuezva, S.; Fernandez-Cortes, A.; Benavente, D.; Rogerio-Candelera, M. A.; Reyes, J.; Cañaveras, J. C.; Sanchez-Moral, S.; Saiz-Jimenez, C. Recolonization of Mortars by Endolithic Organisms on the Walls of San Roque Church in Campeche (Mexico): A Case of Tertiary Bioreceptivity. *Construct. Build. Mater.* **2014,** *53*, 348–359.
318. Akatova, E.; Laiz, L.; Gonzalez, J. M.; Saiz-Jimenez, C. Natural Re-colonization of Restored Mural Paintings. In *International Conference on Heritage, Weathering and Conservation*; Fort, R., Alvarez de Buergo, M., Gomez-Heras, M., Vazquez-Calvo, C., Eds.; 2006; Vol. 1, pp 381–386.
319. Kigawa, R.; Mabuchi, H.; Sano, C.; Miura, S. Biological Issues in Kitora Tumulus During Relocation Work of the Mural Paintings. *Hozon kagaku* **2008,** *47*, 129–134.
320. Nascimbene, J.; Salvadori, O. Lichen Recolonization on Restored Calcareous Statues of Three Venetian Villas. *Int. Biodeteriorat. Biodegradat.* **2008,** *62*(3), 313–318.
321. Nascimbene, J.; Salvadori, O.; Nimis, P. L. Monitoring Lichen Recolonization on a Restored Calcareous Statue. *Sci. Total Environ.* **2009,** *407*, 2420–2426.
322. Nascimbene, J.; Thüs, H.; Marini, L.; Nimis, P. L. Early Colonization of Stone by Freshwater Lichens of Restored Habitats: A Case Study in Northern Italy. *Sci. Total Environ.* **2009,** *407*, 5001–5006.
323. Bracci, S.; Melo, M. J.; Tiano, P. Comparative Study on Durability of Different Treatments on Sandstone After Exposure in Natural Environment. In *The Silicates in Conservative Treatments. Tests, Improvements and Evaluation of Consolidating Performance*, Fondazione per le Biotecnologie and Associazione Villa dell'Arte: Torino, 2002; pp 129–135.

324. Lee, G.; Wilson, P. Cawthorn Camps (North Yorkshire): Damage and Survival. In *Preserving archaeological remains in situ?*; Nixon, T. J. P., Ed.; Museum of London. Archaeology Service: London, 2004; pp 173–178.
325. Sasse, H. S.; Snethlage, R. Methods for the Evaluation of Stone Conservation Treatments. In *Dahlem Workshop on Saving our Architectural Heritage*; Baer, N. S., Snethlage, R., Eds.; John Wiley & Sons Ltd: Berlin, Germany, 1996.
326. Quaresima, R.; Di Giuseppe, E.; Volpe, R. Impiego di biocidi per la rimozione della microflora algale. The Use of Biocides to Remove Algae. In *La pulitura delle superfici dell'architettura*; Biscontin, G., Driussi, G., Eds.; Libreria Progetto Editore: Bressanone, Italy, 1995; pp 267–275.
327. De Muynck, W.; Maury Ramirez, A.; De Belie, N.; Verstraete, W. Evaluation of Strategies to Prevent Algal Fouling on White Architectural and Cellular Concrete. *Int. Biodeteriorat. Biodegradat.* **2009,** *63,* 679–689.
328. Holah, J. T., CEN/TC 216: Its Role in Producing Current and Future European Disinfectant Testing Standards. *Int. Biodeteriorat. Biodegradat.* **2003,** *51*, 239–243.
329. Fortune, I. S.; Alakomi, H. L.; Young, M. E.; Gorbushina, A. A.; Krumbein, W. E.; Maxwell, I.; McCullagh, C.; Robertson, P.; Saarela, M.; Valero, K.; Vendrell, M. Assessing the Suitability of Novel Biocides for use on Historic Surfaces. In *Heritage Microbiology and Science;* May, E., Jones, M., Mitchell, J., Eds.; RSC Publishing, London, 2008; pp 51–61.
330. Fraise, A. P. Historical Introduction. In *Russel, Hugo & Ayliffe's Priciples and Pratice of Disinfection, Preservation and Sterilization*, Wiley-Blackwell: Hoboken, New Jersey, 2013; pp 1–4.
331. American Chemical Society. New 'green' Pesticides are First to Exploit Plant Defenses in Battle of the Fungi. *ScienceDaily.* **2009**.
332. Salta, M.; Wharton, J. A.; Dennington, S. P.; Stoodley, P.; Stokes, K. R. Anti-biofilm Performance of Three Natural Products Against Initial Bacterial Attachment. *Int. J. Mol. Sci.* **2013,** *14*, 21757–21780.
333. Borges, A.; Simões, L. C.; Saavedra, M. J.; Simões, M. The Action of Selected Isothiocyanates on Bacterial Biofilm Prevention and Control. *Int. Biodeteriorat. Biodegradat.* **2014,** *86*, 25–33.
334. Cuzman, O. A.; Faraoni, C.; Turchetti, T.; Camaiti, M.; Tiano, P. Influence of Some Natural Compounds on Freshwater Microfoulants. In *International Conference on Antimicrobial Research*; Mendez-Vilas, A., Ed.; World Scientific Publishing Co.: Valladolid, Spain, 2010; pp 104–108.
335. Cuzman, O. A.; Camaiti, M.; Sacchi, B.; Tiano, P. Natural Antibiofouling Agents as New Control Method for Phototrophic Biofilms Dwelling on Monumental Stone Surfaces. *Int. J. Conserv. Sci.* **2011,** *2*(1), 3–16.
336. Stupar, M.; Grbić, M. L.; Džamić, A.; Unković, N.; Ristić, M.; Jelikić, A.; Vukojević, J. Antifungal Activity of Selected Essential Oils and Biocide Benzalkonium Chloride against the Fungi Isolated from Cultural Heritage Objects. *South Afr. J. Bot.* **2014,** *93*, 118–124.
337. Sasso, S.; Miller, A. Z.; Rogerio-Candelera, M. A.; Cubero, B.; Coutinho, M. L.; Scrano, L.; Bufo, S. A. Potential of Natural Biocides for Biocontrolling Phototrophic Colonization on Limestone. *Int. Biodeteriorat. Biodegradat.* **2016,** *107*, 102–110.

338. Thakkar, K. N.; Mhatre, S. S.; Parikh, R. Y. Biological Synthesis of Metallic Nanoparticles. *Nanomed.: Nanotechnol. Biol. Med.* **2010,** *6*, 257–262.
339. Ditaranto, N.; Loperfido, S.; Van der Werf, I.; Mangone, A.; Cioffi, N.; Sabbatini, L. Synthesis and Analytical Characterization of Copper-Based Nanocoatings for Bioactive Stone Artworks Treatment. *Analyt. Bioanalyt. Chem.* **2011,** *299*, 473–481.
340. Hashimoto, K.; Irie, H.; Fujishima, A. TiO_2 Photocatalysis: A Historical Overview and Future Prospects. *Jpn. J. Appl. Phys.* **2005,** *Part 1*(44), 8269–8285.
341. Fujishima, A.; Zhang, X.; Tryk, D. A. Heterogeneous Photocatalysis: From Water Photolysis to Applications in Environmental Cleanup. *Int. J. Hydrogen Energy* **2007,** *32*, 2664–2672.
342. Lee, S. H.; Pumprueg, S.; Moudgil, B.; Sigmund, W. Inactivation of Bacterial Endospores by Photocatalytic Nanocomposites. *Colloids Surf. B – Biointerfaces* **2005,** *40*, 93–98.
343. Gopal, J.; George, R. P.; Muraleedharan, P.; Khatak, H. S. Photocatalytic Inhibition of Microbial Adhesion by Anodized Titanium. *Biofouling* **2004,** *20*, 167–175.
344. Colangiuli, D.; Calia, A.; Bianco, N. Novel Multifunctional Coatings with Photocatalytic and Hydrophobic Properties for the Preservation of the Stone Building Heritage. *Construct. Build. Mater.* **2015,** *93*, 189–196.
345. Pinna, D.; Lega, A. M.; Mazzotti, V. Monitoraggio e manutenzione di manufatti ceramici situati all'aperto: gli aspetti relativi al degrado biologico nel caso di alcuni monumenti faentini. *Arkos* **2009,** *21*, 18–25.
346. Eyssautier-Chuine, S.; Gommeaux, M.; Moreau, C.; Thomachot-Schneider, C.; Fronteau, G.; Pleck, J.; Kartheuser, B. Assessment of New Protective Treatments for Porous Limestone Combining Water-Repellency and Anti-Colonization Properties. *Quart. J. Eng. Geol. Hydrogeol.* **2014,** *47*, 177–187.
347. Martinez, T.; Bertron, A.; Escadeillas, G.; Ringot, E. Algal Growth Inhibition on Cement Mortar: Efficiency of Water Repellent and Photocatalytic Treatments Under UV/VIS Illumination. *Int. Biodeteriorat. Biodegradat.* **2014,** *89*, 115–125.
348. Hu, H.; Ding, S.; Katayama, Y.; Kusumi, A.; Li, S. X.; de Vries, R. P.; Wang, J.; Yu, X. Z.; Gu, J. D. Occurrence of *Aspergillus allahabadii* on Sandstone at Bayon Temple, Angkor Thom, Cambodia. *Int. Biodeteriorat. Biodegradat.* **2013,** *76*, 112–117.
349. Valentini, F.; Diamanti, A.; Palleschi, G. New Bio-cleaning Strategies on Porous Building Materials Affected by Biodeterioration Event. *Appl. Surf. Sci.* **2010,** *256*, 6550–6563.
350. Warscheid, T. Heritage Research and Practice: Towards a Better Understanding? In *Heritage Microbiology and Science: Microbes, Monuments and Maritime Materials*; May, E., Jones, M., Mitchell, J., Eds.; RSC Publishing: London, 2008; pp 11–26.
351. Maekawa, S.; Toledo, F. Sustainable Climate Control for Historic Buildings in Hot and Humid Regions. *International Preservation News: A Newsletter of the IFLA Programme on Preservation and Conservation* **2008,** *44*, 24–29.
352. Napp, M.; Kalamees, T. Energy Use and Indoor Climate of Conservation Heating, Dehumidification and Adaptive Ventilation for the Climate Control of a Mediaeval Church in a Cold Climate. *Energy Build.* **2015,** *108*, 61–71.
353. Ruga, L.; Orlandi, F.; Romano, B.; Fornaciari, M. The Assessment of Fungal Bioaerosols in the Crypt of St. Peter in Perugia (Italy). *Int. Biodeteriorat. Biodegradat.* **2015,** *98*, 121–130.

354. Gu, J. D., Microbiological Deterioration and Degradation of Synthetic Polymeric Materials: Recent Research Advances. *Int. Biodeteriorat. Biodegradat.* **2003,** *52*(2), 69–91.
355. Cappitelli, F.; Zanardini, E.; Sorlini, C. The Biodeterioration of Synthetic Resins Used in Conservation. *Macromol. Biosci.* **2004,** *4*(4), 399–406.
356. Gysels, K.; Delalieux, F.; Deutsch, F.; Van Grieken, R.; Camuffo, D.; Bernardi, A.; Sturaro, G.; Busse, H. J.; Wieser, M. Indoor Environment and Conservation in the Royal Museum of Fine Arts, Antwerp, Belgium. *J. Cult. Heritage* **2004,** *5*(2), 221–230.
357. Nugari, M. P.; Roccardi, A.; Cacace, C. La Cripta di San Magno ad Anagni: indagini aerobiologiche e microclimatiche. *Bollettino Istituto Centrale per il Restauro Nuova Serie* **2007,** *14*, 76–80.
358. Saiz-Jimenez, C.; Gonzalez, J. M. Aerobiology and Cultural Heritage: Some Reflections and Future Challenges. *Aerobiologia* **2007,** *23*, 89–90.
359. Mandrioli, P.; Pasquariello, G.; Roccardi, A. Prevention of Cultural Heritage. Aerobiological Monitoring. In *Plant Biology for Cultural Heritage*; Caneva, G., Nugari, M. P., Salvadori, O., Eds.; The Getty Conservation Institute: Los Angeles, 2008; pp 298–302.
360. Horner, E. W. Assessment of the Indoor Environment: Evaluation of Mold Growth Indoors. *Immunol. Allergy Clin. N. Am.* **2003,** *23*, 519–553.
361. Abe, K.; Murata, T. A Prevention Strategy Against Fungal Attack for the Conservation of Cultural Assets Using A Fungal Index. *Int. Biodeteriorat. Biodegradat.* **2014,** *88*, 91–96.
362. Borderie, F.; Tête, N.; Cailhol, D.; Alaoui-Sehmer, L.; Bousta, F.; Rieffel, D.; Aleya, L.; Alaoui-Sossé, B. Factors Driving Epilithic Algal Colonization in Show Caves and New Insights into Combating Biofilm Development with UV-C Treatments. *Sci. Total Environ.* **2014,** *484*, 43–52.
363. Cuzman, O. A.; D., T.; Fratini, F.; Mazzei, B.; C., R.; Tiano, P. Assessing and Facing the Biodeteriogenic Presence Developed in the Roman Catacombs of Santi Marco, Marcelliano and Damaso, Italy. *Eur. J. Sci. Theol.* **2014,** *10*(3), 185–197.
364. Bartolini, M.; Nugari, M. P. Resistance to Biodeterioration od Some Products Used for Rising Damp Barrier. In *Protection and Conservation of the Cultural Heritage of the Mediterranean Cities*; Zezza, G. a., Ed.; Swets & Zeitlinger, Lisse: 2002; pp 397–400.
365. Altieri, A.; Pinna, D. Prevention of Bioderioration. Outdoor Environments. In *Plant Biology for Cultural Heritage*; Caneva, G., Nugari, M. P., Salvadori, O., Eds.; The Getty Conservation Institute: Los Angeles, 2008; pp 197–198.
366. Laurenti, M. C. Il progetto delle coperture in aree archeologiche. In *Le coperture delle aree archeologiche. Museo aperto*, Gangemi Editore: Roma, 2006; pp 15–18.
367. Rizzi, G. Sheltering the Mosaics of Piazza Armerina: Issues of Conservation and Presentation *Site Preservation Program Heritage, Conservation, and Archaeology* [Online], 2008, p. 1–3.
368. Santopuoli, N.; Curuni, S. A.; Maietti, F.; Vanacore, S.; Seccia, L.; Troiani, E.; Virgili, V.; De Vincenzo, D.; Concina, E.; Tapini, L. Il consolidamento degli apparati decorativi mediante dispositivi a memoria di forma: il progetto di ricerca sui dipinti murali di via dell'Abbondanza a Pompei. The consolidation of decorations by means of shape memory alloys: the research project on the wall paintings in the Street of

Abundance in Pompeii. In *Il consolidamento degli apparati architettonici e decorativi: conoscenze, orientamenti, esperienze*; Biscontin, G., Driussi, G., Eds.; Arcadia Ricerche: Bressanone, Italy, 2007; pp 439–448.
369. Lithgow, K.; Curteis, T.; Bullock, L. Managing External Environments through Preventive Conservation: The Investigation and Control of Environmentally-Caused Deterioration of Decorative Surfaces in the Marlborough Pavilion, Chartwell, Kent. In *Museum Microclimates*; Padfield, T., Borchersen, K., Christensen, M. C., Eds.; Nationalmuseet (Denmark): Copenhagen, 2007; pp 175–184.
370. Wessel, D. P. Case study: Field Observations on the Effectiveness of Zinc Strips to Control Biocolonization of Stone. In *Biocolonization of Stone: Control and Preventive Methods*; Charola, A. E., McNamara, C., Koestler, R. J., Eds.; Smithsonian Institute Scholarly Press: Washington, 2011; pp 109–112.
371. Wessel, D. P. The Use of Metallic Oxides in Control of Biological Growth on Outdoor Monuments. In *Art, Biology, and Conservation: Biodeterioration of Works of Art*; Koestler, R. J., Koestler, V. H., Charola, A. E., Nieto-Fernandez, F. E., Eds.; Metropolitan Museum of Art: New York, 2003; pp 536–551.
372. Magadán, M. L.; Korth, G. M. A.; Cedrola, M. L.; Charola, A. E.; Pozzobon, J. L. Case Study: Biocontrol Testing at the San Ignacio Miní Jesuit-Guaraní Mission, Misiones, Argentina. In *Biocolonization of Stone: Control and Preventive Methods*; Charola, A. E., McNamara, C., Koestler, R. J., Eds.; Smithsonian Institute Scholarly Press: Washington, 2011; pp 91–98.
373. Papida, S.; Dionysis Garbis, D.; Papakonstantinou, E.; Karagouni, A. D. Biodeterioration Control for the Athens Acropolis Monuments: Strategy and Constraints. *Stud. Conserv.* **2010,** *55*(Supp. 2), 74–79.
374. Prieto, B.; Sanmartín, P.; Silva, C.; Vázquez-Nion, D.; Silva, B. Deleterious Effect Plastic-Based Biocides on Back-Ventilated Granite Facades. *Int. Biodeteriorat. Biodegradat.* **2014,** *86*, 19–24.
375. Gladis, F.; Schumann, R. A Suggested Standardised Method for Testing Photocatalytic Inactivation of Aeroterrestrial Algal Growth on TiO_2-Coated Glass. *Int. Biodeteriorat. Biodegradat.* **2011,** *65*, 415–422.
376. von Werder, J.; Venzmer, H. The potential of Pulse Amplitude Modulation Fluorometry for Evaluating the Resistance of Building Materials to Algal Growth. *Int. Biodeteriorat. Biodegradat.* **2013,** *84*, 227–235.
377. Sedlbauer, K.; Krus, M.; Fitz, C.; Künzel, H. M. Reducing the Risk of Microbial Growth on Insulated Walls by PCM Enhanced Renders and IR Reflecting Paints. In *12th International Conference on Durability of Building Materials and Components*, Porto, Portugal, 2011; pp 93–99.
378. Cheng, M. D.; Pfiffner, S. M.; Miller, W. A.; Berdahl, P. Chemical and Microbial Effects of Atmospheric Particles on the Performance of Steep-Slope Roofing Materials. *Build. Environ.* **2011,** *46*, 999–1010.
379. Henriques, F.; Charola, A. E.; Moreira Rato, V.; Fraria Rodriguez, P. Development of Biocolonization Resistant Mortars: Preliminary Results. *Restorat. Build. Monuments,* **2007,** *13*(6), 389–400.
380. Ivanov-Omskii, V. I.; Panina, L. K.; Yastrebov, S. G. Amorphous Hydrogenated Carbon Doped with Copper as Antifungal Protective Coating. *Carbon* **2000,** *38*(4), 495–499.

381. Munafò, P.; Goffredo, G. B.; Quagliarini, E. TiO_2-Based Nanocoatings for Preserving Architectural Stone Surfaces: An Overview. *Construct. Build. Mater.* **2015,** *84*, 201–218.
382. Franzoni, E.; Fregni, A.; Gabrielli, R.; Graziani, G.; Sassoni, E. Compatibility of Photocatalytic TiO_2-Based Finishing for Renders in Architectural Restoration: A Preliminary Study. *Build. Environ.* **2014,** *80*, 125–135.
383. Graziani, L.; Quagliarini, E.; D'Orazio, M., TiO_2-Treated Different Fired Brick Surfaces for Biofouling Prevention: Experimental and Modelling Results. *Ceram. Int.* **2016,** *42*, 4002–4010.
384. Graziani, L.; Quagliarini, E.; Osimani, A.; Aquilanti, L.; Clementi, F.; Yéprémian, C.; Lariccia, V.; Amoroso, S.; D'Orazio, M. Evaluation of Inhibitory Effect of TiO_2 Nanocoatings Against Microalgal Growth on Clay Brick Façades under Weak UV Exposure Conditions. *Build. Environ.* **2013,** *64*, 38–45.
385. La Russa, M. F.; Macchia, A.; Ruffolo, S. A.; De Leo, F.; Barberio, M.; Barone, P.; Crisci, G. M.; Urzì, C. C. Testing the Antibacterial Activity of Doped TiO_2 for Preventing Biodeterioration of Cultural Heritage Building Materials. *Int. Biodeteriorat. Biodegradat.* **2014,** *96*, 87–96.
386. Villa, F.; Cappitelli, F. Plant-Derived Bioactive Compounds at Sub-lethal Concentrations: Towards Smart Biocide-free Antibiofilm Strategies. *Phytochem. Rev.* **2013,** *12*(1), 245–254.
387. Davies, J. How to Discover New Antibiotics: Harvesting the Parvome. *Curr. Opin. Chem. Biol.* **2011,** *15*(1), 5–10.
388. Pagliaro, M.; Ciriminna, R.; Palmisano, G. Silica-based Hybrid Coatings. *J. Mater. Chem.* **2009,** *19*, 3116–3126.
389. Malagodi, M.; Nugari, M. P.; Altieri, A.; Lonati, G. Effects of Combined Application of Biocides and Protectives on Marble. In *9th International Congress on Deterioration and Conservation of Stone*, Venice, 2000; pp 225–233.
390. Ariño, X.; Canals, A.; Gomez-Bolea, A.; Saiz-Jimenez, C. Assessment of the Performance of a Water-Repellent/Biocide Treatment after 8 Years. In *Protection and Conservation of the Cultural Heritage of the Mediterranean Cities*, Swets & Zeitlinger, Lisse: 2002; pp 121–125.
391. Urzì, C.; De Leo, F. Evaluation of the Efficiency of Water-Repellent and Biocide Compounds Against Microbial Colonization of Mortars. *Int. Biodeteriorat. Biodegradat.* **2007,** *60*, 25–34.
392. Moreau, C.; Verges-Belmin, V.; Leroux, L.; Orial, G.; Fronteau, G.; Barbin, V. Water-Repellent and Biocide Treatments: Assessment of the Potential Combinations. *J. Cult. Heritage* **2008,** *9*(4), 394–400.
393. Majumdar, P.; Lee, E.; Gubbins, N.; Christianson, D. A.; Stafslien, S. J.; Daniels, J.; VanderWal, L.; J., B.; Chisholm, B. J. Combinatorial Materials Research Applied to the Development of New Surface Coatings XIII: An Investigation of Polysiloxane Antimicrobial Coatings Containing Tethered Quaternary Ammonium Salt Groups. *J. Comb. Chem.* **2009,** *11*(6), 1115–1127.
394. Cuzman, O. A.; Tiano, P.; Ventura, S. New Control Methods against Biofilms' Formation on the Monumental Stones. In *11th International Congress on Deterioration and Conservation of Stone*; Lukaszewicz, J. W., Niemcewicz, P., Eds.; Wydawnictwo Naukowe Uniwersytetu Mikolaja Kopernika: Torun, Poland, 2008; pp 837–846.

395. Eyssautier-Chuine, S.; Vaillant-Gaveau, N.; Gommeaux, M.; Thomachot-Schneider, C.; Pleck, J.; Fronteau, G. Efficacy of Different Chemical Mixtures Against Green Algal Growth On Limestone: A Case Study with *Chlorella vulgaris*. *Int. Biodeteriorat. Biodegradat.* **2015,** *103*, 59–68.
396. MacMullen, J.; Zhang, Z.; Dhakal, H. N.; Radulovic, J.; Karabela, A.; Tozzi, G.; Hannant, S.; Alshehri, M. A.; Buhé, V.; Herodotou, C.; Totomis, M.; Bennett, N. Silver Nanoparticulate Enhanced Aqueous Silane/Siloxane Exterior Facade Emulsions and Their Efficacy Against Algae and Cyanobacteria Biofouling. *Int. Biodeteriorat. Biodegradat.* **2014,** *93*, 54–62.
397. Essa, A. M. M.; Khallaf, M. K. Biological Nanosilver Particles for the Protection of Archaeological Stones Against Microbial Colonization. *Int. Biodeteriorat. Biodegradat.* **2014,** *94*, 31–37.
398. Bellissima, F.; Bonini, M.; Giorgi, R.; Baglioni, P.; Barresi, G.; Mastromei, G.; Perito, B. Antibacterial Activity of Silver Nanoparticles Grafted on Stone Surface. *Environ. Sci. Pollut. Res.* **2013**, *21*, 13278–13286.
399. Zhang, Z.; MacMullen, J.; Dhakal, H. N.; Radulovic, J.; Herodotou, C.; Totomis, M.; Bennett, N. Biofouling Resistance of Titanium Dioxide and Zinc Oxide Nanoparticulate Silane/Siloxane Exterior Facade Treatments. *Build. Environ.* **2013,** *59*, 47–55.
400. Ditaranto, N.; van der Werf, I. D.; Picca, R. A.; Sportelli, M. C.; Giannossa, L. C.; Bonerba, E.; Tantillo, G.; Sabbatini, L. Characterization and Behaviour of ZnO-Based Nanocomposites Designed for the Control of Biodeterioration of Patrimonial Stoneworks. *New J. Chem.* **2015,** *39*, 6836.
401. Gómez-Ortíz, N. M.; González-Gómez, W. S.; De la Rosa-García, S. C.; Oskam, G.; Quintana, P.; Soria-Castro, M.; Gómez-Cornelio, S.; Ortega-Morales, B. O. Antifungal Activity of $Ca[Zn(OH)_3]_2{\cdot}2H_2O$ Coatings for the Preservation of Limestone Monuments: An in Vitro Study. *Int. Biodeteriorat. Biodegradat.* **2014,** *91*, 1–8.
402. Pistone, A.; Visco, A. M.; Galtieri, G.; Iannazzo, D.; Espro, C.; Marino Merlo, F.; Urzì, C.; De Leo, F. Polyester Resin and Carbon Nanotubes Based Nanocomposite as New Generation Coating to Prevent Biofilm Formation. *Int. J. Polym. Anal. Character.* **2016,** *21*(4), 327–336.
403. Stroganov, V. F.; Kukoleva, D. A.; Akhmetshin, A. S.; Stroganov, I. V. Biodeterioration of Polymers and Polymer Composite Materials. *Polym. Sci. – Ser. D* **2009,** *2*(3), 164–166.
404. Koestler, R. J. Polymers and Resins as Food for Microbes. In *Of Microbes and Art – The Role of Microbial Communities in the Degradation and Protection of Cultural Heritage*, Kluwer Academic: Firenze, 1999; pp 153–167.
405. Flemming, H. C. Relevance of Biofilms for the Biodeterioration of Surfaces of Polymeric Materials. *Polym. Degradat. Stability* **1998,** *59*(1–3), 309–315.
406. Lugauskas, A.; Levinskaitė, L.; Pečiulytė, D. Micromycetes as Deterioration Agents of Polymeric Materials. *Int. Biodeteriorat. Biodegradat.* **2003,** *52*, 233–242.
407. Heyn, C.; Petersen, K.; Krumbein, W. E. Mikrobieller Angriff auf Synthetische Polymere. Microbial Attack on Synthetic Polymers. In *Schimmel: Gefahr für Mensch und Kulturgut durch Mikroorganismen*; Rauch, A., Miklin-Kniefacz, S., Harmssen, A., Eds.; Konrad Theiss Verlag GmbH & Co.: Stuttgart, 2004; pp 208–214.

408. Tarnowski, A.; Xiang, Z.; McNamara, C.; Martin, S. T.; Mitchell, R. Biodeterioration and Performance of Anti-graffiti Coatings on Sandstone and Marble. *J. Can. Assoc. Conserv.* **2007,** *32*, 3–16.
409. Cappitelli, F.; Nosanchuk, J. D.; Casadevall, A.; Toniolo, L.; Brusetti, L.; Florio, S.; Principi, P.; Borin, S.; Sorlini, C. Synthetic Consolidants Attacked by Melanin-producing Fungi: Case Study of the Biodeterioration of Milan (Italy) Cathedral Marble Treated with Acrylics. *Appl. Environ. Microbiol.* **2007,** *73*(1), 271–277.
410. Pinna, D.; Salvadori, O. Biological Growth on Italian Monuments Restored with Organic or Carbonatic Compounds. In *Of Microbes and Art: The Role of Microbial Communities in the Degradation and Protection of Cultural Heritage*; Ciferri, O., Mastromei, G., Tiano, P., Eds.; Kluwer Academic: Firenze, 1999; pp 149–154.
411. Viles, H. Field and Laboratory Results. In *Soft Capping Historic Walls. A Better Way of Conserving Ruins?*; Zoë Lee, H. V., Chris Wood, Ed.; English Heritage: 2009; pp 37–48.
412. Roby, T.; Alberti, L.; Ben Abed, A. A Preliminary Assessment of Mosaic Reburials in Tunisia. In *Conservation and the Eastern Mediterranean*; Rozeik, C., Roy, A., Saunder, D., Eds.; International Institute for Conservation of Historic and Artistic Works: Istanbul, 2010; pp 207–213.
413. Cappitelli, F.; Toniolo, L.; Sansonetti, A.; Gulotta, D.; Ranalli, G.; Zanardini, E.; Sorlini, C. Advantages of Using Microbial Technology Over Traditional Chemical Technology in the Removal of Black Crusts from Stone Surfaces of Historical Monuments. *Appl. Environ. Microbiol.* **2007,** *73*, 5671–5675.
414. Troiano, F.; Gulotta, D.; Balloi, A.; Polo, A.; Toniolo, L.; Lombardi, E.; Daffonchio, D.; Sorlini, C.; Cappitelli, F. Successful Combination of Chemical and Biological Treatments for the Cleaning of Stone Artworks. *Int. Biodeteriorat. Biodegradat.* **2013,** *85*, 294–304.
415. Sanmartin, P.; Cappitelli, F.; Mitchell, R. Current Methods of Graffiti Removal: A Review. *Construct. Build. Mater.* **2014,** *71*, 363–374.
416. Polo, A.; Cappitelli, F.; Brusetti, L.; Principi, P.; Villa, F.; Giacomucci, L.; Ranalli, G.; Sorlini, C. Feasibility of Removing Surface Deposits on Stone Using Biological and Chemical Remediation Methods. *Microbial Ecol.* **2010,** *60*(1), 1–14.
417. Gioventù, E.; Lorenzi, P. F.; Villa, V.; Sorlini, C.; Rizzi, M.; Cagnini, A.; Griffo, G.; Cappitelli, C. Comparing the Bioremoval of Black Crusts on Colored Artistic Lithotypes of the Cathedral of Florence with Chemical and Laser Treatment. *Int. Biodeteriorat. Biodegradat.* **2011,** *65*, 832–839.
418. Alfano, G.; Lustrato, G.; Belli, C.; Zanardini, E.; Cappitelli, F.; Mello, E.; Sorlini, C.; Ranalli, G. The Bioremoval of Nitrate and Sulfate Alterations on Artistic Stonework: The Case Study of Matera Cathedral after Six Years from the Treatment. *Int. Biodeteriorat. Biodegradat.* **2011,** *65*, 1004–1011.
419. Lustrato, G.; Alfano, G.; Andreotti, A.; Colombini, M. P.; Ranalli, G. Fast Biocleaning of Mediaeval Frescoes Using Viable Bacterial Cells. *Int. Biodeteriorat. Biodegradat.* **2012,** *69*, 51–61.
420. Webster, A.; May, E. Bioremediation of Weathered-Building Stone Surfaces. *Trends Biotechnol.* **2006,** *24*(6), 255–260.
421. De Muynk, W.; De Belie, N.; Verstraete, W. Microbial Carbonate Precipitation in Construction Materials: A Review. *Ecol. Eng.* **2010,** *36*, 118–136.

422. De Muynck, W.; Verbekenc, K.; De Belie, N.; Verstraete, W. Influence of Urea and Calcium Dosage on the Effectiveness of Bacterially Induced Carbonate Precipitation on Limestone. *Ecol. Eng.* **2010,** *36*, 99–111.
423. De Belie, N.; De Muynck, W.; Verstraete, W. A Synergistic Approach to Microbial Presence on Concrete: Cleaning and Consolidating Effects. *FIB Struct. Concrete* **2006**, *7*(3), 1–14.
424. De Belie, N.; Wang, J.; De Muynck, W.; Manso Blanco, S.; Seguro Pérez, I. Microbial Interactions with Mineral Building Materials. *Kuei Suan Jen Hsueh Pao/Journal of the Chinese Ceramic Society* **2014,** *42*(5), 563–567.
425. Shirakawa, M. A.; John, V. M.; De Belie, N.; Alves, J. V.; Pinto, J. B.; Gaylarde, C. Susceptibility of Biocalcite-Modified Fiber Cement to Biodeterioration. *Int. Biodeteriorat. Biodegradat.* **2015,** *103*, 215–220.
426. Tiano, P. E.; Cantisani, I.; Sutherland, J. M. Paget Biomediated Reinforcement of Weathered Calcareous Stones. *J. Cult. Heritage* **2006,** *7*, 49–55.
427. Perito, B.; Marvasi, M.; Barabesi, C.; Mastromei, G.; Bracci, S.; Vendrell, M.; Tiano, P. A *Bacillus subtilis* Cell Fraction (BCF) Inducing Calcium Carbonate Precipitation: Biotechnological Perspectives for Monumental Stone Reinforcement. *J. Cult. Heritage* **2014,** *15*, 345–351.
428. Albini, M.; Comensoli, L.; Brambilla, L.; Domon Beuret, E.; Kooli, W.; Mathys, I.; Letardi, P.; Joseph, E. Innovative Biological Approaches for Metal Conservation. *Mater. Corrosion* **2016,** *67*(2), 200–206.
429. Delgado Rodrigues, J. Defining, Mapping and Assessing Deterioration Patterns in Stoneconservation Projects. *J. Cult. Heritage* **2015,** *16*(3), 267–275.
430. Di Tullio, V.; Proietti, N.; Gobbino, M.; Capitani, C.; Olmi, R.; Priori, S.; C., R.; Giani, E. Non-destructive Mapping of Dampness and Salts in Degraded Wall Paintings in Hypogeous Buildings: The Case of St. Clement at Mass Fresco in St. Clement Basilica, Rome. *Anal. Bioanal. Chem.* **2010,** *396*(5), 1885–1896.
431. Capitani, D.; Di Tullio, V.; Proietti, N. Nuclear Magnetic Resonance to Characterize and Monitor Cultural Heritage. *Prog. Nucl. Magnetic Resonance Spectr.* **2012,** *64*, 29–69.
432. Herrera, L. K.; Videla, H. A. Surface Analysis and Materials Characterization for the Study of Biodeterioration and Weathering Effects on Cultural Property. *Int. Biodeteriorat. Biodegradat.* **2009,** *63*(7), 813–822.
433. Herrera, L. K.; Le Borgne, S.; Videla, H. A. Modern Methods for Materials Characterization and Surface Analysis to Study the Effects of Biodeterioration and Weathering on Buildings of Cultural Heritage. *Int. J. Architect. Heritage* **2009,** *3*(1), 74–91.
434. Gazzano, C.; Favero-Longo, S. E.; Matteucci, E.; Piervittori, R. Image Analysis for Measuring Lichen Colonization on and Within Stonework. *The Lichenologist* **2009,** *41*(3), 299–313.
435. Welton, R. G.; Silva, M. R.; Gaylarde, C.; Herrera, L. K.; Anleo, X.; De Belie, N.; Modry, S. Techniques Applied to the Study of Microbial Impact on Building Materials. *Mater. Struct. Matériaux et constructions* **2005,** *38*(284), 883–893.
436. Mihajlovski, A.; Seyer, D.; Benamara, H.; Bousta, F.; Di Martino, P. An Overview of Techniques for The Characterization and Quantification of Microbial Colonization on Stone Monuments. *Ann. Microbiol.* **2015,** *65*(3), 1243–1255.

437. Urzì, C.; De Leo, F. Sampling with Adhesive Tape Strips: An Easy and Rapid Method to Monitor Microbial Colonization on Monument Surfaces. *J. Microbiol. Methods* **2001,** *44*, 1–11.
438. Urzì, C.; De Leo, F.; Bruno, L.; Krakova, L.; Pangallo, D.; Albertano, P. How to Control Biodeterioration of Cultural Heritage: An Integrated Methodological Approach for the Diagnosis and Treatment of Affected Monuments. In *Works of Art and Conservation Science Today*, Thessaloniki, Greece, 2010.
439. Gonzalez, J. M.; Portillo, M. C.; Saiz-Jimenez, C. Molecular Studies for Cultural Heritage: State of the Art. In *Heritage Microbiology and Science. Microbes, Monuments and Maritime Materials*; May, E., Jones, M., Mitchell, J., Eds.; RSC Publishing: Portsmouth, 2005; pp 97–107.
440. Saad, D. S.; Gaylarde, C. C. The Potential of DGGE for Analysis of Fungal Biofilms on Historic Buildings. In *Molecular Biology and Cultural Heritage: Proceedings of the International Congress on Molecular Biology and Cultural Heritage*; Saiz-Jimenez, C., Ed.; A.A. Balkema: Sevilla, Spain, 2003; pp 145–148.
441. Portillo, M. C.; Gonzalez, J. M. Microbial Community Diversity and the Complexity of Preserving Cultural Heritage. In *Biocolonization of Stone: Control and Preventive Methods*; Charola, A. E., McNamara, C., Koestler, R. J., Eds.; Smithsonian Institute Scholarly Press: Washington, 2011; pp 19–28.
442. Urzì, C. Fluorescent In Situ Hybridisation (FISH) as Molecular Tool to Study Bacteria Causing Biodeterioration. In *Heritage Microbiology and Science. Microbes, Monuments and Maritime Materials*, May, E., Jones, M., Mitchell, J., Eds.; RSC Publishing: Portsmouth, 2005; pp 143–150.
443. Cardinale, M., Scanning a Microhabitat: Plant-Microbe Interactions Revealed by Confocal Laser Microscopy. *Front. Microbiol.* **2014,** *7 March.*
444. Müller, E.; Drewello, U.; Drewello, R.; Weißmann, R.; Wuertz, S. In situ Analysis of Biofilms on Historic Window Glass Using Confocal Laser Scanning Microscopy. *J. Cult. Heritage* **2001,** *2*, 31–42.
445. Sanmartín, P.; Villa, F.; Silva, B.; Cappitelli, F.; Prieto, B. Color Measurements as a Reliable Method for Estimating Chlorophyll Degradation to Phaeopigments. *Biodegradation* **2011,** *22*, 763–771.
446. Sanmartín, P.; Vázquez-Nion, D.; Silva, B.; Prieto, B. Spectrophotometric Color Measurement for Early Detection and Monitoring of Greening on Granite Buildings. *Biofouling* **2012,** *28*(3), 329–338.
447. Mabuchi, H.; Kigawa, R.; Sano, C. Application of ATP Swab Test in Conservation Facilities for Cultural Properties. *Sci. Conserv.* **2010,** *49*, 1–11.
448. Moses, C.; Robinson, D.; Barlow, J. Methods for Measuring Rock Surface Weathering and Erosion: A Critical Review. *Earth-Sci. Rev.* **2014,** *135*, 141–161.
449. de los Rios, A.; Ascaso, C. Contributions of In Situ Microscopy to the Current Understanding of Stone Biodeterioration. *Int. Microbiol.* **2005,** *8*(3), 181–188.
450. Gorbushina, A. A.; Diakumaku, E.; Muller, L.; Krumbein, W. E. Biocide Treatment of Rock and Mural Paintings: Problems of Application, Molecular Techniques of Control and Environmental Hazards. In *Molecular Biology and Cultural Heritage*; Saiz-Jimenez, C., Ed.; A.A. Balkema: Sevilla, Spain, 2003; pp 61–71.

451. Marques, A. P.; Freitas, M. C. Cell-Membrane Damage and Element Leaching in Transplanted *Parmelia sulcata* Lichen Related to Ambient SO_2, Temperature, and Precipitation. *Environ. Sci. Technol.* **2005,** *39*, 2624–2630.
452. Paoli, L.; Loppi, S. A Biological Method to Monitor Early Effects of the Air Pollution Caused by the Industrial Exploitation of Geothermal Energy. *Environ. Pollut.* **2008,** *155*, 383–388.
453. Cecchi, G.; Pantani, L.; Raimondi, V.; Tomaselli, L.; Lamenti, G.; Tiano, T.; Chiari, R. Fluorescence Lidar Technique for the Remote Sensing of Stone Monuments. *J. Cult. Heritage* **2000,** *1*, 29–36.
454. Lognoli, D.; Lamenti, G.; Pantani, L.; Tirelli, D.; Tiano, P.; Tomasetti, L. Detection and Characterization of Biodeteriogens on Stone Cultural Heritage by Fluorescence Lidar. *Appl. Opt.* **2002,***41*(9), 1780–1787.
455. Grönlund, R.; Hallllström, J.; Svanberg, S.; Barup, K. Fluorescence Lidar Multispectral Imaging for Diagnosis of Historical Monuments, Ovedskloser: A Swedish Case Study. In *LACONA VI Lasers in the Conservation of Artworks*; Nimmrichter, J., Kautek, W., Schreiner, M., Eds.; Springer: Vienna, 2005; pp 583–591.
456. Hällström, J.; Barup, K.; Gronlund, R.; Johansson, A.; Svanberg, S.; Palombi, L.; Lognoli, D.; Raimondi, V.; Cecchi, G.; Conti, C. Documentation of Soiled and Biodeteriorated Facades: A Case Study on the Coliseum, Rome, Using Hyperspectral Imaging Fluorescence Lidars. *J. Cult. Heritage* **2009,** *10*, 106–115.
457. Raimondi, V.; Cecchi, G.; Lognoli, D.; Palombi, L.; Gronlund, R.; Johansson, A.; Svanberg, S.; Barup, K.; Hällström, J. The Fluorescence Lidar Technique for the Remote Sensing of Photoautotrophic Biodeteriogens in the Outdoor Cultural Heritage: A Decade of In Situ Experiments. *Int. Biodeteriorat. Biodegradat.* **2009,** *63*, 823–835.
458. Andreotti, A.; Cecchi, A.; Cecchi, G.; Colombini, M. P.; Cucci, C.; Guzman, D.; Fornacelli, C.; Galeotti, M.; Gambineri, F.; Gomoiu, J.; Lognoli, D.; Mohanu, D.; Palombi, P.; Penoni, S.; Picollo, M.; Pinna, D.; Raimondi, V.; Tiano, P. An Investigation of the Potential of Hyperspectral Fluorescence Lidar Imaging of Biodeteriogens on Mural Paintings. In *Lasers in the Conservation of Artworks IX*; Saunders, D., Strlic, M., Korenberg, C., Luxford, N., Birkholzer, K., Eds.; Archetype Publications: London, UK, 2013; pp 140–145.
459. Delgado Rodrigues, J.; Valero Congil, J.; Wakefield, R.; Brechet, E.; Larrañaga, I.; Rizzo, S. Monitoraggio della biocolonizzazione e valutazione dell'efficacia di un biocida. *Arkos: scienza e restauro dell'architettura* **2004,** *5* (7), 52–58.
460. Stibal, M.; Elster, J.; Šabacká, M.; Kaštovská, K. Seasonal and Diel Changes in Photosynthetic Activity of the Snow Alga Chlamydomonas Nivalis (Chlorophyceae) from Svalbard Determined by Pulse Amplitude Modulation Fluorometry. *FEMS Microbiol. Ecol.* **2007,** *59*(2), 265–273.
461. Vandevoorde, D.; M., P.; Schalm, O.; Vanhellemont, Y.; Cnudde, V.; Verhaeven, E. Contact Sponge Method: Performance of a Promising Tool for Measuring the Initial Water Absorption. *J. Cult. Heritage* **2009,** *10*(1), 41–47.
462. Tiano, P.; Pardini, C. Valutazione in situ dei trattamenti protettivi per il materiale lapideo. Proposta di una nuova semplice metodologia. *Arkos* **2004,** *5*, 30–36.
463. Salvadori, O.; Nugari, M. P. Aspetti biologici del degrado: la normativa in ambito nazionale ed europeo. *Kermes* **2008,** *21* (71), 109–111.

APPENDIX

This appendix contains information on most of the active ingredients listed and discussed in the text. The selected biocides are not those recommended for use. The reader finds their pros and cons in the text where the results of the literature are discussed.

This appendix does not include chemical agents that pose evident risks to health, show negative interactions with the materials or low efficacy on organisms. It neither contains products that are no more on the market (see Section 4.6).

This appendix provides data on *chemical and physical properties, toxicity, potential health effects of the active ingredients, along with the manufactures, and/or suppliers of trade products containing the active ingredients*. The information on the stability and incompatibility of the biocides with other compounds contribute to a correct use of them. The trade products' description and the recommendations by the manufactures refer to the efficiency and the application of the product.

The toxicity section provides information on acute toxicity caused by the compounds. Many chemicals have not been thoroughly studied for their toxic effects. The manufacturers can keep confidential some data that are consequently not available to the public. It was even difficult to find the technical data sheets of some commercial products. Moreover, human toxicity data do not exist for many compounds. For all these reasons, some chemicals listed in this appendix do not have such information.

Ecotoxicity values are reported for some active ingredients. EC_{50} (effective concentration at 50%) is the concentration that causes adverse effects in 50% of the test organisms. The ecotoxicology is the study of the effects of toxic chemicals on populations, communities, and terrestrial, freshwater and marine ecosystems. It focuses on mechanisms and processes by which chemicals exert their effects on ecosystems, and examines the impact caused at the population or community level.

The data sources have been:

- The book *Directory of microbicides for the protection of materials: A handbook*, edited by Wilfried Paulus, published by Springer, 2008;

- The book *Russell, Hugo & Ayliffe's: Principles and Practice of Disinfection, Preservation and Sterilization* (editors Adam P. Fraise, Jean-Yves Maillard, Syed A. Sattar) published by Wiley-Blackwell, 2013
- The book *Il Controllo del Degrado Biologico–I Biocidi nel Restauro dei Materiali Lapidei* (authors G. Caneva, M.P. Nugari, D. Pinna, O. Salvadori) published by Nardini Editore, 1996
- The website http://www.epa.gov
- The Pesticide Action Network (PAN) Pesticide Database
- The website https://www.ulprospector.com
- The United States National Library of Medicine, Toxicology Data Network (TOXNET), Hazardous Substance Data Bank (HSDB), 2013, http://toxnet.nlm.nih.gov
- The technical data sheets provided by manufactures.

The description of each active ingredient includes just the data that are relevant for operators and professionals involved in the field of conservation of cultural heritage.

The data are listed according to the chemical classes (in alphabetic order) which the biocides belong to.

Except where noted otherwise, the data are given for materials in their standard state (at 25°C and 101 kPa).

ALCOHOLS

Ethanol

Synonym ethyl alcohol

Chemical and physical properties

Chemical formula C_2H_6O
Molecular weight 46.07
Appearance colorless liquid
Boiling point 78.2–80°C
Melting point −114°C
Density 0.79 g/ml (20°C)
Vapor pressure 59.3 mmHg
Solubility soluble in water, alcohols, acetone, ketones, ether
Stability volatile

Toxicity data

LD_{50} oral 13,700 mg/kg (rat)
LD_{50} oral 9500 mg/kg (rabbit)
Irritant to skin

Environmental fate

Vapor-phase ethanol is degraded in the atmosphere by reaction with photochemically produced hydroxyl radicals; the half-life for this reaction in air is estimated to be 36 h.

Protective equipment and clothing

Wear appropriate personal protective clothing to prevent skin contact. Wear appropriate eye protection to prevent eye contact.

Storage conditions

Keep tightly closed, cool, and away from flame.

2-Propanol

Synonym isopropyl alcohol, isopropanol

Propanol is an aliphatic alcohol hydrocarbon. It is used in antifreeze; as an industrial solvent; as a solvent for gums, shellac, essential oils, quick drying oils, and resins; in quick drying inks; in denaturing ethyl alcohol; in hand lotions, after shave lotions, cosmetics, and pharmaceuticals; in manufacture of acetone, glycerol, isopropyl acetate.

Chemical and physical properties

Chemical formula C_3H_8O
Molecular weight 60.10
Appearance colorless liquid
Boiling point 82–83°C
Melting point −88°C
Density 0.785 g/ml (20°C)
Vapor pressure 45.4 mmHg
Solubility soluble in water, alcohols, acetone, ketones, ether, benzene, and chloroform; it is insoluble in salt solutions.
Stability volatile

Toxicity and ecotoxicity data

LD_{50} *oral* 5850 mg/kg (rat)
LD_{50} *oral* 4475 mg/kg (mouse)
LC_{50} 8970 mg/l (fish)

Isopropyl alcohol is a potent eye and skin irritant. Not classifiable as a human carcinogen.

Protective equipment and clothing

Wear appropriate chemical protective gloves, boots, and goggles.

Storage conditions

Keep the container tightly closed in a dry and well-ventilated place. Store at room temperature in the original container. As it is sensitive to light, store in light-resistant containers. Keep away from heat and sources of ignition.

ALDEHYDES

Formaldehyde

Formaldehyde is used as an antimicrobial agent and disinfectant for industrial processes and some household purposes. It is used to make resins for building materials, coatings for paper and clothing fabrics, and synthetic fibers. A second major use is as a starting chemical to make other chemicals. Building materials with formaldehyde include certain insulation materials, glues, and pressed wood products like particleboard, plywood, and fiberboard. Products containing urea and formaldehyde are used as slow-release nitrogen fertilizers in farming and gardening. Uses of formaldehyde in medicine include disinfecting hospital wards and preserving specimens. In the United States, formaldehyde is also registered as a materials preservative for consumer products such as laundry detergents, general-purpose cleaners and wallpaper adhesives.

Formaldehyde is expected to biodegrade readily in soil and water.

Chemical and physical properties

Chemical formula CH_2O
Molecular weight 30.03

Appearance reactive, inflammable gas
Melting point −92°C
Boiling point −19°C
Vapor pressure 3890 mmHg
Solubility soluble in water up to 55% forming formalin, soluble in alcohols and in polar solvents.
Stability the dry gas tends to polymerize slowly; it is easily oxidized; it reacts with ammonia and proteins under inactivation.

Toxicity data

LC_{50} inhalation 0.48 mg/l/4 h (rat)
LC_{50} inhalation 0.414 mg/l/4 h (mouse)
EU classification oxidant, corrosive, harmful (Xn)

Contact with the skin causes irritation, tanning effect, and allergic sensitization. Contact with eyes causes irritation, itching, and lacrimation. Formaldehyde can provoke skin reactions in sensitized individuals, not only by contact but also by inhalation. It is suspected of causing genetic defects. Harmful to aquatic life.

Acute effects of airborne formaldehyde exposure: odor detection, 0.05–1.0 ppm; eye irritation, 0.01–2 ppm; upper respiratory tract irritation (e.g., irritation of the nose or throat), 0.10–11 ppm; lower airway irritation (e.g., cough, chest tightness, and wheezing), 5–30 ppm; pulmonary edema, inflammation, pneumonia, 50–100 ppm; death >100 ppm.

The International Agency for Research on Cancer determined that formaldehyde is carcinogenic to humans, based on evidence for myeloid leukemia or cancer of the nose or pharynx in formaldehyde-exposed workers and nasal tumors in laboratory animals.

Absorption and distribution

Formaldehyde is absorbed readily from the respiratory and oral tracts and, to a much lesser degree, from the skin. It is the simplest aldehyde and reacts readily with macromolecules, such as proteins and nucleic acids. Inhalational exposure has been reported to result in almost complete absorption. Dermal absorption due to contact with formaldehyde-containing materials (e.g., textiles, permanent-press clothing, cosmetics, or other materials) is of low order of magnitude.

Environmental fate

Gas-phase formaldehyde is degraded in the atmosphere by reaction with photochemically produced hydroxyl radicals; the half-life for this reaction in air is 45 h. Laboratory tests classified the compound as readily biodegradable.

Solutions containing formaldehyde are unstable, oxidizing slowly to form formic acid and polymerizing to form oligomers. In the presence of air and moisture, polymerization readily takes place in concentrated solutions at room temperatures to form paraformaldehyde, a solid mixture of linear polyoxymethylene glycols containing 90–99% formaldehyde.

Protective equipment and clothing

Avoid breathing vapors. Do not handle broken packages unless wearing appropriate personal protective equipment. Wash away any material which may have contacted the body with copious amounts of water or soap and water. Wear appropriate personal protective clothing to prevent skin contact. Wear appropriate eye protection to prevent eye contact. Handle with gloves. Eye/face protection: Tightly fitting safety goggles.

Storage conditions

Store in a cool, dry, well-ventilated location. Special temperature control may be required. Separate from oxidizing materials, alkalis, acids, amines.

Formaldehyde solution

Common name formalin

Chemical and physical properties

Appearance the 37% standard solution is a clear, colorless liquid with the pungent odor of formaldehyde
Content 37%. It contains 10–15% methanol to avoid polymerization
Density 1.083 g/ml (20°C)
Flash point 56°C
Solubility miscible in water and alcohols
Boiling point −19.1°C
Stability sensitive to strong oxidizing and reducing agents; it reacts readily with amines, proteins, phenols.
pH 2.8–4.0

Toxicity data

LD_{50} dermal 300 mg/kg (mouse)
LD_{50} percutaneous 270 mg/kg (rabbit)

Glutaraldehyde

Glutaraldehyde is allowed as a preservative in cosmetics in Europe at concentrations up to 0.1%. It is not allowed in aerosols and sprays. It is commonly used in a 2% concentration for cold sterilization of surgical and dental equipment. It is used as algaecide, bactericide, and fungicide.

Glutaraldehyde is used as a disinfectant in hospitals, agriculture and aquaculture, food handling and food storage establishments, and water treatment plants. It is used as a preservative in the manufacture of several consumer products, including cosmetics, cleaners, adhesives, paper, textiles and leathers, paints and coatings, and inks and dyes. Glutaraldehyde is also used as a tissue fixative in laboratories and embalming fluid and in photographic and X-ray development fluids. Moreover, it is used in hydraulic fracturing and offshore oil operations.

Chemical and physical properties

Chemical formula $C_5H_8O_2$
Molecular weight 100.12
Appearance colorless oily liquid
Boiling point 187–189°C (decomposes)
Melting point −14°C
Density 1.066 g/ml (20°C) (25% dilution in water), 1.131 g/ml (20°C) (50% dilution in water)
Vapor pressure 0.6 mmHg at 30°C
pH 3.1–4.5 (respectively, 25% and 50% dilutions)
Solubility miscible in water, lower in alcohols and glycols
Stability volatile with water vapor; tends to polymerize in water; stable in acidic solutions; stable to a limited extent in alkaline solutions; sensitive to strong oxidizing agents; inactivation by ammonia and primary amines at neutral and higher pH values. Incompatible materials: strong bases, strong oxidizing agents, strong acids.

Toxicity and ecotoxicity data

LD_{50} *oral* 246 mg/kg (rat male); 154 mg/kg (rat female)
LD_{50} *oral* 100 mg/kg (mouse)
LD_{50} *dermal* >2000 mg/kg (rat)
LD_{50} *dermal* 1800 mg/kg (rabbit)
LC_{50} *inhalation* (4h) 350 mg/l (rat male), 280 mg/l (rat female)
LD_{50} species: *Anas platyrhynchos* (Mallard duck) oral 820 mg/kg for 14 days
LC_{50} species: *Daphnia magna* (Water flea); conditions: static, closed system; concentration: 0.35 mg/l for 48 h/50% glutaraldehyde
Toxic to aquatic life with long lasting effects.

Exposure to concentrations < 1 ppm by inhalation or skin contact may cause irritation of the skin and/or mucous membranes. The critical effects of glutaraldehyde exposure are eye, skin, and respiratory irritation, skin sensitization, and occupational asthma.

Glutaraldehyde is not classifiable as a human carcinogen.

General population exposure may occur by breathing in air and skin contact with consumer products containing glutaraldehyde.

Environmental fate

Vapor-phase glutaraldehyde is degraded in the atmosphere by reaction with photochemically produced hydroxyl radicals; the half-life for this reaction in air is estimated to be 16 h. If released to water or soil, glutaraldehyde is expected to bind to soil particles or suspended particles. Glutaraldehyde is not expected to move into air from wet soils or water surfaces, but may move to air from dry soils.

Results of biodegradation screening tests indicate that glutaraldehyde is readily biodegradable.

Protective equipment and clothing see formaldehyde

Preventive measures

Avoid contact with skin and eyes. Avoid inhalation of vapor or mist.

Storage conditions

Keep container tightly closed in a dry and well-ventilated place.

AZOLES

Triazoles are heterocyclic compounds with molecular formula $C_2H_3N_3$, having a five-membered ring of two carbon atoms and three nitrogen atoms.

Tebuconazole

Chemical and physical properties

Chemical formula α-[2-(4-chlorophenyl)ethyl]-α-(1,1-dimethylethyl)-1*H*-1,2,4-triazole-1-ethanol
Molecular weight 307.83
Appearance colorless crystals
Density 1.202 g/ml (20°C)
Melting point approx. 103°C
Ignition approx. 468°C
Vapor pressure 1.7×10^{-6} Pa (1.3×10^{-8} mmHg) at 20°C
Solubility (20°C) in water, 36 mg/l at pH 5–9; in ethyl glycol, approx. 27%; in white spirit, approx. 0.3%

Toxicity data

LD_{50} *oral* >5000 mg/kg (rat)
LD_{50} *oral* 1615 mg/kg (mouse)
LD_{50} *oral* >1000 mg/kg (rabbit)
LC_{50} *inhalation* 0.82 mg/l/4 h (rat)

Environmental fate

Tebuconazole's use as a fungicide will result in its direct release to the environment. If released to air, its vapor pressure indicates it will exist solely in the particulate phase in the atmosphere. Particulate-phase tebuconazole will be removed from the atmosphere by wet or dry deposition.

Biodegradation data are not available.

Hazardous decomposition

When heated to decomposition, it emits toxic vapors of hydrogen chloride and nitrogen oxides.

MANUFACTURERS AND PRODUCTS

LANXESS

PREVENTOL A8 → tebuconazole 95.37%
Stability resistant to hydrolysis in acidic up to strong alkaline formulations.

Imidazoles are organic compounds that contain an imidazole ring.
Imidazole
Chemical formula $C_3H_4N_2$
Appearance white or pale yellow solid
Density 1.23 g/cm^3
Melting point 89–91°C
Boiling point 256°C
Solubility soluble in water

MANUFACTURERS AND PRODUCTS

Schulke & Mayr

Parmetol DF 12 → blend of tetrahydro-1,3,4,6-tetrakis(hydroxymethyl) imidazo[4,5-*d*]imidazole-2,5(1*H*,3*H*)-dione, octylisothiazolinone, methylchloroisothiazolinone, and methylisothiazolinone

Appearance light yellow liquid
Density 1.035–1.047 g/ml
Solubility emulsion in water
Recommended use pH range 3–9.5
EU classification Xi (Irritant)

CARBAMATES

Zinc dimethyldithiocarbamate

Synonym/common name Ziram, zinc bis (dimethyldithiocarbamate)

The compound has a broad spectrum of effectiveness covering fungi, yeasts, bacteria, algae; additionally, it is nonvolatile, nonleaching and relatively heat resistant.

Chemical and physical properties

Chemical formula $C_6H_{12}N_2S_4Zn$
Molecular weight 305.83
Appearance white to odorless powder
Melting point 250–252°C
Vapor pressure 7.5×10^{-9} mmHg at 0°C
Density 1.66g/ml
Solubility 0.065 g/l in water; scarcely soluble in organic solvents; soluble in carbon disulfide, chloroform, diluted alkalis, and concentrated HCl.
Stability stable in pH range 5–10; decomposition in acid media with release of CS_2 and dimethylamine; causes coloration when in contact with traces of heavy metals, for example, Cu, Fe.

Noncorrosive except to iron and copper.

Toxicity data

LD_{50} *oral* 320 mg/kg (rat)
LD_{50} *dermal* >6000 mg/kg (rat)
LD_{50} *oral* 100–150 mg/kg (Guinea pig)
LD_{50} *oral* 100–300 mg/kg (rabbit)
LC_{50} *inhalation* 81 mg/cu m/4 h (rat)

Ziram is not classifiable as carcinogen to humans. Numerous laboratory trials showed mutagenic activity.

It does not cause irritation, but irritates mucous membranes and eyes. It is a severe irritant to eyes, nose, and throat.

When heated to decomposition it emits very toxic fumes of nitrogen oxides and sulfur oxides.

Environmental fate

Ziram's production and use as a rubber accelerator and biocide in water treatment, paper sizing, and adhesives may result in its release to the environment through various waste streams. Its use as an agricultural fungicide will result in its direct release to the environment. If released to air, its vapor pressure indicates that ziram will exist solely in the particulate phase in the ambient atmosphere. Particulate-phase ziram will be removed from the atmosphere by wet and dry deposition.

In soils with medium to high content of organic matter, ziram will be moderately bound. A field half-life of 30 days has been estimated for ziram, indicating a low to moderate persistence. Because the compound is toxic to bacteria, biodegradation in sediments may be rather slow, or occur only at very low concentrations.

MANUFACTURERS AND PRODUCTS

Vanderbilt

VANCIDE 51 → aqueous solution of sodium dimethyldithiocarbamate 27.6% and sodium mercaptobenzothiazole 2.4%

3-Iodopropynylbutylcarbamate

Synonyms 3-iodine-2-propylbutylcarbamate, iodopropynyl butylcarbamate

Chemical and physical properties

Chemical formula $C_8H_{12}INO_2$
Molecular weight 281.07
Appearance crystalline off-white powder
Content (%)>90
Melting point 65–66°C
Density 450g/ml (20°C)
Solubility 0.168 g/l in water; soluble in organic solvents
Stability stable under normal conditions; decomposition at 180°C; hydrolyses in strong alkaline media.

Toxicity data

LD_{50} *oral* 1400 mg/kg (rat)
LD_{50} *dermal* >2000 mg/kg (rabbit)
LC_{50} *inhalation* 689 mg/l (4h) (rat)
Severe eye irritant, skin irritant, not a sensitizer.

MANUFACTURERS AND PRODUCTS

LAMIRSA

Mirecide-TF/580.ECO → concentrated aqueous dispersion of 3-iodine-2-propylbutylcarbamate and 2-*N*-octyl-4-isothiazolin-3-one

CTS

Biotin R → mixture of iodopropynyl butylcarbamate (10–25%), 2-*n*-ctyl-4-isothiazolin-3-one(2.5–10%) and 2-2′-oxydiethanol

CHLORINE-CONTAINING COMPOUNDS

Calcium hypochlorite dihydrate

Calcium hypochlorite is widely used as a disinfectant in swimming pools and drinking water supplies, treatment of industrial cooling water for slime control of bacteria, algal, and fungal growths and for disinfection and odor control in sewage and wastewater effluents. It is employed as a sanitizer in households, schools, hospitals, and public buildings, microbial control in restaurants and other public eating places. It is used for general sanitation in dairies, wineries, breweries, canneries, food processing plants, and beverage bottling plants. The largest use of calcium hypochlorite is for water treatment. It is also used to bleach textiles in commercial laundries and textile mills.

Chemical and physical properties

Chemical formula $CaCl_2O_2$
Molecular weight 142.98
Appearance white/gray powder; white crystalline solid
Boiling point decomposes at 100°C
Melting point 100°C
Solubility 21% in water
Density 2.35 g/cm^3 (20°C)
pH (1% solution) 10.4–10.8

Stability Not heat resistant (to be stored at temperatures < 32°C). It forms explosive compounds with ammonia and amines. It is a strong oxidizer. Other incompatible materials include organics, nitrogen containing compounds, combustible or flammable materials. It reacts vigorously with acids to generate heat and toxic chlorine gas. Dry calcium hypochlorite reacts with organics such as fuels, oils, greases, solvents, lotions, cosmetics, food, dead vegetation, cardboard, soap, and many other organics-containing materials to start spontaneously a fire.

Toxicity and ecotoxicity data

LD_{50} *oral* 790 mg/kg (rat male)
LD_{50} *dermal* >2000 mg/kg (rabbit)
LC_{50} Species: *Daphnia* sp. (Water flea); conditions: freshwater, static, 29.5°C; concentration: 4270 μg/l for 2 h

EU classification oxidant (O), corrosive (C), harmful (Xn), dangerous for the environment (N).

Corrosive to skin, mucous membranes, and eyes. Hypochlorite salts are not classifiable as carcinogens to humans. Calcium hypochlorite releases chlorine gas when added to water and can cause respiratory effects. Dermal and eye injury can occur because of the caustic nature of chlorine.

Protective equipment and clothing

Wear face shield or eye protection in combination with breathing protection.

Storage conditions

Store in a cool, dry, well-ventilated location at a temperature below 50°C to avoid slow decomposition and to prevent formation of dangerous concentration of chlorine gas. Separate from oxidizing materials, acids, ammonia, amines, and other chlorinating agents.

Formulations. Chlorinated lime (bleaching powder) consists of a mixture of calcium hypochlorite, calcium chloride, and calcium hydroxide; it should not be confused with pure calcium hypochlorite.

Impurities. Commercial products normally contain various impurities such as $CaCl_2$, $CaCO_3$, $Ca(OH)_2$, and water.

Sodium hypochlorite

Household and laundry bleaching agent; bleaching agent in paper, pulp, textile industries; disinfectant for glass, ceramics and water; algaecide and molluscicide in cooling water for power stations.

Chemical and physical properties

Chemical formula ClNaO
Molecular weight 74.44
Appearance yellowish liquid with disagreeable, sweetish odor
Content of active chlorine ≥13%
Boiling point 111°C
Density 1.11 g/cm^3
Solubility soluble in water in any portion
Stability Relatively stable in alkaline solution; unstable at lower pH values; sensitive to light; not heat resistant (to be stored at temperatures <15°C). It releases chlorine when heated above 35°C. Anhydrous sodium hypochlorite is very explosive.

Toxicity data

LD_{50} *oral* 5800 mg/kg (mouse)
LD_{50} *oral* 8910 mg/kg (rat)

EU classification oxidant, corrosive, harmful (Xn)

It has a pronounced irritant effect on the skin. Hypochlorite salts are not classifiable as carcinogens to humans.

Storage Conditions

Store at a temperature not exceeding 20°C, away from acids, in well-fitted air-tight bottles. Protect from light.

MANUFACTURERS AND PRODUCTS

Angelini

Amuchina → sodium hypochlorite 1.02% w/w
This biocide is registered for unrestricted use.

Croda

Thickened Sodium Hypochlorite Bleach Solution → sodium hypochlorite 96.6% w/w, sodium hydroxide 0.4% w/w, Crodasinic LS30 (sodium lauroyl sarcosinate 1.0% w/w, myristyl dimethyl amine oxide 2.0% w/w). Sodium hypochlorite bleach solutions are widely used as the basis for many household products for their dual cleaning and disinfectant properties. Crodasinic LS30 is a bleach-stable surfactant with surface wetting properties. The addition of Crodasinic LS30 increases the cleaning, foaming, and wetting properties of the bleach solution. Combinations of Crodasinic LS and a myristyl/cetyl amine oxide, as in this formulation, provide a thickened bleach system.

COMPOUNDS WITH ACTIVATED HALOGEN ATOMS

Bronopol

Synonyms 2-nitro-2-bromo-1,3-propanediol, 2-bromo-2-nitropropane-1,3-diol, 2-bromo-2-nitro-1,3-propanediol

The product is suitable for fungicidal and bactericidal preservation of a broad range of water-based products.

Chemical and physical properties

Chemical formula $C_3H_6BrNO_4$
Molecular weight 199.99
Appearance white crystalline powder
Melting point 131.5°C
Solubility soluble in alcohol, ethyl acetate; slightly soluble in chloroform, acetone, ether, benzene; insoluble in ligroin; 28% soluble in water.
Stability Incompatible materials: strong oxidizing agents, strong bases, strong reducing agents. The substance decomposes on heating or on burning producing toxic and corrosive fumes, including hydrogen bromide and nitrogen oxides.
Vapor pressure 1.26×10^{-5} mmHg at 20°C

Toxicity data

LD_{50} oral 350 mg/kg (mouse)
LD_{50} oral 342 mg/kg (rat female)
LD_{50} oral 307 mg/kg (rat male)
LD_{50} dermal 64–160 mg/kg (rat male)
LC_{50} inhalation >5 mg/l air/6 h (rat)

The active ingredient Bronopol is severely acutely toxic by the dermal route. It irritates the eyes, the skin, and the respiratory tract.

Protective equipment and clothing

Eye/face protection: face shield and safety glasses. Skin protection: handle with gloves. Avoid contact with skin and eyes. Avoid formation of dust and aerosols.

Storage conditions

Provide appropriate exhaust ventilation at places where dust is formed. Keep away from heat and sources of ignition.

Environmental fate

Bronopol's use as a preservative for cosmetics and pharmaceuticals may result in its release to the environment through various waste streams; its use as an agricultural bacteriostat will result in its direct release to the environment. If released to air, its vapor pressure indicates bronopol will exist in both the vapor and particulate phases in the atmosphere. Vapor-phase bronopol is degraded in the atmosphere by reaction with photochemically produced hydroxyl radicals; the half-life for this reaction in air is estimated to be 11 days. Particulate-phase bronopol is removed from the atmosphere by wet and dry deposition.

MANUFACTURERS AND PRODUCTS

Schülke Inc.

Parmetol N 20 → 2-*n*-octyl-4-isothiazolin-3-one + bronopol (2-bromo-2-nitropropane-1,3-diol)

Thor

Acticide MBL5515 → blend of 2-methyl-4-isothiazolin-3-one (5%), 1,2-benzisothiazolin-3-one (5%), and bronopol (15%).

GLYPHOSATE

N-(phosphonomethyl)glycine

Other name 2-[(phosphonomethyl)amino] acetic acid

Glyphosate is a nonselective herbicide registered for use on many food and nonfood field crops as well as noncrop areas where total vegetation control is desired. When applied at low rates, glyphosate is also a plant growth regulator. It enters plants through the foliage and moves throughout the plant and into the root system. Therefore, all parts of the plant treated with glyphosate may contain the herbicide. Glyphosate is applied to crops before emergence as otherwise crop destruction would result. Uptake through the root system is precluded by soil inactivation.

Glyphosate is marketed worldwide under many trade names: Roundup Pro (41% of a.i.), Roundup PROMAX (49% of a.i.), Roundup Pro Concentrate (50.2% of a.i.), Aquamaster (53.8%) by Monsanto; Buccaneer Plus (41% of a.i.) by Tenkoz; Rodeo (53.8% of a.i.) by Dow AgroSciences; Aquaneat (53.8% of a.i.) by Nufarm. The products contain inert ingredients such as water and surfactants that are added to aid penetration of the active ingredient through leaf surfaces. The type of surfactant can affect the performance of the herbicide. Glyphosate is registered for use in over 100 countries. The original glyphosate formulation concentrate (Roundup Herbicide, Monsanto) contained the isopropylamine salt of glyphosate (41% w/v), a polyethoxylated tallow amine surfactant (15.4% w/v), and water. Over 100 different brand names comprised of many formulations have been sold worldwide by Monsanto. In addition, other companies may now market the glyphosate molecule, and it is estimated to appear in thousands of other herbicide products. Some products contain the original isopropylamine salt, but some contain the monoammonium, sodium sesqui-, or other salts of glyphosate. Some products are dry formulations; some contain different surfactant systems, or greater or lesser amounts of total surfactant, or no surfactant at all. Glyphosate may also be sold in combination with other herbicides including various chorophenoxy compounds, simazine, linuron, and picloram or with a fertilizer to boost plant growth, thereby enhancing lethality.

Chemical and physical properties

Chemical formula $C_3H_8NO_5P$
Molecular weight 169.07
Appearance white crystalline powder
Melting point 230°C
Density 1.704 g/cm^3 (20°C)
Boiling point decomposition at 187°C. When heated to decomposition it emits very toxic fumes of nitrogen and phosphorus oxides.
Solubility in water 1.01 g/100 ml (20°C), insoluble in common organic solvents, for example, acetone, ethanol and xylene. Incompatible materials: strong oxidizing agents, metals, bases.
Vapor pressure 9.8×10^{-8} mmHg /1.31×10^{-2} mPa
Flash point nonflammable

Toxicity and ecotoxicity data

LD_{50} *oral* 1568 mg/kg (mouse)
LD_{50} *oral* 3800 mg/kg (rabbit)
LD_{50} *percutaneous* >5000 mg/kg (rabbit)
LC_{50} *inhalation* >4.98 mg/l air/4 h (rat)
EC_{50} species: *Daphnia magna* (Water flea) age <24 h; Conditions: freshwater, static; Concentration: >22,000 μg/l for 48 h; Effect: intoxication, immobilization.

EU classification irritant (Xi), dangerous for the environment (N)

Glyphosate in the formulated products can cause eye and skin irritation.

Glyphosate is an active ingredient of the most widely used herbicide, and it is believed to be less toxic than other pesticides. However, several studies showed its potential adverse health effects to humans as it may be an endocrine disruptor. Concentrated solutions of glyphosate can also cause dermal irritation. Most intoxicated cases are from ingestion, inhalation, and skin exposure. Commercial formulations were more cytotoxic than the active component alone, supporting the concept that additives in commercial formulations play a role in the toxicity attributed to glyphosate-based herbicides. Glyphosate was found nongenotoxic in human lymphocytes with or without metabolic activation.

Glyphosate was originally classified as possible human carcinogen because of increased incidence of renal tumors in mice. Following independent review, the classification was changed for a lack of statistical significance and uncertainty as to a treatment-related effect. Therefore, it is not classifiable as a human carcinogen.

Protective equipment and clothing

Eye/face protection: face shield and safety glasses. Use equipment for eye protection tested and approved under appropriate government standards such as NIOSH (US) or EN 166(EU). Skin protection: handle with gloves. Body protection: complete suit protecting against chemicals. The type of protective equipment must be selected according to the concentration and amount of the dangerous substance at the specific workplace.

Storage conditions

Always store pesticides in their original containers, complete with labels that list ingredients, directions for use, and first aid steps in case of accidental poisoning. Never store pesticides in cabinets with or near food, animal feed, or medical supplies. Do not store pesticides in places where flooding is possible or in places where they might spill or leak into wells, drains, ground water, or surface water.

Keep container tightly closed in a dry and well-ventilated place.

HYDROGEN PEROXIDE

It is marketed as a solution in water in concentrations of 3–90% by weight. Hydrogen peroxide is used as a 6% solution for bleaching hair. Some disinfectant solutions for contact lenses contain a 3% hydrogen peroxide. Chlorine free bleaches contain 6% hydrogen peroxide. Some newer fabric stain removers and bleaches contain 5–15% of hydrogen peroxide.

Chemical and physical properties

Chemical formula H_2O_2
Molecular weight 34.02
Appearance clear, colorless liquid
Boiling point 152°C

Melting point −0.43°C
Ignition temperature 150.5°C
Solubility miscible in water
Stability It is an unstable compound and decomposes slowly to H_2O and oxygen if not stabilized. Solutions of hydrogen peroxide are usually stabilized by addition of acetanilide or similar organic compounds. Incompatibilities: alkalies, ammonia and their carbonates, albumin, balsam peru, phenol, charcoal, chlorides, alkali citrates; ferrous, and gold salts; hypofosfites, iodides, lime water, permanganates, sulfites, and organic matter in general.
pH weak acid; H_2O_2 concentration wt% = 35, 50, 70, 90; corresponding pH: 4.6, 4.3, 4.4, 5.1

Toxicity data

LC_{50} for rats and guinea pigs 100 ppm after inhalation.
Corrosive to skin, mucosa, and eyes.
EU classification oxidant, corrosive, harmful (Xn).
Hydrogen peroxide is not classifiable as carcinogen to humans.

The dissociation of hydrogen peroxide is a violent and exothermic reaction. Ingestion results in gastrointestinal irritation, the severity of which depends on the concentration of the solution.

Protective equipment and clothing

Wear appropriate personal protective clothing to prevent skin contact. Wear appropriate eye protection to prevent eye contact.

Storage conditions

Storage containers should be properly ventilated and kept away from direct heat and sun, and combustible materials. Once removed from the original container, the hydrogen peroxide must not be returned to it.

IMIDAZOLINONE HERBICIDES

Nonselective herbicides for control of weeds, broadleaved herbs, and woody species.

Imazapyr

Synonym 2-(4-isopropyl-4-methyl-5-oxo-2-imidazolin-2-yl) nicotinic acid

In plants, imazapyr acts as a meristem inhibitor through inhibition of amino acid branched chain biosynthesis. It is a systemic, nonselective, contact, and residual herbicide, absorbed by the foliage and roots, with rapid translocation in the xylem and phloem to the meristematic regions, where it accumulates. It is used for the pre and postemergence control of a broad range of weeds, including terrestrial annual and perennial grasses, broad-leaved herbs, woody species, and riparian and emergent aquatic species. Imazapyr may be used in a variety of agricultural, commercial, and residential settings. Use sites include corn, forestry sites, rights-of-way, fence rows, hedge rows, drainage systems, outdoor industrial areas, outdoor buildings and structures, domestic dwellings, paved areas, drive-ways, patios, parking areas, walkways, various water bodies (including ponds, lakes, streams, swamps, wetlands, and stagnant water), and urban areas. Imazapyr may also be used on recreation areas, athletic fields, and golf course roughs.

Chemical and physical properties

Chemical formula $C_{13}H_{15}N_3O_3$
Molecular weight 261.276
Appearance clear, slightly viscous, pale yellow to dark green aqueous liquid
Melting point 171°C
Corrosivity corrosive to iron, mild steel, and brass, but not to stainless steel
Density 0.34 g/ml
Vapor pressure 1.79×10^{-11} mmHg

Toxicity and ecotoxicity data

LD_{50} *oral* >5000 mg/kg (rat)
LD_{50} *percutaneous* >2000 mg/kg (rabbit)
LD_{50} *oral* >2000 mg/kg (mouse, female)
LC_{50} >100 mg/l(bluegill sunfish)
LC_{50} >100 mg/l (rainbow trout)

Irritating to eyes and skin. Evidence of noncarcinogenicity for humans.

Aquatic animal toxicity: low toxicity to invertebrates and practically nontoxic to fish, birds, and mammals.

Protective equipment and clothing

Herbicides should be handled and applied only with full attention to safety measures that minimize personal contact. Many formulations contain adjuvants (stabilizers, penetrants, surfactants) that may have significant irritating and toxic effects.

MANUFACTURERS AND PRODUCTS

Cyanamid

Arsenal → Imazapyr 27.8%, other ingredients 72.2%
Harmful if swallowed. Avoid contact with eyes or clothing.
Arsenal Herbicide Technical (Basf Corporation) Imazapyr 98.5%
Arsenal 75 SG Herbicide (Basf Corporation) Imazapyr 75%

ISOTHIAZOLONES

2-Methyl-4-isothiazolin-3-one

Synonym/common name 2-methyl-3(2*H*)-isothiazolone; 2-methyl-1,2-thiazol-3-one; 2-methyl-2*H*-isothiazol-3-one, 2-methyl-3(2*H*)-isothiazoline; 2-methylisothiazol-3-one; 3(2*H*)-isothiazolone; methylisotiazolinone; *N*-methylisothiazolone; 2-methyl-2,3-dihydroisothiazol-3-one Methylisothiazolinone (MIT) is used as a biocide in cooling water systems, pulp, and paper mill water systems, plastics and marine paint. It is commonly found in a 3:1 ratio with 5-chloro-2-methyl-4-isothiazolinone (CMIT) in biocide products. It is also an ingredient in latex paint, air fresheners, household cleaners, laundry products, and dish soaps. It is used as a preservative in personal care and cosmetics products such as leave-on products, namely hand and body lotions and moisturizers, sun tanning lotions and some rinse-off products like shampoos, surfactants, and conditioners. In addition, MIT is an antimicrobial used to control slime-forming bacteria, fungi, and algae in fuel storage tanks, oil extraction systems, and other industrial settings. It is also used to control the growth of mold on wood products.

Chemical and physical properties

Chemical formula C_4H_5NOS
Molecular mass 115.16
Appearance colorless, extremely hygroscopic crystals
Boiling point 93°C
Melting point 50–51°C
Density 1.35 g/ml
Solubility 30 g/l in water, highly soluble in organic solvents.
Stability At exposure to air, it is converted into an oily compound. Decomposition on heating starts at 55°C. Incompatibilities: strong oxidizing agents, amines, mercaptans, reducing agents.
Vapor pressure 0.062 mmHg

Toxicity and ecotoxicity data

LD_{50} *oral* 285 mg/kg (rat)
LD_{50} *dermal* >2000 mg/kg (rat)
EC_{50} species: *Daphnia magna* (Water flea) age <24 h; Conditions: freshwater, flow through; Concentration: 180 μg/l for 48 h; Effect: intoxication, immobilization.

Caustic effect on skin and mucous membranes: can cause severe eye corrosion.

Protective equipment and clothing

Hand protection: Handle with gloves. Eye protection: Tightly fitting safety goggles. Avoid contact with skin, eyes and clothing. Avoid inhalation of vapor or mist.

Chemical and physical properties of an aqueous solution

Appearance colorless to pale yellow liquid
Boiling point 100°C
Density 1.2 g/ml (20°C)
pH 4–5
Stability stable in the presence of light, over the pH range 2–10 and up to 80°C; sensitive to oxidizing and reducing agents.

MANUFACTURERS AND PRODUCTS

Sigma-Aldrich

ProClin 950 → aqueous solution containing 9.5% of 2-methyl-4-isothiazoline-3-one

2-n-Octyl-4-isothiazolin-3-one

Synonym/common name 2-*n*-octyl-3(2*H*)-isothiazolone; 2-Octyl-3(2*H*)-isothiazolone; 2-*n*-octyl-4-isothiazolin-3-one; octhilinone; octyl-3(2*H*)-isothiazolone; 2,3-octylisothiazolone

It is mainly used as a fungicide being highly toxic for fungi (paint film protection, decorative wood stains, leather industry, adhesives and sealants, paper and cardboard, etc.).

Chemical and physical properties

Chemical formula $C_{11}H_{19}NOS$
Molecular weight 213.34
Appearance clear pale yellow liquid
Boiling point 188°C
Density 1.038 g/ml
pH (10% in water) 2.4
Solubility 0.480 g/l in water, soluble in alcohols, acetone, ether, ethyl acetate; partly soluble in oils.
Stability stable in the pH range 2–10 at room temperature; sensitive to oxidizing and reducing agents.

Toxicity data

LD_{50} *oral* 760 mg/kg (rat)
LD_{50} *dermal* >690 mg/kg (rabbit)
LC_{50} on aerosol inhalation 1.25 mg/l (rat)
It may cause skin sensitization. Corrosive.

MANUFACTURERS AND PRODUCTS

Schülke Inc.

Parmetol N 20 → 2-*n*-octyl-4-isothiazolin-3-one + bronopol (2-bromo-2-nitropropane-1,3-diol)

Rohm and Haas

ROCIMA 103 → Concentrated, stable blend of 2-*n*-octyl-4-isothiazolin-3-oneand a quaternary ammonium compound.

It is a liquid preparation of active substances, which are effective on fungi and algae, and used for the preservation and remedial treatment of masonry. Masonry preservation: very effective on fungi and algae, long-term remedial effect, improves the adhesion of dispersion paint and plasters. It prevents blisters and flaking of the new paint.

Lamirsa

Mirecide-TF/580.ECO → concentrated aqueous dispersion of 3-iodine-2-propylbutylcarbamate and 2-*N*-octyl-4-isothiazolin-3-one). Supplied by Axioma (Milan, Italy).

Thor

Acticide MBL5515 → blend of 2-methyl-4-isothiazolin-3-one (5%), 1,2-benzisothiazolin-3-one (5%), and bronopol (15%).

5-Chloro-2-methyl-4-isothiazolin-3-one

CMIT is used as an antimicrobial preservative in household cleaners, paints, shampoos, car maintenance products, and water storage units. It is commonly found in biocide products that also contain methylisothiazolinone at a 3:1 ratio. CMIT is an ingredient in over 101 pesticide formulations.

CMIT's use as a preservative in cosmetics such as shampoos, conditioners, gels, and surfactants may result in its release to the environment through various waste streams. Its use as a biocide in industrial water treatment, wood treatment, oil drilling muds, hydraulic fracturing fluids and in household products will result in its direct release to the environment. If released to air, its vapor pressure indicates that it will exist solely as a vapor in the atmosphere. Vapor-phase CMIT is degraded in the atmosphere by reaction with photochemically produced hydroxyl radicals; the half-life for this reaction in air is estimated to be 18 h.

Chemical and physical properties

Chemical formula C_4H_4ClNOS
Molecular weight 149.60
Appearance pale yellow crystals
Melting point 54–55 °C
Solubility 5 g/l in water, moderately soluble in organic solvents
Stability sensitive to sodium bisulfite, SH compounds, amines, alkaline solutions
Vapor pressure 0.018 mmHg

Ecotoxicity values

EC_{50} Species: *Daphnia magna* (Water Flea) age <24 h; Conditions: freshwater, flow through; Concentration: 180 μg/l for 48 h; Effect: intoxication, immobilization.

MANUFACTURERS AND PRODUCTS

Troy Corporation

Mergal S97 → Blend of 5-chloro-2-methylisothiazolinone, 2-methylisothiazolinone, methyl-benzimidazol-2-ylcarbamate and 2-*n*-octylisothiazolinone.

Dry film preservative, fungicide, algaecide and bactericide for paints and lacquers. It offers compatibility with aqueous polymer emulsions such as pure acrylic, styrene acrylic, polyvinyl acetate and copolymers, polyvinyl alcohol and vinyl acetate-ethylene pressure polymers. Use concentration: 0.3–2%.

Chemical and physical properties

Appearance liquid
pH stability 3–9

Heat stability 60°C
Density 1.05 g/ml

Rohm and Haas

Kathon CG → 5-chloro-2-methyl-4-isothiazolin-3-one 1.15% + 2-methyl-4-isothiazolin-3-one 0.35%. Inert ingredients: magnesium salts (chloride and nitrate) 23%, water to 100.

Chemical and physical properties

Appearance clear liquid
pH (as manufactured) 1.7–3.7
Stability stable at least 1 year at ambient temperatures and at least 6 months at 50°C

Toxicity data (product as sold)

LD_{50} *oral* 2630 mg/kg (rat male), 3350 mg/kg (rat female)
LC_{50} *inhalation* exposure 4h, 0.33 mg a.i./l air (rat)
Skin irritation corrosive
Eye irritation corrosive, severe corneal damage
Sensitization skin sensitizer

Applications the mixture is one of the most important preservatives for the in-can/in-tank protection of aqueous functional fluids including detergents and cosmetics.

1,2-Benzisothiazolin-3-one

The compound has a broad spectrum of effectiveness that covers fungi, yeasts, algae, bacteria, lichens. It is used as a preservative and antimicrobial agent in industrial and consumer products such as cosmetics and paints. It is a component of air fresheners, printer inks, household cleaners, and laundry products. It is also used in private area and public health area disinfectants and other biocidal products, in-can preservatives, fiber, leather, rubber, and polymerized materials preservatives, masonry preservatives, preservatives for liquid-cooling and processing systems, metalworking-fluid preservatives.

Chemical and physical properties

Chemical formula C_7H_5NOS
Molecular weight 151.19
Appearance off-white to yellowish solid brown, almost odorless oil
Boiling point 327.6°C
Melting point 156.6°C
Solubility 1.1g/l in water at 20°C
Vapor pressure 2.78×10^{-6} mmHg
Stability Incompatible materials: strong oxidizing agents. When heated to decomposition it emits toxic vapors of NO_x and SO_x.

Toxicity and ecotoxicity data

LD_{50} *oral* 1020 mg/kg (rat)
LD_{50} *oral* 1150 mg/kg (mouse)
EC_{50} species: *Daphnia magna* (Water flea) age <24 h; Conditions: freshwater, flow through; Concentration: 3700 μg/l for 48 h; Effect: intoxication, immobilization.
It is a severe eye irritant.

Protective equipment and clothing

Skin protection: handle with gloves. Eye/face protection: face shield and safety glasses. Precautions for safe handling: avoid contact with skin and eyes. Avoid formation of dust and aerosols.

Environmental fate

1,2-Benzisothiazoline-3-one's production and use as a preservative and antimicrobial agent in industrial and consumer products such as cosmetics and paints may result in its release to the environment through various waste streams. Its use as a fungicide, microbicide, and disinfectant and as an inert ingredient in pesticide products may result in its direct release to the environment. If released to air, its vapor pressure indicates that it will exist in both the vapor and particulate phases in the atmosphere. Vapor-phase 1,2-benzisothiazoline-3-one will be degraded in the atmosphere by reaction with photochemically produced hydroxyl radicals; the half-life for this reaction in air is estimated to be 23 days. The particulate-phase of 1,2-benzisothiazoline-3-one is removed from the atmosphere by wet and dry deposition.

PRODUCTS

See Acticide MBL5515

n-Butyl-1,2-benzisothiazolin-3-one

The compound has a spectrum of effectiveness that covers bacteria, algae, and fungi.

Chemical and physical properties

Chemical formula $C_{11}H_{13}NOS$
Molecular weight 207.29
Appearance brown oily liquid
Density approx. 1.17 g/ml (20°C)
Solubility <0.5 ppm in water; highly soluble in ethanol and most other organic solvents.
Vapor pressure 1.3×10^{-4} mmHg
Stability decomposition at ~ 300°C; stable over a wide pH range (2–12).
pH 2.77

Toxicity data

LD_{50} *oral* >2000 mg/kg (rat)
LD_{50} *dermal* >2000 mg/kg (rat)

It causes burns on the skin and it is corrosive to eyes and mucous membranes. It may cause sensitization by skin contact.

SUPPLIER PHASE RESTAURO

ALGOPHASE→ *n*-butil-1,2 benzoisotiazolin-3-one (20% w/w) and organic solvents (80% w/w)

Chemical and physical properties

Appearance pale yellow liquid
Boiling point 202°C
Density 1.027 g/ml

Solubility 25 ppm in water; highly soluble in ethyl methyl ketone, cyclohexane, tetrahydrofuran.

Toxicity data

LD_{50} oral 700 mg/kg

Causes irritation to the respiratory system. Causes eye and skin irritations. May cause skin sensitization.

Not carcinogenic. Not mutagenic.

PICLORAM

Chemical name 4-Amino-3,5,6-trichloro-2-pyridinecarboxylic acid

Chemical and physical properties

Chemical formula $C_6H_3Cl_3N_2O_2$
Appearance colorless to white crystalline solid
Molecular weight 241.45
Melting point 218.5°C
Vapor pressure 6.0×10^{-16} mmHg
Solubility Water 430 mg/l. Organic solvents g/100 ml: acetone 1.98; acetonitrile 0.16; benzene 0.02; carbon disulfide less than 0.005; diethyl ether 0.12; ethanol 1.05; isopropanol 0.55; kerosene 0.001; methylene chloride 0.06.

pH of saturated solution 3.0 (24.5°C)

Toxicity data

LD_{50} oral 2000 mg/kg (rabbit)
LD_{50} oral 1922 mg/kg (Guinea pig)
LD_{50} oral 1061 mg/kg (mouse)
LD_{50} percutaneous >2000 mg/kg (rabbit)
Irritant to eyes, skin, and respiratory system.

Environmental fate

Picloram's use as an herbicide will result in its direct release to the environment. If released to air, its vapor pressure indicates picloram will exist solely in the particulate phase in the atmosphere. Particulate-phase picloram is removed from the atmosphere by wet or dry deposition. It does not contain chromophores that absorb at wavelengths >290 nm and, therefore, is not expected to be susceptible to direct photolysis by sunlight. If released to soil, picloram is expected to have very high to high mobility. Aerobic degradation half-lives for picloram at various application rates in soil ranged from 18 days at 0.0025 ppm to 300 days at 2.5 ppm. Based on these half-lives, picloram is expected to biodegrade in soil.

Protective equipment and clothing

Wear appropriate personal protective clothing to prevent skin contact. Wear appropriate eye protection to prevent eye contact.

MANUFACTURERS AND PRODUCTS

Dow AgroSciences

Tordon 101→ 2,4-Dichlorophenoxyacetic acid, triisopropanolamine salt 39.2 % + picloram 10.2 % + Isopropanol 5.0% + alkylphenol alkoxylate 5.2% + triisopropanolamine 1.3% + other ingredients 39.1 %

Chemical and physical properties

Appearance yellow liquid
pH 7.0 (10% aqueous solution)
Boiling point > 82°C
Solubility soluble in water
Stability thermally stable at typical use temperatures. Avoid contact with strong acids, strong bases, and strong oxidizers.

Toxicity data

LD_{50} *oral* 2598 mg/kg (female, rat)
LD_{50} *dermal* >5000 mg/kg (rabbit)
LC_{50} *inhalation* 4h >1.38 mg/l (rat)

The product is moderately toxic to aquatic organisms on an acute basis. It is practically nontoxic to birds on a dietary basis.

Combustible liquid and vapor. May cause allergic skin reaction. May cause skin and eye irritation. May be harmful if inhaled. Vapor explosion hazard. Harmful if swallowed. Toxic fumes may be released in fire situations.

QUATERNARY AMMONIUM COMPOUNDS

Benzalkonium chloride

Chemical name *N*-alkyl (C8-C18)-*N*,*N*-dimethyl-*N*-benzylammonium chloride

The number of carbon atoms in the alkyl group may differ between *n*-C8 and *n*-C18 and benzalkonium chloride may contain several corresponding alkyl groups.

Chemical and physical properties

Average molecular weight 408
Appearance white-yellow powder
Melting point 61°C
Ignition temperature 365°C
pH (1% aqueous solution) 5–8 at 20°C
Solubility 250 g/l (20°C) in water, soluble in alcohols and ketones
Stability stable at normal conditions and at pH 1–12. Compatible with nonionic surfactants. Incompatibilities with anionic compounds, strong-oxidizing agents, strong acids, strong bases, metals.
Hazardous decomposition products Hydrogen chloride, nitrogen oxides, carbon monoxide

Toxicity data

LD_{50} *oral* 600 mg/kg (rat)

Potential Health Effects. Causes severe eye irritation and burns. Irritant to skin and mucosa. Harmful if absorbed through the skin. Ingestion: causes gastrointestinal tract burns. May be harmful if inhaled. Prolonged or repeated skin contact may cause dermatitis.

MANUFACTURERS AND PRODUCTS

Thor

Acticide BAC 50 → 50% benzalkonium chloride (for preparation of disinfectants and hard surface cleaners)

Acticide BAC 80 → 80% benzalkonium chloride (for preparation of disinfectants, cleansers, and sterility products and for production of fungicidal/algaecidal washes)

Acticide LV706 → blend of benzalkonium chloride 10% + 2-*n*-octyl-3 (2*H*)-isothiazolone in distilled water

Sigma-Aldrich sells pure benzalkonium chloride.

Lonza

Hyamine 3500, Barquat MB-50, Barquat MB-80, and Barquat 50–65.

Lanxess

Cementone Fungicidal Solution → a solution of benzalkonium chloride for the removal and prevention of algae, moss and molds growing on roofs, walls, fences, and monuments. Areas where the product is applied must be surface dry. Do not apply to exposed surfaces when rain is imminent or during frosty or windy weather. On vertical surfaces apply by brush or coarse low-pressure spray; air assisted sprays should not be used. The rate of application strongly depends on the porosity of the surface and the amount of biological growth present at the time of treatment. On absorbent surfaces, such as cement render, concrete stone, and on surfaces heavily contaminated with growth, fungicidal solution should be applied at a rate of 1–2 l/m^2. On smooth, dense surfaces and previous paintwork where organic growth is light, rates of up to 4 l/m^2 may be adequate. Where growth is very thick, two separate applications may be necessary. Remove growth killed by the first treatment after 7–10 days before applying second treatment.

PREVENTOL RI 80 → Solution of approx. 80% benzalkonium chloride and 2% isopropyl alcohol in distilled water. For the formulation of disinfectants or disinfecting cleaning agents for the medical, veterinary and cosmetic sectors, the food processing industry, agriculture, and households. For water treatment (algae prevention in cooling water and swimming pools) and for the elimination of bacteria, yeasts, fungi, algae, lichens.

Langlow

ABICIDE FUNGICIDAL WASH → aqueous solution of 5% w/w benzalkonium chloride. It is a surface fungicidal wash designed to prevent and eradicate the growth of mold, algae, and fungi on most common masonry surfaces. It is suitable for use on paving, stone, and concrete. It can also be used on most common types of masonry wall surfaces. A small test area application is recommended to verify compatibility. Depending on the texture and porosity of the masonry surface, 1 liter of diluted (1/2 or 1/5) product is effective over 8–20 m^2. Repeated applications may be necessary on areas where mold and algae has been previously well established, or where the substrate is particularly porous and friable.

Sunshine makers (distributor Prosoco, Inc.)

D/2 Architectural Antimicrobial → A proprietary combination of octyl decyl dimethyl ammonium chloride (0.30%), dioctyl dimethyl ammonium chloride (0.12%), didecyl dimethyl ammonium chloride (0.18), and alkyl dimethyl benzyl ammonium chloride (0.40%) with surfactants, wetting agents, and buffers. Its pH is 9.5.

Huckert's International

Umonium 38 → isopropyl-tridecyl-dimethyl ammonium chloride 32g/100ml. The molecule is obtained by combining isopropyl alcohol and tridecyl ceteth alcohol with benzalkonium chloride in carefully controlled proportions. The molecule exhibits a remarkable synergistic effect respect to the single components in terms of spectrum of activity and contact time required to kill microorganisms.

Chemical and physical properties

pH 7
Solubility soluble in water and alcohols
Stability chemically stable molecule

Toxicity data

LD_{50} *oral* 5840 mg/kg (rat)

Noncorrosive. Irritant in undiluted form, 2.5% solution is nonirritant. Biodegradable > 81%.

Di-n-decyl-dimethylammonium chloride

Synonym didecyldimonium chloride, didecyl–dimethyl ammonium chloride

Chemical and physical properties

Chemical formula $C_{22}H_{48}NCl$
Molecular weight 362.08
Appearance clear, colorless to pale yellow liquid
Boiling point >180°C
Density 0.87 g/ml (20°C)
pH (29.5% aqueous solution) 6.8
Solubility soluble in water and alcohols
Stability stable over the pH range 2–10 and up to 120°C. Not compatible with anionic compounds.

Toxicity and ecotoxicity data

LD_{50} *oral* 84 mg/kg (rat)
LD_{50} *oral* 268 mg/kg (mouse)
EC_{50} Species: *Daphnia magna* (Water flea) age <20 h; Conditions: freshwater, static; Concentration: 18 μg/l for 48 h; Effect: intoxication, immobilization.

Strongly irritant to skin mucosa and eyes. Not mutagenic nor teratogenic. *Biodegradability* Biologically well degradable; test period 28d; method OECD 302 B.

MANUFACTURERS AND PRODUCTS

Kimia (Perugia, Italy)

Kimistone Biocida → a ready-to-use blend of didecyl–dimethyl ammonium chloride and benzalconiumchloride in water.

CTS

Biotin T → mixture of didecyl–dimethyl ammonium chloride (40–60%), 2-*N*-ottil-2*H*-isotiazol-3-one (7–10%), isopropanol (15–20%) and formic acid (1–2.5%).

New Des 50 → didecyl–dimethyl ammonium chloride (50%) in water
Appearance pale yellow or clear liquid
Density 20°C 0.9 ± 0.02 kg/l
pH 6.5–8 aqueous solution at 10% concentration

TRIAZINES

Atrazine

2-Chloro-4-(ethylamine)-6-(isopropylamine)-*s*-triazine
$C_8H_{14}ClN$

Its use is controversial due to its effects on nontarget species, such as amphibians, and because of widespread contamination of waterways and drinking water supplies. There are also concerns for implications in human birth defects, low birth weights and menstrual problems. Like many commercial products, it is sold under numerous trade names. Its use has been banned in the European Union since 2004 because of its persistent groundwater contamination, its long-term persistence in the environment together with toxicity for wildlife and possible link to effects on human health. It is still one of the most widely used herbicides in the United States.

Chemical and physical properties

Molecular weight 215.684
Appearance colorless powder
Melting point 173°C
Boiling point 205°C
Vapor pressure 2.89×10^{-7} mmHg
Solubility 183 g/kg dimethyl sulfoxide; 52 g/kg chloroform; 28 g/kg ethyl acetate; 18 g/kg methanol; 12 g/kg diethyl ether; 0.36 g/kg pentane. In water, 34.7 mg/l at 26°C.
Stability stable in neutral, slightly acidic or basic media

Toxicity and ecotoxicity data

LD_{50} *oral* 1480–5100 mg/kg (rat)
LD_{50} *oral* 750 mg/kg (rabbit)
LD_{50} *oral* 850 mg/kg (mouse)
LD_{50} *percutaneous* >3100 mg/kg (rat)
LC_{50} *inhalation* >0.71 mg/l air/1 h (rat)
EC_{50} species: *Daphnia magna* (Water flea); Conditions: freshwater, flow through; Concentration: 49,000 μg/l for 48 h; Effect: intoxication, immobilization.

A skin and severe eye irritant.

Protective equipment and clothing

Wear appropriate protective gloves, boots, and googles.

Storage conditions

Store the material in a well ventilated, secure area.

MANUFACTURERS AND PRODUCTS

Sigma-Aldrich

Irgarol 1051 → 2-(*tert*-butylamino)-4-(cyclopropylamino)-6-(methylthio)-*s*-triazine

Synonym Cybutryne
Chemical formula $C_{11}H_{19}N_5S$
Molecular weight 253
The compound inhibits the growth of marine organisms.

Simazine

6-Chloro-*N,N′*-diethyl-1,3,5-triazine-2,4-diamine

Chemical and physical properties

Chemical formula $C_7H_{12}ClN_5$
Appearance off-white crystalline compound
Melting point 225–227°C (decomposition)
Solubility 6.2 mg/l at 20°C in water. Soluble in many organic solvents.

Toxicity data

LD_{50} *oral* 971 mg/kg (rat)
LD_{50} *oral* >5000 mg/kg (rabbit)
LC_{50} *inhalation* 4h >9800 mg/l (rat)

Simazine is an herbicide of the triazine class. It is used to control broad-leaved weeds and annual grasses. It is now banned in the European Union (EU directive 91/414/EEC). It remains active in the soil for 2–7 months after application.

Hexazinone

3-Cyclohexyl-6-dimethylamino-1-methyl-1,3,5-triazine-2,4(1*H*,3*H*)-dione

Chemical and physical properties

Chemical formula $C_{12}H_{20}N_4O_2$
Appearance white crystalline solid
Density 1.25 g/cm^3
Melting point 116°C
Vapor pressure 2.25×10^{-7} mmHg
Solubility soluble in water
Stability stable in aqueous media between pH 5 and 9 and below 37°C. Decomposed by strong acids and strong bases. Stable to light.

Toxicity and ecotoxicity data

LD_{50} *oral* 1690 mg/kg (rat)
LD_{50} *dermal* 5278 mg/kg (rat)
LD_{50} *oral* 860 mg/kg (Guinea pig)
LC_{50} inhalation >7.48 mg/l/1 h exposure (rat)
LC_{50} 370 mg/l (bluegill sunfish)

Environmental Fate

Hexazinone's use as an herbicide will result in its direct release to the environment. If released to air, its vapor pressure indicates it will exist in both the vapor and particulate phases in the atmosphere. Vapor-phase hexazinone is degraded in the atmosphere by reaction with photochemically produced hydroxyl radicals; the half-life for this reaction in air is estimated to be 1.4 h. Particulate-phase hexazinone is removed from the atmosphere by wet and dry deposition. Aqueous solutions of this compound have been shown to degrade when exposed to sunlight. If released to soil, hexazinone is expected to have high mobility. It is not expected to volatilize from dry soil surfaces based upon its vapor pressure. Studies of its biodegradation indicate that the process is slow.

Protective equipment and clothing

Wear appropriate personal protective clothing to prevent skin contact. Wear appropriate eye protection to prevent eye contact.

MANUFACTURERS AND PRODUCTS

Du Pont

Velpar L → hexazinone 25%, other ingredients 75%
Water-dispersible liquid.
Corrosive, causes irreversible eye damage. Harmful if swallowed. Do not get in eyes or on clothing.

URACIL HERBICIDES

Bromacil

Chemical name 5-bromo-6-methyl-3-(1-methylpropyl)-2,4(1*H*,3*H*) pyrimidinidione
Synonym 5-bromo-3-sec-butyl-6-methyluracil

Chemical and physical properties

Chemical formula $C_9H_{13}BrN_2O_2$
Molecular weight 261.10
Appearance odorless, colorless to white, crystalline solid
Density 1.46 g/cm^3
Melting point 158°C
Vapor pressure 4.1×10^{-2} mPa /3.07×10^{-7} mmHg
Solubility 815 mg/l in water; 201 g/kg in acetone; 77 g/kg in acetonitrile; 155 g/kg in ethyl alcohol; 33 g/kg xylene
Stability stable in water except under strongly acidic conditions and elevated temperatures

Toxicity data

LD_{50} oral 5200 mg/kg (rat, male)
LD_{50} oral 3998 g/kg (rat, female)

LD_{50} percutaneous >5000 mg/kg (rabbit)
LC_{50} inhalation >4.8 mg/l air/4h (rat)
LD_{50} oral 3040 mg/kg (mouse)

Liquid formulations of bromacil are moderately toxic, while dry formulations are practically nontoxic. The herbicide is irritating to the skin, eyes, and respiratory tract.

Environmental fate

Bromacil's use as an herbicide can result in its direct release to the environment. If released to air, its vapor pressure indicates it will exist in both the vapor and particulate phases in the atmosphere. Vapor-phase bromacil is degraded in the atmosphere by reaction with photochemically produced hydroxyl radicals; the half-life for this reaction in air is estimated to be 6.6 h. Particulate-phase bromacil is removed from the atmosphere by wet or dry deposition. If released to soil, bromacil is expected to have very high to high mobility.

Protective equipment and clothing

Wear appropriate personal protective clothing to prevent skin contact. Wear appropriate eye protection to prevent eye contact.

Storage conditions

Do not contaminate water, other pesticides, fertilizer, food, or feed in storage. Store in original container. Keep tightly closed in a dry, cool, and well-ventilated place.

MANUFACTURERS AND PRODUCTS

DuPont

HyvarX→ bromacil 80%, other ingredients 20%

Chemical and physical properties

Appearance off-white solid powder
Density 0.53 g/cm^3

pH 8.5–9.5, 20 g/l at 20°C
Solubility dispersible in water
Stability stable at normal temperatures and storage conditions; incompatible with amines.

Toxicity data

LD_{50} *oral* 2000 mg/kg (male, rat)
LD_{50} *oral* 1300 mg/kg (female, rat)
LD_{50} *dermal* >2000 mg/kg (rabbit)
LC_{50} *inhalation* 4h > 5.2 mg/l (rat)

UREA HERBICIDES AND ALGAECIDES

Diuron

3-(3,4-Dichlorophenyl)-1,1-dimethylurea
Other names DCMU, Dichlorfenidim, Herbatox

Diuron is a photosynthesis inhibitor used mainly for general weed control on noncrop areas. It has also been used in the selective control of germinating broadleaf and grass weeds in sugarcane, citrus, pineapples, cotton, asparagus, and temperate climate tree and bush fruits.

Environmental Fate

Diuron's use as a preemergence herbicide and sugarcane flowering suppressant will result in its direct release to the environment. If released to air, the low vapor pressure indicates diuron will exist solely in the particulate phase in the atmosphere. In other words, its volatilization is considered insignificant. Particulate-phase diuron will be removed from the atmosphere by wet or dry deposition. If released to soil, diuron is expected to have moderate to low mobility. Volatilization from moist soil surfaces is not expected to be an important fate process. Diuron is not expected to volatilize from dry soil surfaces based upon its vapor pressure. If released into water, diuron is expected to adsorb to suspended solids and sediment. The rate of diuron hydrolysis is negligible at neutral pH but increases as the conditions become strongly acidic or alkaline, leading to its principle derivative, 3,4-dichloroaniline; a hydrolysis half-life of 4 months has been reported. Occupational exposure to diuron may occur through inhalation of dust and dermal contact with this compound at workplaces.

Chemical and physical properties

Chemical formula $C_9H_{10}Cl_2N_2O$
Appearance white, odorless, crystalline solid
Molecular weight 233.09
Boiling point decomposes at 180–190°C
Melting point 158–159°C
Density 1.48g/cm^3
Vapor pressure 6.9×10^{-8} mmHg
Solubility very low in hydrocarbon solvents; in acetone 53 g/kg at 27°C; very slightly soluble in water.
Stability stable under normal conditions. It decomposes on heating to 180–190°C emitting highly toxic fumes of chlorine and nitrooxides (dimethylamine and 2,4-dichlorophenyl isocyanate). Incompatible with strong acids.

Toxicity and ecotoxicity data

LD_{50} *oral* >5000 mg/kg (rat)
LD_{50} *dermal* >1000 mg/kg (rat)
LD_{50} *oral* >2500 mg/kg (mouse)
LD_{50} *oral* >1000 mg/kg (rabbit)
LC_{50} *inhalation* for 4h >0.265 mg/l (rat)
LC_{50} *inhalation* for 4h >0.335 mg/l (mouse)
LC_{50} Species: *Daphnia magna* (Water flea); Concentration: 1.4 mg/l for 2 days; Conditions: temperature 20°C.

Not classifiable as a human carcinogen. Contact with diuron (particularly in concentrated form) may irritate the eyes, nose, throat, and skin. Exposure to diuron may even prove fatal if sufficient quantities are inhaled, swallowed, or absorbed through the skin. The primary hazard of diuron is the threat to the environment. It is very toxic to aquatic life with long lasting effects.

Protective equipment and clothing

Wear appropriate personal protective clothing to prevent skin contact. Wear appropriate eye protection to prevent eye contact.

Storage Conditions

Store in tightly closed containers in a cool, well-ventilated area. Avoid freezing liquid suspension. Dry formulations are stable under normal storage.

MANUFACTURERS AND PRODUCTS

Diachem

Toterbane 50F → diuron 600 g/l + monoethylene glycol, water, coformulants

Chemical and physical properties

Appearance off-white aqueous suspension
Density 1200 g/l
pH (1% in water) 7.4
Solubility 36.4 mg/l in water, 53 g/l at 27°C in acetone (pure diuron)
Stability stable under normal conditions

Toxicity data

LD_{50} *oral* 3400 mg/kg (rat)
LD_{50} *dermal* >2000 mg/kg (rabbit)
LC_{50} >5 mg/l (4h) (rat)
Irritant to eyes.

EU classification Harmful (Xn), dangerous for the environment (N).

Lanxess

Preventol A6 → 98% diuron

Preventol A6 is a broad-spectrum algaecide highly effective on algae. It is nonvolatile and not light sensitive. The overall fungicidal effect of the product is not very strong.

Chemical and physical properties

Appearance powder
Bulk density approx. 450 kg/m^3

pH (aqueous suspension) approx. 5.8
Melting point approx. 154–159°C
Solubility 0.035 g/l in water; 50 g/l in acetone; 1.2 g/l in xylene; 62 g/l in butyl glycol
Stability It hydrolyses in strong acids and alkalis.

DuPont

Karmex DF → 80% diuron, 20% other ingredients

INDEX

C

D

H

I

N

O

P

T

U

V

W

X

Y

Z